Clinical Microbiology
Diagnosis, Treatment and
Prophylaxis of Infections

THE MICROBIOLOGY OF CENTRAL NERVOUS SYSTEM INFECTIONS

Clinical Microbiology Diagnosis, Treatment and Prophylaxis of Infections

THE MICROBIOLOGY OF CENTRAL NERVOUS SYSTEM INFECTIONS

Edited by

KATERYNA KON

Associate Professor, Department of Microbiology, Virology and Immunology, Kharkiv National Medical University, Kharkiv, Ukraine

MAHENDRA RAI

Professor and Head, Department of Biotechnology, SGB Amravati University, Maharashtra, India

ELSEVIER

ACADEMIC PRESS

An imprint of Elsevier

Academic Press is an imprint of Elsevier
125 London Wall, London EC2Y 5AS, United Kingdom
525 B Street, Suite 1800, San Diego, CA 92101-4495, United States
50 Hampshire Street, 5th Floor, Cambridge, MA 02139, United States
The Boulevard, Langford Lane, Kidlington, Oxford OX5 1GB, United Kingdom

Library of Congress Cataloging-in-Publication Data
A catalog record for this book is available from the Library of Congress

British Library Cataloguing-in-Publication Data
A catalogue record for this book is available from the British Library

ISBN: 978-0-12-813806-9
ISSN: 2451-9006

For information on all Academic Press publications
visit our website at https://www.elsevier.com/books-and-journals

Publisher: Shirley Decker-Lucke
Acquisition Editor: Linda Versteeg-buschman
Editorial Project Manager: Timothy Bennett
Production Project Manager: Poulouse Joseph
Cover Designer: Matthew Limbert

Typeset by SPi Global, India

CONTENTS

CONTRIBUTORS

Philippe V. Afonso
Institut Pasteur; Centre National de la Recherche Scientifique (CNRS), Paris, France

Talevi Alan
University of La Plata, La Plata, Argentina

Rabia Cakir-Koc
Yildiz Technical University, Istanbul, Turkey

Tapash Chakraborty
Dibrugarh University, Dibrugarh, India

Céline Curis
Institut Pasteur; Centre National de la Recherche Scientifique (CNRS); Université Paris Diderot, Sorbonne Paris Cité, Paris, France

Malay K. Das
Dibrugarh University, Dibrugarh, India

Carlos Franco-Paredes
University of Colorado Denver, Denver, CO, United States; Hospital Infantil de México, Federico Gómez, Mexico City, Mexico

Mohammadhassan Gholami-Shabani
Pasteur Institute of Iran, Tehran, Iran

Indarchand Gupta
SGB Amravati University, Amravati; Government Institute of Science, Aurangabad, India

Andrés F. Henao-Martínez
University of Colorado Denver, Denver, CO, United States

Avinash P. Ingle
SGB Amravati University, Amravati, India

Hafiz M.N. Iqbal
Tecnologico de Monterrey, School of Engineering and Sciences, Campus Monterrey, Ave. Eugenio Garza Sada 2501, Monterrey, N. L., CP 64849, Mexico

Fatemehsadat Jamzivar
Pasteur Institute of Iran, Tehran, Iran

Benard W. Kulohoma
University of Nairobi, Nairobi, Kenya; University of Oxford, Oxford, United Kingdom

Sbaraglini Maria Laura
University of La Plata, La Plata, Argentina

Bellera Carolina Leticia
University of La Plata, La Plata, Argentina

Hoda Moosa
Pasteur Institute of Iran, Tehran, Iran

Dipali Nagaonkar
SGB Amravati University, Amravati, India

Burak Özdemir
Yildiz Technical University, Istanbul, Turkey

Raksha Pandit
SGB Amravati University, Amravati, India

Priti Paralikar
SGB Amravati University, Amravati, India

Roberto Parra-Saldívar
Tecnologico de Monterrey, School of Engineering and Sciences, Campus Monterrey, Ave. Eugenio Garza Sada 2501, Monterrey, N. L., CP 64849, Mexico; Microsystems Technologies Laboratories, MIT; Division of Engineering in Medicine, Department of Medicine, Brigham and Women's Hospital, Harvard Medical School; Harvard-MIT Division of Health Sciences and Technology, Massachusetts Institute of Technology, Cambridge, MA02139, USA

Ali Parsaeimehr
Tecnologico de Monterrey, School of Engineering and Sciences, Campus Monterrey, Ave. Eugenio Garza Sada 2501, Monterrey, N. L., CP 64849, Mexico

Mahendra Rai
SGB Amravati University, Amravati, India

Mehdi Razzaghi-Abyaneh
Pasteur Institute of Iran, Tehran, Iran

Manish Sadarangani
University of British Columbia, Vancouver, BC, Canada; University of Oxford, Oxford, United Kingdom

Masoomeh Shams-Ghahfarokhi
Tarbiat Modares University, Tehran, Iran

Rewati R. Sharma
NMC (Atlas) Hospitals Ruwi and Al Ghoubra, Muscat, Sultanate of Oman

Apollina Sharma
McGill University, Montreal, QC, Canada

Sudhir Shende
SGB Amravati University, Amravati, India

Gwenn L. Skar
University of Nebraska Medical Center, Omaha, NE, United States

Jessica N. Snowden
University of Nebraska Medical Center, Omaha, NE, United States

Saeed Soleiman-Meigooni
AJA University of Medical Sciences, Tehran, Iran

Praveen Sudhindra
Cooper University Hospital, Camden, NJ, United States

Daniel Vela-Duarte
University of Colorado Denver, Denver, CO, United States

Samira Zamani
Pasteur Institute of Iran, Tehran, Iran

PREFACE

Infections of the central nervous system (CNS) include a diverse group of bacterial, viral, and fungal infections of the brain and meninges. Moreover, the number of cases of multidrug-resistant microbes is increasing. Despite the presence of advanced hospital techniques and the discovery of new antimicrobial drugs, CNS infections are associated with significant mortality and morbidity worldwide. The healthcare importance of this group of diseases makes it necessary to have a well-structured source of updated scientific information that discusses existent clinical and diagnostic guidelines as well as new and perspective trends in diagnosis, treatment, and prophylaxis of CNS infections.

The present book has been divided into three sections. The first section contains general reviews on the current trends in diagnosis, treatment, and prophylaxis of CNS infections. It includes chapters on the overall management of meningitis, CNS catheter infections, the molecular diagnosis of CNS viral infections, and vaccine prevention of CNS infections. This section also incorporates the strategic role of metal nanoparticles and natural products for the management of CNS infections.

The second section is devoted to CNS infections caused by bacterial and viral pathogens, which represent the majority of cases of CNS infections. This section includes chapters on CNS infections caused by pneumococci and tuberculosis as well as nontuberculosis mycobacteria. Viral infections covered in this book include tick-borne and retroviral infections.

The third section discusses rarer CNS infections such as infections of fungal and protozoal etiology. Special attention is paid to CNS infections caused by true pathogenic fungi, such as blastomycosis, coccidioidomycosis, paracoccidioidomycosis, histoplasmosis, and sporotrichosis; opportunistic fungal infections of the CNS, particularly aspergillosis, candidiasis, and cryptococcosis; and the melanized fungi causing CNS infections, such as *Cladophialophora* spp., *Rhinocladiella mackenziei*, and others. Among protozoal infections, particular focus has been made on sleeping sickness and cerebral toxoplasmosis.

The book would be very useful for graduate and postgraduate students, researchers, teachers at universities, scientists, medical practitioners, and specialists from pharmaceutical and laboratory diagnostic companies.

SECTION I

Central Nervous System Infections: Current Trends in Diagnosis, Treatment and Prophylaxis

CHAPTER 1

Meningitis: Current Understanding and Management

Rewati R. Sharma*, Apollina Sharma†
*NMC (Atlas) Hospitals Ruwi and Al Ghoubra, Muscat, Sultanate of Oman
†McGill University, Montreal, QC, Canada

1. INTRODUCTION

Meningitis is an inflammation of the meninges. It is a serious, life-threatening emergency with high morbidity and mortality despite aggressive management. Even after achieving effective clinical control, central nervous system (CNS) infections leave either transient or permanent neurological deficits in a significant number of cases.[1]

The meningeal irritation irrespective of its etiological factors is referred to as meningism. Significant meningitis due to noninfective causes is called aseptic meningitis; when it is caused by microorganisms, it is referred to as infective meningitis. In complicated meningitis, the extension of infection leads to cerebritis, ventriculitis, vasculitis, cranial neuritis, meningoencephalitis, encephalomyelitis, cerebral abscesses, granulomas, etc.[1,2]

2. CLASSIFICATION

Based on the wide variety of causative factors, meningitis covers a wide spectrum that can be classified in the following broad groups:

1. Viral meningitis
2. Bacterial meningitis
3. Fungal meningitis
4. Parasitic meningitis
5. Chemical meningitis
6. Carcinomatous meningitis

2.1 Viral Meningitis[1–4]

2.1.1 Introduction

The meningeal inflammation due to viruses is called viral meningitis.[3] Viral meningitis is ubiquitous with an approximate incidence of 11–20 individuals per 100,000 people in

The Microbiology of Central Nervous System Infections
https://doi.org/10.1016/B978-0-12-813806-9.00001-9

the United States. Its occurrence depends on many factors such as the virulence of the virus, the inoculums level, the immunological status of the patient, and the tropism of the virus for a specific CNS cell type.[2–4]

2.1.2 Clinical Characteristics

(a) *Etiology*: Infants are generally infected with *enteroviruses*; young children with *arboviruses*; school-going children with the polio virus, measles, and mumps; young adults with mumps and measles; adults and middle-aged people with the *herpes virus*; and older people with *enteroviruses*. There is no bar to age and sex. However, the younger the age, the greater the incidences of viral infection.[3,4]

Males are more commonly affected with *mumps and enteroviruses. Viral* meningitis is commonly (80%–85%) caused by *enteroviruses* (*Coxsackie viruses A and B, Echoviruses, polio viruses*); less commonly (10%–15%) by the *arboviruses, herpes viruses, and mumps*; and rarely (5% or less) by the *measles, lymphocytic choriomeningitis viruses, HIV*, and *adenoviruses*.[2–4]

The *arboviruses* families are: (1) Alphaviruses (*Eastern, Western*, as well as *Venezuelan equine encephalitic viruses*); (2) Bunyaviruses (*California encephalitic viruses and Jamestown Canyon viruses*); (3) Flaviviruses (*Japanese B encephalitis virus, Colorado tick fever virus, West Nile virus, St. Louis encephalitis virus, Murray valley viruses*, etc.). The two most common viruses are *St. Louis encephalitis virus* and *Japanese B encephalitis virus* for most cases worldwide.

The Herpes virus family: *Herpes simplex viruses (HSV-1 and HSV-2), Varicella-zoster virus* (usually causing chickenpox and shingles), *Human herpes virus-6, Epstein-Barr virus (EBV)*, and *cytomegalovirus (CMV)*. *Herpes simplex virus-2* meningitis occurs in less than 5% of cases. In this family, *cytomegalovirus* causes congenital intracranial infections in the fetus with long-term disabilities. *Paramyxoviruses* such as *mumps* and *measles* are uncommon causes of viral meningitis due to vaccination programs.

The *influenza virus* rarely causes meningitis. *Adenoviruses, especially Ad 3 and Ad 7,* are also rare causes of meningitis in immunocompetent subjects. They are a major cause of meningitis in AIDS cases, patients with severe combined immunodeficiency (SCID), and immune-compromised hosts as well as in bone marrow transplant recipients. Retroviruses, *Human Immunodeficiency Virus (HIV)*, and *Human T-cell Lymphotrophic Virus (HTLV)* cause atypical meningitis, at the time of seroconversion, in about 4%–10% of HIV infections.

(b) *Viral transmission occurs via various routes*[1–4]: *Enteroviruses* transmit via the orofecal route; *arboviruses* via vectors; *herpes viruses* through personal contact; and the *herpes family viruses (HSV-1, HSV-2, VZV-B)* and *Rabies virus* get to the CNS from the mucocutaneous regions via retrograde extension up the peripheral nerves. Reactivated *Herpes simplex virus (HSV) type-1* most frequently travels up the olfactory nerves to the temporal lobes, causing necrotizing-hemorrhagic encephalitis. LCMV is contracted from a rodent's fresh urine, saliva, droppings, and nesting materials.

Potentially, the viral infection is also caused by an infected blood transfusion and infected donor organs. *The incubation period* for *enteroviral* meningitis is about 2 days, and it is different for other causes of viral meningitis.

Initial viral infection and rapid replication occur in the primary foci, then spread to the reticulo-endothelial system; secondary viremia reaches the CNS by the hematological route. Infective viruses also reach the CNS via direct neuronal penetration, retrograde transportation, and migration along the craniospinal nerves from adjacent structures such as the nasal mucosa, cutaneous tissues via spinal ganglia, etc.

(c) *Meningeal inflammation*: It occurs within 24–48 h following viral inoculums. The initial response is polymorphonuclear leukocytosis, followed later by progressively increasing monocytosis and T-lymphocytosis. The CSF study confirms the active meningitic process. If the infective process involves the cerebrum or spinal cord, there will be parenchymal edema, vasculitis, leucocytic/lymphocytic infiltration, glial cells proliferation, and neuronal degenerations. The severity of meningitis depends on the type of virus and its virulence as well as the penetration into the craniospinal axis with attendant degree of morbidity and mortality.

2.1.3 Clinical Presentation and Differential Diagnoses[2–4]

Initially, patients present with prodromal symptoms (fever, chills and rigors, general malaise, lethargy, muscle pains) and extraneural symptoms (sore throat, upper respiratory infection, skin rashes, joint pains, nausea, vomiting, abdominal pains, and lymphadenopathy). This is followed by meningeal irritation that results in neck stiffness, headaches, vomiting, and photophobia. There is increasing resistance to the passive flexion of the neck (Brudzinski's sign) and inability to extend the knee when the hip (thigh) is flexed at 90 degrees with the trunk (Kernig's sign). Further clinical deterioration manifests as confusion, disorientation, irritability, agitation, drowsiness, seizures, focal neurological deficits, or coma. One should look for mucocutaneous vesicles in *Coxsackie viruses* and *Herpes viruses* as well as rashes in HIV seroconversion.

Uncomplicated viral meningitis is often benign and self-limiting with good recovery in about fortnight in the majority of cases. Seizures, altered sensorium, focal or general cerebrospinal deficits, and raised ICP are carefully managed to avoid complications.

Asymptomatic or mildly symptomatic cases of LCM are common. It has a biphasic clinical course: the initial course (febrile systemic extraneural symptoms) lasting for about 1 week is mainly marked with prodromal and constitutional symptoms along with upper respiratory infection, polyarthropathy, and glandular (parotid and testicular) pains. This is usually followed by recovery. The second phase presents with obvious symptoms of meningitis, less commonly with encephalitis and raised intracranial pressure.

Differential diagnosis is done with the other causes of meningitis (bacterial, fungal, parasitic), postvaccinal states (especially following measles and rubella vaccines), vascular episodes (intracranial subarachnoid hemorrhage, migraine), neoplastic conditions (carcinomatosis), drugs (NSAIDs, allopurinol, azathioprine etc.), systemic disorders (vasculitis, sarcoidosis), trauma, etc.

2.1.4 Investigations[2–4]

The lumbar CSF analysis usually shows mild lymphocytic pleocytosis with normal sugar and moderately raised proteins. Tissue culture, animal inoculation, serological tests, viral culture, PCR, and typing are used in establishing the diagnosis. The CT brain scan is performed, which usually shows normal findings unless there is an acute hydrocephalus or evolving encephalitis. In these cases, periodic CT/MRI scans are done to help in management.

2.1.5 Management and Prognosis[3,4]

The management is mainly supportive: adequate bed rest, good rehydration, antipyretics, analgesics, and nutritional support. Significant symptoms warrant lumbar CSF analysis.

Viral meningitis is treated with antiviral therapy. If there is hypogammaglobulinemia, then the needed immunoglobulins must be instituted. In cases of *HSV-1* encephalitis and *HSV-2* meningitis, a full course of intravenous acyclovir must be started immediately for a period of 5–7 days. Similarly, ganciclovir should be given in cases of *CMV* infections or AIDS-related infections. *Enteroviruses* are usually self-limiting and need no specific antiviral therapy. For the prevention of meningitis due to the *mumps, measles, varicella, the polio virus, the influenza virus, Japanese-B encephalitis virus,* etc., specific vaccinations are given.

Intensive care management of the complicated meningitis and meningoencephalitis is undertaken. Periodic neuroimaging studies, close ICP monitoring, anticerebral edema medications (mannitol, 3% hypertonic saline, acetazolamide), anticonvulsant drugs, ventilation support, CSF diversion, and precautions *to prevent the spread and recurrence* while receiving treatment are mandatory. Family members and friends should be evaluated for obvious reasons.

The overall prognosis in viral meningitis is better than the other infective causes except in infants and young children. Usually there is complete recovery within 2–4 weeks. However, the complicated viral meningitis/meningoencephalitis is the fifth-leading cause of significant infantile morbidities and mortality. Prompt recognition and treatment will reduce symptoms, lessen morbidity, and reduce mortality. The role of preventive measures therefore cannot be over emphasized.

2.2 Bacterial Meningitis[4,5]

2.2.1 Introduction[5,6]

Bacterial meningitis is the inflammation of the cerebrospinal meninges due to the bacterial infection. It is a life-threatening illness with high morbidity and mortality. Bacterial meningitis is ubiquitous with worldwide distribution and is the second-leading cause of CNS infections. The impact of the HIV infection and the availability of antibacterial vaccines have changed the epidemiology of bacterial meningitis. Currently in the United States, *Streptococcus pneumoniae* is the leading cause of meningitis followed by the *Neisseria meningitidis* and then *Hemophilus influenzae*. For bacterial meningitis, there is no bar for age, sex, race, and geographical locations. In the United States, bacterial meningitis is in the range of 0.5–4.0 cases per 100,000 adult populations and about 0.25–1 case per 1000 live births, being 10 times more prevalent in premature than term neonates. Incidence is slightly more in males, immigrants, Africans and Asians, and patients with HIV and tuberculosis (*Micobacterium avium intracellulare*) with a relatively poor prognosis.

2.2.2 Etiopathogenesis[5–7]

(a) *Community-acquired bacterial meningitis* in adults is due to *N. meningitidis*, *S. pneumoniae* (encapsulated bacteria), and *H. influenzae type-B* or *HIB*.[8] In neonates and early infancy, *Escherichia coli, Streptococcus agalactiae (group B streptococci)*, and *Listeria monocytogenes* are the main culprits. In the elderly as well as immune-compromised patients, the *L. monocytogenes* (an aerobic Gram-negative bacilli) join the group of organisms (except *H. influenzae*) causing meningitis in adults.

S. pneumonia[9] is a lancet-shaped, Gram-positive diplococci and *N. meningitidis* is a Gram-negative intracellular cocci; however, *H. influenzae type B* is a Gram-negative pleomorphic bacillus. These infections can be transmitted from person to person; their incubation period ranges from 4 to 7 days.[5–9] Asymptomatic carriers harbor *N. meningitidis* and *S. pneumoniae* in their nasopharynx and *H. influenzae* in their paranasal sinuses.[10,11]

(b) *Nosocomial bacterial meningitis*: In postoperative neurosurgical cases (craniotomies, shunt surgeries, external ventricular drains, lumbar punctures, intrathecal medication infusions, spinal anesthesia, and posttraumatic compound fractures of the calvaria), Gram-positive cocci (*Staphylococcus aureus, Staphylococcus epidermidis*, and *coagulase-negative staphylococci*), aerobic Gram-negative bacilli (*Pseudomonas aeruginosa*) as well as *Propionibacterium acnes* are predominantly responsible.

The majority of infections manifest in the first month following surgery (90%–95%). A craniotomy is associated with a 1%–3% rate of nosocomial infections whereas the internal shunts are about 5%–10%, external ventricular drains 10%, and lumbar punctures/invasive techniques ~1%–2%.

(c) *Chronic meningitis* is commonly caused by *Tubercular bacilli* and *Treponema pallidum* worldwide. For tuberculosis and spirochete infections, humans are the main reservoir.[11] The *Tubercular bacilli* is an aerobic Gram-positive acid-fast bacillus. The CNS tuberculosis is usually secondary to pulmonary tuberculosis by hematogenous dissemination. Initially, the bacilli seed the meninges and the brain parenchyma to form subpial and subependymal tiny caseous lesions known as Roch foci. These subpial foci usually enlarge and rupture in the subarachnoid spaces causing meningitis, adhesions, and vasculitis. The parenchymal Roch foci extremely rarely rupture in the ventricular system and more commonly grow to form tuberculomas and abscesses.

2.2.3 The Predisposing Risk Factors

These are naso-oro-pharyngeal and respiratory infections, middle ear and mastoid infections, contaminated objects, compound head injuries, HIV, immune-compromised states, malnutrition, alcoholism, drug abuse, diabetes mellitus, corticosteroid use, etc. The common mode of transfer of bacterial infections is via person-to-person contact and spread of the respiratory droplets during coughing, sneezing, kissing, and using shared objects. Use of contaminated milk and cheese (*L. monocytogenes*) can also cause meningitis.

Dissemination occurs by a hematogenous, contiguous route, and direct implantation during the medical-surgical procedures. Bacteria produce intense inflammation of the leptomeninges and result in an immediate polymorphonuclear response as well as high protein and low sugar contents with presence of bacteria and the pleocytosis in the CSF.

The host defenses destroy the bacteria and, in turn, the products of the bacterial cell wall destruction excite cytokine activation. Interleukin-1 (IL-1), tumor necrosis factors (TNF), and nitric acid cause further enhancement in inflammatory responses and excite production of platelet-activating factors (prostaglandins, thromboxanes, and leukotrienes) in the arachidonic acid pathway. These platelet-activating factors promote leukocyte adhesions via the activation of the receptors on the endothelial cells. Leukocyte adhesions and destruction with the release of proteolytic enzymes result in increased blood-brain barrier permeability: activation of the coagulation cascade, cerebral edema, multifocal infarction, hemorrhages, and brain swelling. There are mural thrombi and obstructions to the vascular blood flow with resultant multifocal ischemia, cerebral edema, infarctions (neuronal injury and apoptosis), and loss of neurological functions with clinical deterioration.

2.2.4 Clinical Presentation and Differential Diagnosis[12]

(a) *Acute bacterial meningitis presents* initially with high-grade fever, headaches, neck stiffness, photophobia, nausea, and vomiting and later with confusion, disorientation, irritability, drowsiness, seizures, and focal neurological deficits or a coma. Elderly patients present with forgetfulness, confusion, disorientation, changes in personality, etc. *Immune-compromised patients* may present with mild fever, subtle neck stiffness, and mental change. *Chronic meningitis* (onset more than 3 weeks) may present with features of raised intracranial pressure (headaches, vomiting, papilledema, visual obscuration, and drowsiness) and disturbed higher mental faculties.

(b) *Clinical signs of meningitis* are usually neck stiffness, photophobia, Brudzinski's sign, and a positive Kernig's sign. Initially the CNS examination may otherwise be normal. About 10%–15% of patients have cranial nerve signs (usually second, sixth, and eight cranial nerves). The presence of papilledema suggests raised ICP, which may be due to hydrocephalus, subdural effusion/empyema, venous thrombosis, and cerebral abscess. Disturbed mental functions, altered sensorium, and appearance of focal neurological deficits herald the progression to complicated meningitis needing ICU management.

(c) In neonates, medical specialists look for alterations in vital parameters, feeding habits, sound production, bodily status, limb activities and social activities as well as signs of cerebral irritation, increased ICP (irritability, seizures, and bulging fontanelle), features of systemic derangements (pallor, jaundice, hypoglycemia, and metabolic acidosis), *some features of meningism* (photophobia, phonophobia, and nuchal rigidity with opisthotonus posture), constitutional symptoms (lethargy, anorexia, nausea, and vomiting) and signs of cerebral irritation (irritability, excessive crying, seizures, and bulging fontanelle). *Young children may even complain* of headaches, neck pains, photophobia, tiredness, and bodily pains. There may be papilledema. In meningococcal meningitis, apart from the meningeal signs, there may be petechial purpuric rashes, endotoxic shock, and intravascular coagulopathy with a poor prognosis.[13,14]

(d) *Gram-negative meningitis* is more serious and far more common in infants than in adults. The main pathogenic Gram-negative bacteria are *Acineto-bacter baumannii, E. coli, Enterobacter aerogenes, Klebsiella pneumoniae, and Pseudomonas aeroginosa*. The predisposing factors are prematurity, immune-compromised states, urogenital abnormalities and infections, spinal meningomyeloceles, shunt surgery, etc. These infants show rapid clinical deterioration in their level of consciousness with irritability, agitation, and seizures. Recurrent vomiting and fever with a toxic look may be present. The feeding is poor with tachycardia, tachypnea, and bulging fontanelles. These infants may have abnormal postures such as decerebrate, decorticate, and opisthotonos states.

(e) *In the acute stage of tubercular meningitis*,[11] the typical clinic-pathological features are noted with raised CSF proteins and reduced CSF absorption. Communicating hydrocephalus may develop with clinically raised ICP. The partially treated meningitis, subacute meningitis, and chronic tubercular meningitis, are all present with progressive raised intracranial pressure due to the blockage of the subarachnoid spaces by the inflammatory exudates. Tuberculosis produces arachnoiditis, tuberculomas, abscesses, vasculitis, and vascular occlusions. Various clinical syndromes due to tubercular vasculitis of the basilar artery and its branches are described in relation to the brain stem and the carotid arteries in the cerebral hemispheres. Brucellosis mainly produces granulomatous changes. *Treponema pallidum* is a slender, tightly coiled spirochete that is usually acquired through sexual contact; it may present as tertiary syphilis (meningoencephalitis) if effectively untreated.

2.2.5 Differential Diagnosis of Bacterial Meningitis

Differential diagnosis of bacterial meningitis will include other infective causes of meningitis (viral, fungal, and parasitic infections), vascular problems (intracranial subarachnoid hemorrhage, migraine), neoplastic lesions (carcinomatosis), medications (NSAIDs, allopurinol, azathioprine etc.), systemic disorders (vasculitis, sarcoidosis), trauma, etc. *Recurrent meningitis* occurs in patients with CSF fistulae, anatomical defects, persistence of primary foci, immune-compromised states, asplenia, etc.

2.2.6 Investigations[5–8]

(a) *It is imperative that the causative organism is isolated and identified on an urgent basis.*[7–10] Lumbar puncture and CSF analyses are therefore mandatory. If raised ICP is suspected, the CT brain scan is performed prior to the lumbar puncture. The CSF analysis usually shows a high-polymorphonuclear pleocytosis (cell count more than 500 cells/mm^3), with low glucose levels (less than 40% of the blood-sugar levels), and highly raised protein levels (more than 100 mg/dL) as well as bacterial organisms seen on microscopy or in the CSF cultures. The Gram stain is positive in the majority (80%–90%) of meningitis cases and in about 70% of cases due to the Gram-negative bacilli. If needed, bacterial antigens may be conclusively detected in the CSF studies, especially on the latex agglutination for bacterial antigens.

Lactic acid levels are higher (more than 35 mg/dL) in acute bacterial meningitis. A limulus lysate assay for endotoxins is positive in cases of meningitis due to Gram-negative bacilli. For the *tubercular bacilli* (acid fast bacilli), the ZN stain is positive more on spun CSF in tuberculous meningitis. In tubercular meningitis, the positive cultures are obtained in about 40%–70% cases.

(b) *In postneurosurgical infections*, the CSF glucose level (less than the 40% of the blood-sugar levels) is more important than the protein levels and cellular counts. Bursting of a brain abscess in the ventricular system will result in serious fulminant meningitis with the CSF cell counts in the range of 10,000–20,000 cells/Cumm. The CSF is sent periodically for routine analysis, bacterial cultures, and antibiotic sensitivity. In partially treated meningitis where bacterial cultures are negative, the agglutination tests for the detection of bacterial antigens in the bodily fluid (blood, CSF, urine, etc.) are performed to detect the causative organisms, e.g., *S. pneumoniae, H. influenzae-B, N. meningitidis, E. coli,* etc. There is great potential in PCR technology in detecting the causative organism in the partially treated meningitis cases. In tuberculosis, a definitive diagnosis is achieved following a culture on the Lowenstein Jensen medium, the Bactec radiometric system, the mycobacterial growth indicator tube, the luciferase reporter mycobacteriophage assays, the PCR, and the serological tests.

(c) *Tests to assess the general status of the patient*: Laboratory tests for hematological, biochemical, and coagulation parameters, a systemic infective screen including the blood cultures, and assessment of intake and output are performed. A culture may show bacteria involved, and then antibiotics can be instituted appropriately.[8–14]

(d) *Neuroimaging studies* (Figs. 1 and 2): Contrast-enhanced MRI is the investigation of choice for detection of meningitis and its complications. FLAIR and postcontrast

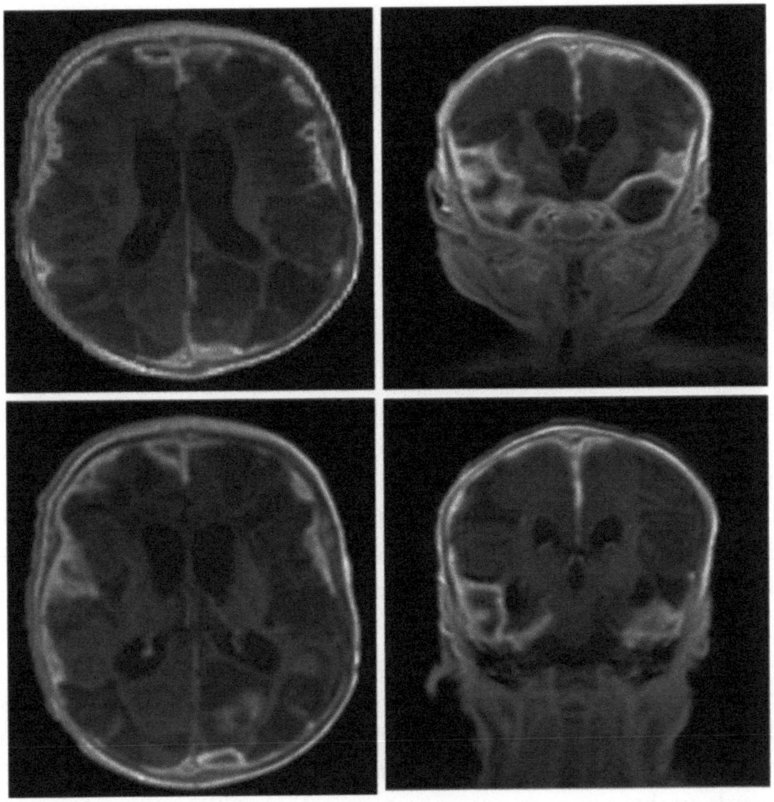

Fig. 1 Neonate with proven B streptococcal meningitis. The axial and coronal contrast enhanced MRI scans show bilateral, extensive meningeal enhancement.

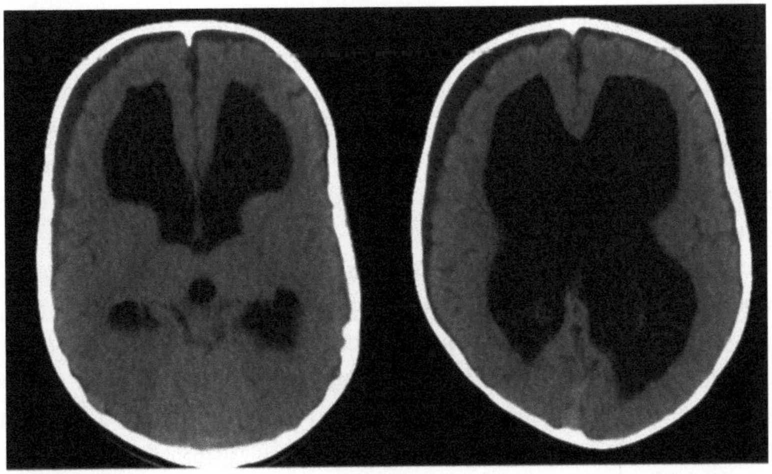

Fig. 2 Axial CT scan with complicated meningitis showing obstructive hydrocephalus and the right frontoparietal subdural collection.

T1W images are most useful and often provide complimentary information. Delayed-contrast enhanced FLAIR images are useful for subtle diagnosis. Diffusion-weighted images are invaluable in detecting the complications of meningitis. The CT is often helpful and a viable alternative to detect hydrocephalus and its effects. An abnormal meningeal enhancement can be seen on postcontrast CT/MRI scans (Fig. 1). The main role of imaging in meningitis is to detect the complications- subdural hygroma and empyema (Fig. 2), hydrocephalus, venous/arterial thrombosis, and ischemia/infarction, cerebritis, abscess, ventriculitis, and ependymitis.

2.2.7 Management[5–15]

Acute bacterial meningitis is treated on an emergency basis in the intensive care unit with full supportive therapy: strict bed rest, monitoring of vital parameters, adequate rehydration with maintenance of electrolytes, empirical antibiotics, antipyretics-analgesics, anticonvulsant therapy, airway management facilities at hand, control of raised ICP (mannitol 20%, hypertonic saline 3%, acetazolamide, dexamethasone, barbiturate coma), and immunoglobulin therapy for hypogammaglobulinemia.

There should be periodic monitoring of hematological, biochemical, and coagulation parameters; arterial blood gas analysis; infection screening; intake/output chart recording; and maintenance of ventilator parameters. Periodic recordings of the clinical assessments, CT/MRI brain scans, and sequential CSF analysis are performed to guide management.

Eradication of the bacterial infection with antimicrobial therapy: The choice of antibiotic therapy depends on three important parameters: its high-CSF penetration, effective bactericidal activity, and proven clinical efficacy.

(a) *The antibiotic therapy in suspected cases of meningitis prior to the CSF analysis*: The empirical antibiotic is chosen based on the predisposing factors in each age group of patients.

The age group ranging from the fourth month to 50 years: the antibiotic regime of choice is Vancomycin (60 mg/kg/day) with one of the cephalosporins (cefotaxime 200–300 mg/kg/day or ceftriaxone 100 mg/kg/day) in their high dosages. It is called a double antibiotic regime. Rifampicin is usually added only in the cases of *cephalosporin resistant pneumococci.*

The vulnerable age groups (newborns, neonates, and elderly patients) as well as immune-compromised patients all need a triple antibiotic regime. The best combination is just adding ampicillin (200 mg/kg/day) to the adult antibiotic regime of choice. *In neonates (age 0–4 weeks),* ampicillin with cefotaxime or an aminoglycoside is used. *In postoperative neurosurgical infections*, vancomycin plus Ceftazidime are commonly used with good results.

In suspected cases of tuberculosis, antitubercular medications (Pyrizinamide, isoniazid, rifampicin, and pyridoxine as a combination regime) are used for 4 months and then only two selected drugs for 18 months to 3 years. Sometimes ethambutol is added; rarely, streptomycin is given in resistant cases.

(b) *The antibiotic therapy after the positive CSF Gram staining but without identifying a specific bacterium*: Usually a double antibiotic policy is used in the interim period.

In the group of patients with cocci: (a) Gram-positive cocci are treated with vancomycin and a cephalosporin (cefotaxime or ceftriaxone); (b) Gram-negative cocci are treated with penicillin-G and if there is penicillin hypersensitivity or resistance, then ceftriaxone is used. In cases of Beta lactam (both penicillins and cephalosporins) hypersensitivity or resistance, antibiotics such as vancomycin, rifampicin (20 mg/kg/day for 4–5 days), carbapenem (imipenem and meropenem), and oxazolidinones (linezolid) are considered.

In the group of patients with bacilli, in general aminoglycoside is used effectively with one other broad-spectrum antibiotic (the double antibiotic policy): (a) *Gram-positive bacilli* are treated with ampicillin, and (b) *Gram-negative bacilli* are treated with cephalosporin (cefotaxime or ceftriaxone; ceftazidime if pseudomonas infection presents).

(c) *The antibiotic therapy following definitive identification of a specific bacterium on culture and sensitivity*: All antibiotics are given intravenously as per sensitivity to achieve adequate blood and CSF levels and to have bactericidal effect at the earliest.

2.2.8 Precautions to Prevent the Spread and Recurrence

Frequent hand washing, barrier nursing, and universal precautions are important. For prevention of meningitis due to *S. pneumoniae, H. influenzae, N. meningitidis,* and *M. tuberculosis* (BCG, or Bacillus Calmette-Guerin vaccination) etc., vaccinations are given. The vaccines are used against *H. influenzae B* (three doses in the first 6 months and the fourth at 12 or 18 months.); *N. meningitidis* (antibiotics to close contacts and vaccines to control outbreaks as well as vaccines to travelers about 1 week prior to their departure to areas where the meningococcal disease is prevalent, such as in African countries); *S. pneumoniae* (for persons prone to recurrent *S. pneumoniae* infections and elderly people); and *M. tuberculosis*. Meningitis cases are reported to the central registry in the local health authority as well as to the state health authority.

2.2.9 Prognosis: Morbidity and Mortality[5–15]

(a) *Prognosis depends on factors* such as age, immune status, type and virulence of the pathogen, dose of the inoculums, and clinical severity. The following facts must be kept in mind.

(b) *Fortunately, with early effective management,* many patients recover without any residual squeal/disability. The overall mortality in bacterial meningitis is about 10%–20% and the morbidity 15%–25%.

(c) *Usually*, the extremes of ages, clinically prolonged seizures, altered state of consciousness/coma, focal neurological deficits, leucopenia, immune compromised state, endotoxic shock, and disseminated intravascular coagulopathy (4%–6% cases), ventriculitis with hydrocephalus, diffuse cerebral edema-hemorrhage-raised intracranial pressure, and cerebral infarction are the factors predicting an overall poor prognosis.

(d) *Among the various age groups*, neonates and infants under the age of one have the highest mortality (10%–20%) and morbidity rates (20%–30%) with long-term disabilities in the survivors. The same is nearly true for the elderly. In adults, the mortality is about 7%–15% and morbidity 12%–15%.

(e) *Currently among the organisms responsible for bacterial meningitis*, *S. pneumoniae* is responsible for one of the highest combined mortality and morbidity rates (30%–40%); followed by *L. monocytogenes* (25%–30%), *N. meningitidis* (15%–20%), *H. influenzae* (5%–10%), etc. Tuberculous meningitis, despite appropriate antituberculous therapy, is associated with appreciable mortality and significant morbidity in surviving cases.

(f) *Transient or permanent morbidities* occur in the higher mental faculties, speech, cranial nerves, motor sensory functions in the limbs, and sphincter disturbances. These conditions are managed on their own individual merits.

2.3 Fungal Meningitis[1,16]

2.3.1 Introduction

In general, fungi are ubiquitous. Although some forms have a restricted geographical distribution. Fungi have low pathogenicity and therefore rarely infect normal subjects. Only, about 10%–15% of the pathogenic fungi produce clinically significant systemic and CNS mycosis. True pathogenic fungi are *Blastomyces, Coccidioides, Paracoccidioides, Histoplasma, Sporothrix*, etc. They produce clinical diseases and provide long-term immunity on recovery, which is highly unlikely with opportunistic fungal infections (*Aspergillus fumigatus, Candida albicans, Cryptococcus neoformans,* and *Rhizopus arrhizus*) in which relapses are common. The CNS fungal infections present with varied clinical syndromes such as meningitis, abscesses, granulomas, parenchymal cyst, hydrocephalus, etc. Among these, meningitis and hydrocephalus are the common ones.[16,17]

Fungi are broadly classified in three major groups: (1) Pseudomycetes/yeast: *Blastomyces, Candida, Coccidioides, Cryptococcus, Histoplasma, Paracoccidioides,* and *Sporotrichum*. (2) Septate mycetes: *Aspergillus, Cephalosporium, Cladosporium, Diplorhinotrichum, Homodendrum Paecilomyces, and Penicillium*. (3) Nonseptate mycetes: *Absidia, Basidiobolus, Cunninghamella, Mucor,* and *Rhizopus*.

2.3.2 Clinical Characteristics[1,16–18]

(a) *The predisposing risk factors* are malnutrition, immune-suppression, organ transplants, HIV-AIDS infection, systemic malignancy, Hodgkin's disease, collagen vascular diseases, uncontrolled diabetes mellitus, alcoholism, intravenous drug abuse, long-term antibiotic therapy, corticosteroid uses, and prolonged parenteral nutrition. In intensive care units, candidiasis and aspergillosis are frequently diagnosed nosocomial infections with high morbidities and mortality.

(b) *Etiopathogenesis*

The blood-brain barrier (BBB) and meninges provide effective barriers to fungal infections. In immune-compromised states, these barriers are easily overcome to produce clinical manifestations. Small pseudomycetes (*Cryptococcus, Blastomyces, Histoplasma, Coccidioidomycetes*, etc.) usually cause meningitis as they gain access to the meningeal microcirculation and infect the CSF, which may pass along the Virchow Robin spaces to the cerebral parenchyma to develop meningoencephalitis. Larger pseudomycetes (*Candida)* produce cerebral abscesses and granulomas. Septate (*aspergillus)* and nonseptate (*zygomycetes*) mycetes infect paranasal sinuses, orbits, and the oral cavity for long periods before invading contiguous cranial bones, meningeal tissues, basal cerebral venous sinuses, and contained intermediate and large arteries. They cause arterial thrombosis and occlusions with resultant extensive cerebral infarctions.

(c) *Clinical Presentations*[1,16,18,19]

Acute fungal meningitis is distinctly rare, while subacute meningitis is less common; the chronic fungal meningitis is a common presentation. Fungal meningitis frequently occurs with Pseudomycetes-*Cryptococcus, Coccidioides, Blastomyces, Paracoccidioides, Sporotrichum, Histoplasma,* and *Candida.* A good clinical assessment to delineate the predisposing factors and the site of the primary focus is of paramount importance.

Clinical features: The fungemia, meningitis, and meningoencephalitis usually present with headaches, nausea, vomiting, visual impairment, and papilledema, then later with neck stiffness, fever, and personality changes. Further worsening may occur with seizures, cranial nerve palsies, hydrocephalus, altered sensorium, and a coma. In many patients, there are no focal or general physical signs.

2.3.3 Differential Diagnosis

Fungal meningitis should be differentiated from other forms of meningitis (viral, bacterial, and parasitic infections), postvaccine states, vascular disorders, neoplastic lesions, systemic disorders, etc. Identifying the etiological agent as fungal and not bacterial is vital since antibacterial therapy is not effective against fungi and the fungal meningitis leads to high morbidity and mortality.

(i) *Cryptococcosis:* The inhalation of the airborne spores of *Cryptococcus neoformans* may result in asymptomatic or symptomatic pulmonary infections and later as subacute or chronic meningitis with insidious onset. The symptomatology includes headaches, vomiting, papilledema, seizures, altered sensorium, and cranial nerve deficits (visual impairment, double vision, facial weakness, and hearing loss). Nearly 10% of HIV patients develop cryptococcosis and in a significant number of patients, it manifests as an initial presentation of HIV infection. Cryptococcal meningitis may be complicated by cerebral abscess, cryptococcoma, and intraparenchymal basal ganglia cysts filled with slimy gelatinous material. The CSF analysis shows a typical picture of chronic meningitis; if HIV-AIDS coexists, then the pleocytosis may not be significant. The best method for diagnosis is the India ink preparation of the post-centrifuged CSF sediments, which shows double-refracting spherules. The latex cryptococcal agglutination test is positive for the capsular antigen in about 95% of patients. *Cryptococcus* is well stained with PAS and Gomorimethanamine silver stains. *In acute cryptococcal meningitis*, amphotericin-B and fluconazole achieve good results. Fluconazole is used as maintenance therapy for the prevention of the late relapses of cryptococcal meningitis, especially in AIDS patients after completion of primary therapy.

(ii) *Coccidioides immitis* is a virulent dimorphic fungus producing intense meningeal inflammation (coccidioidomycosis) with an accumulation of exudates, opacification of lepto-meninges, and obliteration of sulci with caseous granulomatous nodules in basal cisterns. Extensive fibrosis causes obstructive hydrocephalus. Invasion of the cranial nerves and basal blood vessels leads to the cranial neurovascular syndromes and, rarely, multiple cerebral aneurysms. The CSF studies usually show the typical features of chronic meningitis. If positive, then the fungi will be well demonstrated on all the routine staining preparations for the fungi: H and E stain, PAS, and Gomorimethanimine silver stains on the biopsy materials with its pathognomonic features of encapsulated spherules with endospores. However, immunological tests initially (up to 3 weeks) for IgM antibodies (latex particle agglutination, immunodiffusion methods) and later on (up to 8 months) the serological tests for compliment-fixing antibodies in the blood and the CSF are routine. The CT head and MRI brain scans demonstrate superolateral and basal granulomatous meningitis. *Coccidioidal meningitis* is treated with oral fluconazole and, in refractory cases, amphotericin-B or alternatively triazoles (voriconazole/posaconazole) are used. Once the patient is stable, then the fluconazole is given for a prolonged period (at least 6 months or more) to prevent relapse.

(iii) *Candidiasis* is a common cerebral mycosis, whether acquired through the community or from a nosocomial infection. Hematogenous spreads to CNS usually occur from the oropharyngo-esophageal regions and commonly produce meningitis and, less commonly, the cerebral abscess. Direct infection from the skull base lesions, intensive care procedures, and neurosurgical operations are also possible. The latter may be caused via colonization of the ventricular drains, shunt tubing, and central venous lines. The CSF study is essential to confirm candidial meningitis. Microscopic examination of the exudates in 10% KOH clearly show yeast-like fungi. The fungal stains show Gram-positive yeast in the CSF, which is more clearly visualized with PAS and Methanamine silver stains. The CNS candidiasis is treated with various medications singly or in combination: amphotericin-B, caspofungin, micafungin, voriconazole, fluconazole, and flucytosin. The drug therapy is continued for about 2 weeks following resolution of symptoms and a negative blood and CSF culture.

(iv) *Histoplasma capsulatum* is also a soil contaminant. About 20% of systemic histoplasmosis cases develop CNS histoplasmosis: meningitis and solitary or multiple cerebral granulomas. Typically, the meningeal inflammation in histoplasmosis excites fibroblastic proliferation with obliteration of the subarachnoid spaces, and vasculopathy with varied presentations. The staining of the CSF sediments with Wright or Giemsa stains shows the organisms clearly. Compliment fixing antibody tests are positive in 80% of patients and the CSF histoplasma polysaccharide antigen is positive in about 90% of patients. *Histoplasma meningitis* is best treated with amphotericin-B for a period of 3 months to start with; then it is better to evaluate the antibody and antigen titers with the progress of the clinical response.

(v) *Aspergillus and zygomycetes* initially colonize paranasal sinuses and the upper respiratory region. This is followed by local tissue and vascular invasion and then these organisms spread by direct extension across the tissues, bones, nerves, blood vessels, and cartilage. They form cerebral granulomas, abscesses, thrombophlebitis, arteritis, and, *rarely, meningitis*. Invasion of cerebral vessels results in cerebral infarction. The cranio-sino-orbital syndrome progresses to cavernous sinus syndrome and orbital apex syndrome. These fungi are basically diagnosed by tissue biopsies from paranasal sinuses. The septate and nonseptate hyphae are best demonstrated on Periodic Acid Schiff and Gomorimethanamine silver stains. The methods of radioimmunoassay for circulating antigens and the PCR for antigens in the CSF are helpful in the diagnosis. A serum immune-diffusion test for antibodies also can be done. Neuroimaging studies (CT/MRI with angiography) show sinusitis, subdural empyema, hydrocephalus, ischemia-infarction, vascular occlusions, filling defects, and cerebral aneurysms. *Very rarely do these fungi demonstrate meningitis*. There is high morbidity and mortality with these infections. Once the patient progresses to a coma, then the mortality is very high.

In CNS Aspergillosis, the high doses of amphotericin-B or its lipid preparations (in systemic medical disorders), voriconazole or caspofungin (if a drug allergy to amphotericin-B and Voriconazole), may be lifesaving. *In patients with cerebral mucormycosis* (Fig. 3), a prolonged course of amphotericin-B (1–1.5 mg/kg/day) or one of its lipid preparations is given. In some cases of allergy, posaconazole may be the next choice.

2.3.4 Management[1,16,18,20]

The management primarily depends on the antifungal drug therapy; in some cases of hydrocephalus, abscesses, and granulomas, neurosurgical intervention may be needed.

The principles of the management of the fungal infections of the CNS are as follows[21,22]:

(1) Identifying and controlling the predisposing risk factors and coexisting medical conditions,

(2) Early empirical broad-spectrum antifungal drug therapy started once the CNS fungal infection is strongly suspected. Amphotericin-B and its lipid formulations, caspofungin, micafungin, voriconazole, fluconazole, posaconazole, and flucytosin are broad-spectrum antifungal medications.[23]

(3) Establishing the diagnosis of a specific fungal infection for an appropriate antifungal drug therapy.

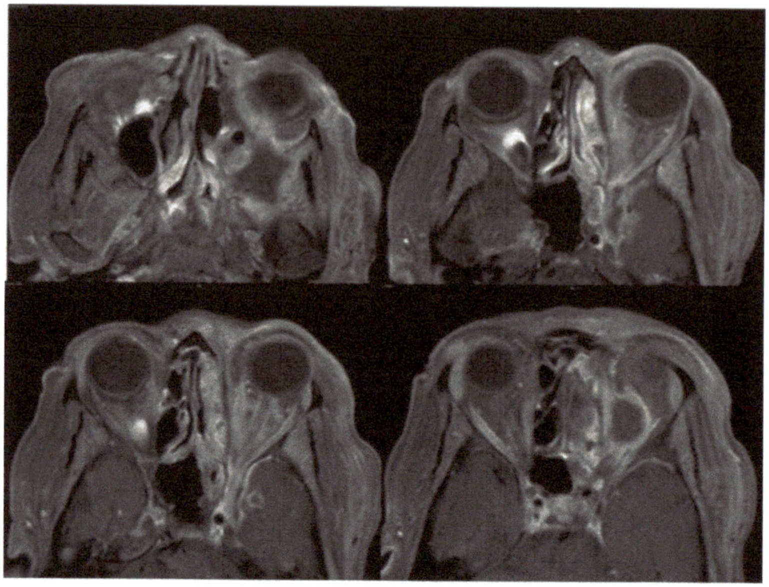

Fig. 3 Diabetic patient with left eye proptosis due to invasive rhinocerebral mucormycosis of ethmoid and sphenoid sinuses with extension in left orbit and periorbital soft tissue as well as in the left temporal lobe with left cavernous sinus thrombosis.

(4) Radical surgical procedures are undertaken to excise the fungal mass in the orbit and paranasal sinuses, and active neurosurgical interventions are performed for intracranial infective lesions or hydrocephalus. Excisions of necrotic infected tissue and the drainage of paranasal sinuses are needed in such patients.[24]

(5) Aggressive antifungal drug therapy including irrigation of aerated paranasal sinuses with antifungal agents is needed for a prolonged period (more than 3 months).

(6) Periodic clinical assessments, laboratory tests, and neuroimaging studies are needed for monitoring the progress and management.

The factors predicting the poor prognosis[23,24] are extremes of ages, organ failures, severity of clinical presentation, type and virulence of the fungal organism, complicated forms of meningitis, cerebral-vascular involvement, infarction, raised ICP, increased antigen titers in the blood and CSF, extensive extracranial mycosis, and coexisting medical disorders such as immune-compromising states.

2.4 Parasitic Meningitis[25,26]

2.4.1 Introduction

Parasitic (protozoal and helminthes) infections of the CNS are rare, causing focal (amoebic meningoencephalitis, cysticercosis), multifocal (toxoplasmosis), and generalized meningoencephalitis (*falciparum malaria*, trypnosomiasis). In the United States, about 6–8 cases of amoebic meningoencephalitis are noted annually, and about 200–300 cases have been recorded so far in the medical literature.

2.4.2 Clinical Etiopathologic Characteristics

(a) *Etiological factors* CNS parasitic infections are common in underdeveloped countries due to the low socioeconomic conditions, poor standard of hygiene, lack of medical facilities, lack of education, and unawareness of preventive measures. In developed countries, factors such as lymphoma, immunosuppression, and diabetes mellitus are frequently responsible. These infections are transmitted and spread via vectors, excretions, contamination of food, and direct contact.

Clinically significant parasites causing meningitis or meningoencephalitis are few, i.e., *Amoebae naegleriafowleri, Acanth amoebata, Malaria, Trypnosomes, Trichinella spiralis, Angiostrongylus cantonensis,* and *Taenia solium.*

(b) *The pathogenesis*: Both the protozoal and helminthetic parasites produce clinicopathological changes due to inflammatory reaction, allergic reaction, foreign body reaction, etc.

Among the protozoans, *Amoebae naegleria-fowleri* and *Acanth-amoebata* are ubiquitous in the soil and warm, stagnant fresh water. When infected water is insufflated in the nose by young swimmers, these organisms migrate through the olfactory nerves to reach intracranial cavity; this causes necrosis of brain tissue very rapidly. Pathologically, widespread necrosis involves the olfactory lobes (parosmia, anosmia, agusea), and frontotemporal lobes. Petechial hemorrhages with *amoeba* in the perivascular spaces and purulent meningeal inflammation with exudates are common with the necrotizing granulomatous meningeal reaction (*Acanth-amoebata*). The CSF examination is positive for the pleocytosis and amoeba. In a majority of helminthes, their larva reaches the meninges via the hematogenous route and produces eosinophilic meningitis with variable meningeal fibrosis, cerebral granulomas, and tissue necrosis.

2.4.3 Clinical Presentations and Their Management[25–27]

(a) *Primary amoebic meningoencephalitis (PAM)*[25]: The main clinical manifestations are headaches, nausea or vomiting, neck stiffness, and an altered state of consciousness. The patients develop repetitive seizures, status epilepticus, anoxia, and an irreversible coma. The brainstem damage leads to respiratory failure and mortality within 2 weeks of infection. The neuroimaging studies show nonspecific meningoencephalitic changes, initially in the medial frontotemporal regions and then in other parts of the brain. The CSF studies, including the microbial cultures (Nelson's medium), are the main basis of the early diagnosis of PAM. *Amoeba* are better seen in a centrifuged CSF deposit stained with a Giemsa stain or Gomaris' trichrome stain after adding a few drops of polyvinyl alcohol. Early medical therapy (including amphotericin-B, rifampicin, azithromycin, azoles, etc.) for the PAM is the only chance (3%–4%) of survival in this highly aggressive and mostly lethal condition.

(b) *American trypnosomiasis* is caused by *Trypanosoma cruzi.*[26] *Trypanosomes* infect meningeal cells and glial cells to cause meningeal inflammation and glial nodules. Acute meningoencephalitis initially presents with fever, headaches, nausea, vomiting, neck stiffness, a confused state, agitation, and recurrent seizures. These lead to altered mental functions and somnolence followed by coma and high mortality. In many survivors, dementia, spastic paralysis, and peripheral neuropathy are commonly present. Blood films prepared during periods of high fever may show trypanosomes. Blood cultures on an NNN medium as well as the Xeno-diagnosis are more sensitive tests for the diagnosis of trypanosomiasis. Serodiagnostics such as indirect fluorescent antibody, CFT, and indirect hemagglutination are also helpful in chronic stages. There is no universally effective chemotherapy available in Trypnosomiasis. In acute cases, Nifurtimox (a nitrofurazone derivative, 8–10 mg/kg/day for 3 months), or benzonidazole (anitroimidazole, 6–8 mg/kg/day for 1 month) are used.

(c) *African trypanosomiasis*[25,26] is caused by *T. brucei-gambiense* and *T. brucei-rhodesiense.* Initially, patients develop lethargy, lassitude, apathy, and depression for months, followed by somnolence and dementia. Some patients develop generalized seizures and extrapyramidal features (tremors, choreiform movements, rigidity) as well as speech and cerebellar disturbances. This may progress to stupor, incontinence, and a coma. This condition is invariably fatal in untreated cases. The diagnosis is established by the presence of trypanosomes in the blood or tissue fluids, including CSF. Melarsoprol (Mel B), Dimercaprol, dexamethasone, and Suramin are the medications used in the medical management.

(d) *Angiostrongylus cantonensis*[25–27] is a neurotropic helminth. Its larvae migrate through the meninges and the brain substances, producing eosinophilic meningitis and cerebral granulomas. The patients develop low-grade fever, nausea, vomiting, severe bitemporal and fronto-occipital headaches, migrating painful paresthesia or dysaesthesia, and neck stiffness. Neuroimaging studies may be normal and the CSF studies via lumbar puncture may show higher CSF pressure, cloudy CSF, and increased proteins but normal glucose levels and pleocytosis with eosinophilia. The management consists of symptomatic treatment, antihelminthic medications (a course of albendazole or mebendazole for 2 weeks), steroid therapy (a course of prednisolone for 2 weeks) and relief of the raised CSF pressure (periodic lumbar punctures).

(e) *Cysticercosis,*[26,27] caused by the *Taenia solium,* is the most common helminthic parasite infecting the CNS. The formation of cysts in various tissues by the larvae of the *Taenia solium* is known as cysticercosis. There are many clinical presentations, including raised ICP, hydrocephalus, seizures, cerebral cysts, strokes, and meningoencephalitis due to allergic-immunological responses. These patients present with features of seizures (60%–70% of cases), focal neurological deficits (20%–25% of cases), and raised ICP (10%–20% of cases).

There is peripheral as well as central (CSF) eosinophilia. The plain X-rays of the muscular tissues in the limbs or body show multiple centimeter-long cysts that are calcified, oval, and oat-shaped. The CT scan of the head and MRI scan of the brain show characteristic features of the cystic or granulomatous lesions as well as hydrocephalus. Immunological tests with variable success are utilized such as indirect hemagglutination, complement fixation, agar gel precipitation, immune-electrophoresis, counter electrophoresis, indirect immune-fluorescence, ELISA, and PCR. Management is basically comprised of antihelminthics (a course of mebendazole, 400–600 mg/day for 4–6 weeks or a course of Praziquantal), anticonvulsants, antiedema measures (prednisolone) where indicated, and surgical treatment in the form of CSF diversion procedures and microsurgical excision of the intracranial or intraspinal cysts with an overall good prognosis. The overall mortality is about 8%–10% of the cases.

(f) *Strongyloidiasis*[26,27] is caused by the *Strongyloides stercoralis* infection in the small intestines, pulmonary system, and bloodstream dissemination. It is ubiquitous in distribution with about 800 million people estimated to be infected the world over. It is endemic in the Southeastern United States and tropical parts of the world. The larval invasion of the bloodstream leads to peritonitis, severe pneumonitis, and meningoencephalitis. Rapidly developing raised intracranial pressure with headaches, vomiting, and neck rigidity progresses to a deterioration in the level of consciousness, coma, and fatality. The worst scenario is the frequent association of the Gram-negative sepsis with this parasitic infection, which leads to very high mortality. The treatment of choice is supportive intensive care management with medical therapy including thiabendazole, 25 mg/kg every 12 h for 2 days, and antibacterial antibiotics for Gram-negative sepsis.

2.4.4 The Prognosis

The prognosis in the protozoal and helminthic infections is variable, being worse in protozoal as compared to the helminthic infections. The overall prognosis is better than bacterial and fungal meningitis. Many times, no diagnosis is made on clinical grounds and autopsy studies are helpful in understanding the severity of the infections, especially in PAM.

2.5 Chemical Meningitis[28,29]

2.5.1 Introduction

Chemical meningitis is reported mainly as anecdotal cases or as reports of a small series of patients. Meningeal inflammation due to chemical substances is called chemical meningitis. Chronic chemical inflammation excites fibrosis and, therefore, results in obliteration of the subarachnoid spaces. This is termed variously as adhesive meningitis, obliterative leptomeningitis, plastic arachnoiditis, or chemical arachnoiditis.

2.5.2 Etiopathogenesis

(a) *The medications* causing chemical meningitis are analgesics (NSAIDs), antibiotics, blood products, cytarabine, immunoglobulins, vaccines, baclofen, methotrexate, methylprednisolone, hydrocortisone, vincristine, radiological contrast mediums, anesthetic medications, preservatives, etc. Among the NSAIDs, ibuprofen results in the maximum number of cases with chemical meningitis, especially in cases of lupus erythematosus and rheumatoid arthritis.

(b) *Intracranial cystic mass lesions* can result in chemical meningitis due to the spillage of their contents in the subarachnoid spaces: cystic craniopharyngiomas, epidermoids, dermoid cysts, pineal cysts, colloid cysts of the third ventricle, etc. Their proteinaceous material excites a strong inflammatory response that proceeds to severe reactive CSF changes. The possible pathological mechanisms may be: (a) direct meningeal irritation/inflammatory response against the medications or chemicals, (b) immunological response to the medications, (c) allergic responses to the preservatives, and (d) chronic infections where the organisms or their components are not identified. Idiosyncratic immunologic meningitis may result especially from NSAIDs and intravenous immunoglobulins.

2.5.3 Clinical Presentations

The patients usually develop generalized headaches, mild fever, nausea, marginal neck stiffness, photophobia, and, rarely, recurrent seizures. The neuroimaging studies are normal in the initial stages but in the late stages there are typical features of arachnoiditis on the CT scans and MRI scans of the brain. The CSF studies show consistently negative culture studies; however, white cell counts are slightly higher, as are the protein estimations, but the CSF glucose may be slightly low.

2.5.4 Management and Prognosis

Chemical meningitis due to drug ingestion usually subsides after the withdrawal of the medications in a few hours to a few days without significant squeal in most patients. However, if it is due to the intrathecal injection of the medications, then its severity will depend upon the concentration of the medication, lipid solubility, drug particle size, duration of the contact with the leptomeninges, etc. Every case of chemical meningitis is treated on its own merit with the treatment strategy most suitable for the type of chemical meningitis the patient has. The management of cystic intracranial mass lesions is done by microsurgical excision of these lesions to prevent spillage in the subarachnoid spaces. These patients usually recover well with good supportive treatment.

The principles of management of chemical meningitis remain as follows: symptomatic supportive therapy including periodic CSF assessments, analgesics, antipyretics, fluid and electrolyte supplements, anticonvulsants where indicated, control of raised ICP (carbonic anhydrase inhibitor-Diamox, mannitol, etc.), checking super infections, and effectively dealing with the causative factors. The overall prognosis in cases of chemical meningitis is good to excellent if the condition is recognized early and effectively treated in time.

2.6 Neoplastic/Carcinomatous Meningitis[30,31]

2.6.1 Introduction

Neoplastic or carcinomatous meningitis is an extremely rare entity. It is more frequently (4%–6% of all cancer patients) recognized and managed to provide palliative therapy. It is usually caused by carcinomatous metastatic meningeal deposits from extracranial primary neoplastic sites (lungs, breast, bowel, bone marrow, and sarcomas) and rarely from intracranial sites such as cerebellar medulloblastoma, lateral ventricular choroid plexus carcinoma, malignant ependymomas, etc.

2.6.2 General Characteristics

The metastatic lesions from the malignancies primarily originating outside the cranial and spinal regions commonly reach the meninges by the hematogenous route, less commonly, the cells from the primary brain tumors in the ventricles (choroids plexus carcinomas) near the ventricular surface (ependymomas, lymphomas) or meningeal surface (astrocytomas, medulloblastomas) invade subarachnoid spaces directly to cause neoplastic meningitis.

The main etiological conditions are solid adenocarcinomas (5%–10% of cases of the breast, lungs, and stomach cancers), melanomas (20% of patients of extracranial melanomas), hematological malignancies (10%–20% of cases of lymphoblastic leukemia and lymphomas), and the primary brain tumors (choroid plexus carcinomas, malignant ependymomas, medulloblastomas and pituitary adenocarcinoma) are more frequently implicated.

Macroscopically, the metastatic nodules are carpeting the meningeal surfaces, especially in the basal CSF cisterns, and acquiring their blood supply. These lesions break the blood-brain barrier and blood-CSF barrier, thereby spreading more rapidly. These malignant cells usually extend in the perivascular spaces and then the pial invasion leads to an intraparenchymal spread. There are beading appearances of the cerebral vessels and, consequently, decreased regional blood flow with ischemia. This results in vascular occlusion, thrombosis, and infarctions along with the multifocal spread in the brain.

2.6.3 Clinical Symptomatology[30–32]

Usually the clinical picture is that of insidious onset with a chronic course, but in a significant number of cases, acute presentations result due to focal radiculopathy, neurological deficits, stroke, and neuropathy. A majority of patients present initially with progressive generalized headaches, nausea, recurrent vomiting, cervicooccipital or frontoorbital headaches, seizures, focal neurological deficits, and cranial nerve palsies (ocular nerve palsies especially the abducent palsy, trigeminal sensory-motor disturbances, hearing impairment, and visual loss). These are followed by general constitutional symptoms such as malaise, lassitude, generalized myalgia, confusion, memory disturbances, etc. The spinal carcinomatous meningitis usually presents with the features of radiculopathy, cauda equina syndrome, and myelopathy.

2.6.4 Diagnosis and Differential Diagnosis[32]

The CT head is usually followed by the lumbar CSF studies. In about 50%–60% of cases, the CSF pressure is elevated with raised proteins and slightly low sugar levels. On the cytological studies, the malignant cells are larger in size with larger, denser nuclei as compared to normal lymphocytes.

The electron microscopic studies and special immunohistochemistry using monoclonal antibodies help in better characterizing the malignant cells. These monoclonal antibodies are basically markers for leucocytes, epithelial cells, neuroblasts, and neuroectodermal tissues. These studies can very well differentiate and categorize these lesions such as carcinomas, lymphomas, and neuroectodermal tumors. Neuroimaging studies (such as CT and MRI scanning (Fig. 4)) are of prime importance in precisely defining the intracranial lesions (subarachnoid carpeting and nodules, intraparenchymal deposits, and hydrocephalus).

2.6.5 Management and Prognosis[30–32]

The management of carcinomatous meningitis comprises chemotherapy, radiotherapy, and surgical therapies. Neurosurgical procedures are primarily the insertion of a ventriculoperitoneal shunt for the hydrocephalus, implantation of an Ommayya reservoir with intraventricular catheter for the delivery of the chemotherapeutic medications, and periodic lumbar punctures. Chemotherapy remains the best modality of treatment where intrathecal drug therapy (via the transventricular route or translumbar route) is used. Among the chemotherapeutic agents (thio-TEPA, arabinoside, cytosine, methotrexate, etc.), the methotrexate (intraventricular via reservoir) is frequently used with success in about 50% of cases. Systemic chemotherapy may be needed either in addition or as a replacement for the intrathecal drug therapy. A course of radiotherapy (2400 rads) over 2 weeks is combined with chemotherapy for greater success, especially in the secondary from breast cancer. The clinical progress is mainly monitored with the CSF cytological studies. The usual survival time in patients with untreated neoplastic meningitis ranges from 6 weeks to 3 or 4 months occasionally.

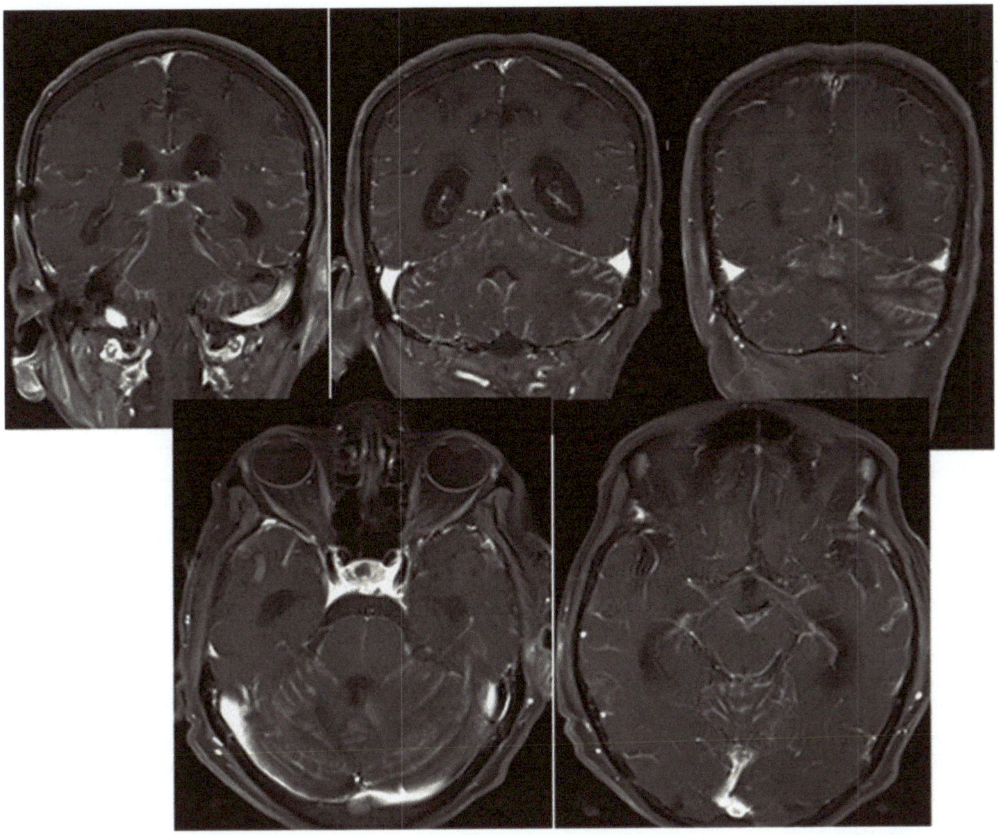

Fig. 4 A 52-year-old female with recurrent headaches having breast cancer and positive CSF findings of carcinomatosis. Postcontrast FAT SAT T1W coronal and axial images reveal leptomeningeal enhancement.

3. CONCLUSIONS[1–4,10,11,16,25,27,28,32]

Meningitis is a serious clinical condition with high morbidity and mortality. Therefore, each case needs to be thoroughly clinically assessed with a firmly established diagnosis. The management must be tailored as per the needs of the patient on a case-to-case basis. Meningitis and meningoencephalitis are caused by a variety of etiological factors within differing demographical situations. Protean, diverse, and perplexing clinical presentations pose special challenges in management. Clinicians must differentiate between a mild form of meningitis (mild aseptic viral meningitis) needing outpatient-based treatment and serious ones (acute Gram-negative bacterial meningitis) needing intensive care management.

Diagnostic armamentarium comprises blood and the CSF analysis and neuroimaging studies as well as microbiological and immunological tests. Medical therapies, intensive care management where indicated, and various surgical procedures form the cornerstone of their management. Timely management significantly improves morbidity and mortality associated with these high-risk conditions. These preventive measures are certainly having far-reaching implications and cannot be overemphasized.

REFERENCES

1. Sharma RR, Pawar SJ, Lad SD, Mishra GP, Netalkar AS, Rege S. Fungal infection of the central nervous system. In: Quinones-Hinojosa A, editor. *Schmidek & sweet: operative neurosurgical techniques. Indications, methods, and results*. 6th ed. vol. 2. Philadelphia, PA: WB Saunder Elsevier Inc.; 2012. p. 1691–732.
2. Canada Communicable Disease Report. *International note: global advisory committee on vaccine safety*. Public Health Agency of Canada Reports & Publications, 11–12 June; 2003.
3. Desmond RA, Accortt NA, Talley L, Vilano SA, Soong S, Whitley RJ. Enteroviral meningitis: natural history and outcome of plecoraril therapy. *Antimicrob Agents Chemother* 2006;**50**(7):2409–14.
4. Fleisher GR. Infectious disease emergencies. In: Fleisher GR, Ludwig S, Henretig FM, editors. *Textbook of pediatric emergency medicine*. 5th ed. The Netherlands: Walters Kluwer, Alphen aan de Rijn; 2006. p. 783.
5. Swartz MN. Bacterial meningitis-a view of the past 90 years. *N Engl J Med* 2004;**351**(18):1826–8.
6. Feigin RD, Cutrer WB. Bacterial meningitis beyond the neonatal period. In: Feigin RD, Cherry JD, Demmler-Harrison GJ, Kaplan SL, editors. *Textbook of pediatric infectious diseases*. 6th ed. Philadelphia, PA: Saunders; 2009. p. 439.
7. Ericson L, DeWals P. Complications and sequalae of meningococcal disease in Quebec, Canada 1990–1994. *Clin Infect Dis* 1998;**26**:1159–64.
8. American Academy of Pediatrics. Haemophilus influenzae infections. In: Pickering LK, editor. *Red book: 2009 report of the committee on infectious diseases*. 28th ed. Elk Grove Village, IL: American Academy of Pediatrics; 2009. p. 455.
9. American Academy of Pediatrics. Pneumococcal infections. In: Pickering LK, editor. *Red book: 2009, report of the committee on the infectious diseases*. 28th ed. Elk Grove Village, IL: American Academy of Pediatrics; 2009. p. 524.
10. Bridy-Pappas AE, Margolis MB, Center KJ, Isaacman DJ. Streptococcus pneumoniae: description of the pathogen, disease epidemiology, treatment, and prevention. *Pharmacotherapy* 2005;**25**(9):1193–212.
11. Shaw JE, Pasipanodya JG, Gumbo T. Meningeal tuberculosis: high long term mortality despite standard therapy. *Medicine (Baltimore)* 2010;**89**(3):189–95.
12. Kaplan SL, Fishman MA. Supportive therapy for bacterial meningitis. *Pediatr Infect Dis J* 1987;**6**(7):670–7.
13. Pomeroy SL, Holmes SJ, Dodge PR. Seizures and other neurological sequalae of bacterial meningitis in children. *N Engl J Med* 1990;**323**(24):1651–7.
14. Yogev R, Guzman-Cottril J. Bacterial meningitis in children: critical review of current concepts. *Drugs* 2005;**65**(8):1097–112.
15. Baraff LJ, Lee SI, Schriger DL. Outcomes of bacterial meningitis in children: a meta-analysis. *Pediatr Infect Dis J* 1993;**12**:389.
16. Sharma RR. Fungal infections of the nervous system; current perspective and controversies in management. *Int J Surg* 2010;**8**(8):591–601.
17. Bazan C, Rinaldi MG, Rauch RR, Jinkins R. Fungal infections of the brain. *Neuroimaging Clin N Am* 1991;**1**:57–88.
18. Bennett JE. Introduction to mycoses. In: Mandell GL, Bennett JE, Dolin R, editors. *Mandell, Douglas, and Bennett's principles and practice of infectious diseases*. 6th ed. Amsterdam, The Netherlands: Elsevier Churchill Livingstone; 2005. p. 2935–8.
19. Sharma RR, Lad SD, Desai AP, Lynch PG. Surgical management of fungal infections of the nervous system. In: Schmidek HH, editor. *Schmidek & Sweet's operative neurosurgical techniques, indications, Methods and Results*. 4th ed. Philadelphia, PA, USA: W.B. Saunders Company; 2000. p. 726–55.

20. Edwards JE. Candida Species. In: Mandell GL, Bennett JE, Dolin R, editors. *Mandell, Douglas, and Bennett's principles and practice of infectious diseases*. 6th ed. Amsterdam, The Netherlands: Elsevier Churchill Livingstone; 2005. p. 2939–57.
21. Sharma RR, Pawar SJ, Ravi RR, Sausa J, Devadas RV, Athale S. A solitary primary aspergillus brain abscess in an immuno-competent host: CT guided stereotaxy with an excellent outcome. *Pan Arab J Neurosurg* 2002;**6**:62–5.
22. Sharma RR, Pawar SJ, Delmendo A, Lad SD. Fatal rhino-orbito-cerebral mucormycosis in an apparently normal host: case report and literature review. *J Clin Neurosci* 2001;**8**(6):583.
23. Sharma RR, Gurusinghe NT, Lynch PG. Cerebral infarction due to aspergillus arteritis following glioma surgery. *Br J Neurosurg* 1992;**6**:485–90.
24. Dismukes WE. Guidelines from the infectious diseases society of America. Introduction to antifungal drugs. *Clin Infect Dis* 2000;**30**:653–7.
25. Cervantes-Sandoval I, Serranno-Luna Jde J, Garcia-Latorre E, Tsutsumi V, Shibayama M. Characterization of the brain inflammation during primary amoebic meningoencephalitis. *Parasitol Int* 2008;**57** (3):307–13.
26. Scheld WM, Whitley RJ, Marra CM, Huang DB, Schantz Pewit CA. Helminthic infections. In: Scheld WM, Whitley RJ, Marra CM, editors. *Infections of the central nervous system*. 3rd ed. Philadelphia, PA: Lippincott Williams &Wilkins; 2004. p. 797–828.
27. Escobedo F, Garcia-Ramos G, Sotelo J. Parasitic disorders and epilepsy. In: Nistico G, Di Perri R, Meinardi H, editors. *Epilepsy: an update on research and therapy*. New York: Alan R Liss; 1983. p. 227–33.
28. Forgacs P, Geyer CA, Freidberg SR. Characterization of chemical meningitis after neurological surgery. *Clin Infect Dis* 2001;**32**(2):179–85.
29. Jolles S, Sewell WA, Leighton C. Drug induced aseptic meningitis: diagnosis and management. *Drug Saf* 2000;**22**(3):215–26.
30. Gleissner B, Chamberlain MC. Neoplastic meningitis. *Lancet Neurol* 2006;**5**(5):443–52.
31. Zachariah B, Zachariah SB, Varghese R, Balducci L. Carcinomatous meningitis: clinical manifestations and managements. *Int J Clin Pharmacol Ther* 1995;**33**(1):7–12.
32. Jimenez Mateos A, Cabrera Naranjo F, Gonzalez Hernandez A, Fabre Pi O, Diaz Nicolas S, Lopez Fernandez JC. Neoplastic meningitis. Review of a clinical series. *Neurologia* 2011;**26**(4):227–32.

CHAPTER 2

Pathogenesis, Diagnosis, and Treatment of Central Nervous System Catheter Infections

Gwenn L. Skar, Jessica N. Snowden
University of Nebraska Medical Center, Omaha, NE, United States

1. INTRODUCTION

Advances in our ability to care for premature and other medically complicated infants have resulted in many lives saved. However, the medical devices used to treat these infants may become infected, requiring additional procedures and prolonged treatment. This occurs frequently in the treatment of hydrocephalus, where life-saving cerebrospinal fluid (CSF) shunts can become infected. Hydrocephalus is the pathologic accumulation of CSF resulting in increased pressure in the brain. CSF is produced in the ventricles of the brain and can accumulate due to overproduction, decreased absorption, or decreased circulation to other brain compartments. The accumulation of CSF creates an increase in the pressure of the brain parenchyma leading to decreased blood flow and altered metabolism in the compressed tissue.[1] There are many etiologies for hydrocephalus, including inherited genetic disorders, injury, infection, intraventricular hemorrhage, or tumors.[2] If left untreated, hydrocephalus will cause cognitive impairment and may even lead to death.[2]

CSF shunts were introduced in the 1950s as treatment for hydrocephalus; they remain the mainstay of treatment today.[3] These devices allow excess CSF to be drained to other body cavities where it may be reabsorbed. The typical components of a shunt are a proximal catheter, a reservoir, a valve to regulate pressure, and a distal catheter.[4] The proximal catheter is inserted into the lateral ventricle of the brain and connected to the reservoir and valve. The distal catheter is then tunneled subcutaneously and inserted into the right atria of the heart, pleural space, or peritoneal space. Ventriculoperitoneal shunts are the most common as the peritoneal space is typically the favored site of insertion.[3,5] While shunts successfully treat thousands of patients each year, they are associated with complications. Most complications can either be classified as malfunctions or infections. The rate of infection in the literature ranges between 1.6% and 33% depending on the definition of infection applied, the study population, and the study design.[5–7] Infection is typically defined as a positive CSF or catheter culture.[8] Additionally, shunts may be considered

The Microbiology of Central Nervous System Infections
https://doi.org/10.1016/B978-0-12-813806-9.00002-0

infected if there is a positive Gram stain from CSF fluid, changes in CSF parameters indicative of infection, or supporting clinical signs and symptoms.[9–11] Unfortunately, diagnosis is frequently delayed or debated due to pretreatment with antibiotics, the influence of postsurgical inflammation, the difficulty of culturing biofilm-predominant infections, and the attenuated inflammatory responses observed in biofilm infections.[12] Identifying more sensitive and specific diagnostic approaches to identifying shunt infection will decrease diagnostic uncertainty, prevent delayed treatment, improve outcomes, and decrease the cost associated with shunt infections.

2. HOSTS AT RISK OF INFECTION

Over the last 60 years, several host factors have been identified that increase the risk of shunt infection. Prematurity and younger age have both been associated with an increased risk of shunt infection.[13,14] This is likely due to the immature skin barrier of preterm infants, which increases the potential for introduction of the bacteria colonizing the skin as well as the immune systems of infants, which are thought to be skewed to be less inflammatory in response to infection.[15] The cause of hydrocephalus may also affect the risk of shunt infection. Those that have hydrocephalus secondary to intraventricular or other hemorrhages, meningitis or other infections, or myelomeningocele are at a higher risk of infection.[4,16] Additionally, CSF leakage and previous EVD placement are risk factors for infection.[4,13] Finally, those with previous shunt infections are at a greatly increased risk of future infections.[17,18] The mechanisms linking the etiology of hydrocephalus with risk of later shunt infection are unknown, but likely reflect a combination of factors including urgency and technical difficulty of the shunt placement and immune skewing of the patient in response to the preceding infection or hemorrhage.

3. MICROBIOLOGY

Many bacterial and fungal species have been reported to cause shunt infections. When thinking about the microbiology of CSF shunt infections, it can be helpful to divide the pathogens into groups: those causing early infection and those causing late infection. Early infection occurs in the first 30 days after shunt placement or revision.[7,9,19] Pathogens causing early infection include Coagulase negative Staphylococci (CoNS), *Staphylococcus aureus*, *Proprionibacterium acnes*, and *Corynebacterium* species.[7,9,15,19] The organisms are typically thought to be introduced at the time of surgery with CoNS, such as *Staphylococcus epidermidis*, causing ~50% of all shunt infections, and *S. aureus,* causing another 25% of all shunt infections.[7,9,20] Late infection occurs several months after shunt placement or revision.[7,9,19] The route of introduction of infecting organisms in late infection in unknown; however, environmental contamination at the time of shunt placement, hematogenous spread, and contiguous infection from wounds or perforated intestines are proposed mechanisms.[21] Other organisms causing later infections include: Enterococci, Gram-negative bacteria, and Candida species.[7,9,19] Recognizing the diversity of organisms associated with shunt infections is important in guiding empiric antibiotic therapy.

3.1 Biofilm

Shunt infections are caused by biofilms forming on the silicone catheter. Biofilms are organized communities of bacteria attached to surfaces that are encased in an extracellular matrix that confer functional bacterial resistance to antibiotics and evasion of host defenses[22,23] (Fig. 1). Of bacteria implicated in shunt infections, *S. epidermidis*, *S. aureus*, *P. acnes*, enterococcus, *Candida* species, *Pseudomonas aeruginosa*, and other Gram-negative organisms are all capable of forming biofilms.[22,24] Understanding the pathophysiology of biofilm formation on an indwelling medical device is crucial for the management of shunt infections and other device-associated infections. The formation of a biofilm is thought of as a three-step process: attachment, proliferation, and detachment[25] (Fig. 2). After attaching to the shunt tubing, bacteria begin to proliferate and mature, leading to the three-dimensional structure of bacteria within an extracellular matrix.[23] Once a mature biofilm is formed, bacterial cells may detach and travel to other sites, leading to severe infection or sepsis.[23] When bacteria have dispersed from the biofilm, they are considered planktonic bacteria and become easier to identify on traditional bacterial cultures.

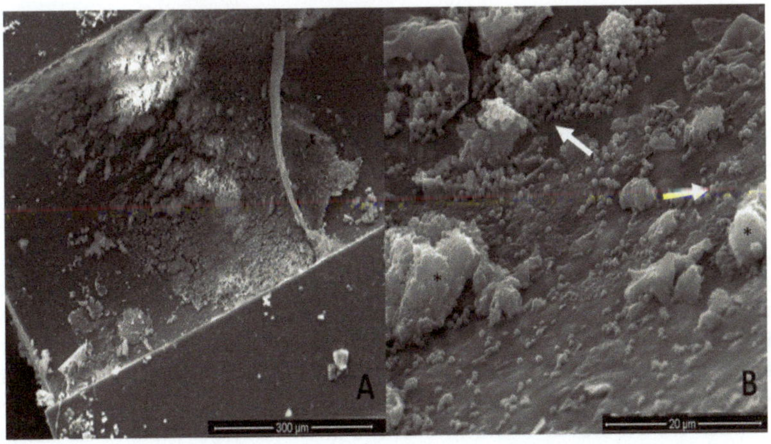

Fig. 1 Visualization of *S. aureus* biofilm growth on explanted CNS catheters. Catheters were removed 10 days postinfection in a mouse model of CNS catheter infection and visualized with SEM. While some biofilm disruption occurred following catheter removal and processing, a large portion of the catheter surface is covered with bacteria (A: 200× magnification). Higher magnification of this area shows bacteria adhering to the catheter surface *(white arrows)*, as well as white blood cells (*) (B: 3000× magnification). *(From Snowden JN, Beaver M, Smeltzer MS, Kielian T. Biofilm-infected intracerebroventricular shunts elicit inflammation within the central nervous system.* Infect Immun *2012;**80**(9):3206–14. https://doi.org/10.1128/IAI.00645-12. Copyright © American Society for Microbiology.)*

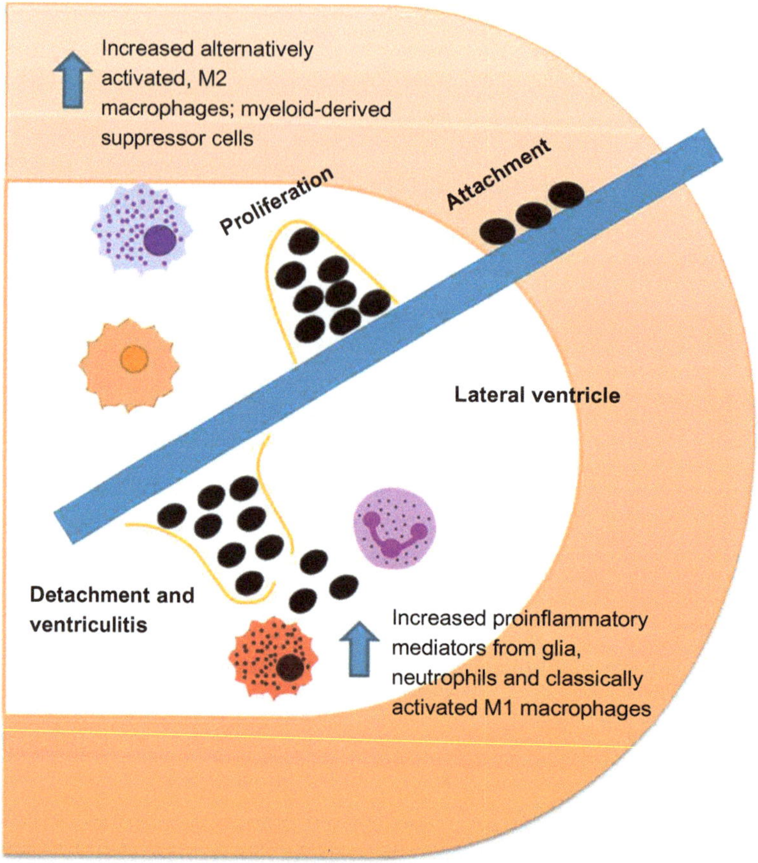

Fig. 2 Overview of biofilm pathogenesis, immune responses, and prevention/treatment options. Biofilm development occurs in three stages with attachment followed by proliferation and then dispersal. Prevention strategies, such as antimicrobial and antibiofilm materials, decrease initial attachment. The proliferation phase is marked by a skewed immune response with decreased inflammation. Strategies currently under investigation for treatment of proliferating biofilm include dispersal agents, classically activated macrophages, and lock therapy with antibiotic or other antibiofilm agents. Patients typically become symptomatic during the dispersal phase with release of planktonic bacteria prompting a more inflammatory immune response. During the dispersal phase, systemic and intraventricular antibiotics can be used to treat the planktonic bacteria but device removal is still needed because of residual biofilm. *(Used with permission from Gutierrez-Murgas Y, Snowden JN. Ventricular shunt infections: immunopathogenesis and clinical management.* J Neuroimmunol *2014;**276**(1–2):1–8. https://doi.org/10.1016/j.jneuroim.2014.08.006. Copyright © Journal of Neuroinflammation.)*

As *S. epidermidis* is the most common cause of shunt infections, we will focus on its virulence factors that lead to biofilm formation. Importantly, parallel mechanisms exist in many other bacteria associated with shunt infections and are under similar study to identify targets for therapeutic manipulation. In the case of *S. epidermidis*, attachment to shunt tubing is through nonspecific, hydrophobic interactions.[26] Bacterial surface proteins aid in this interaction, primarily through teichoic acids and the adhesins/autolysins AtlE and Aae.[25,27] It has also been found that extracellular DNA (eDNA) contributes to attachment.[28] Once the tubing surface becomes coated with human proteins (fibronectin, collagen, fibrinogen, and vitronectin) bacterial microbial surface components recognizing adhesive matrix molecules (MSCRAMMs) aid in attachment.[25] The most well-known of the MSCRAMMs in *S. epidermidis* is SdrG, which binds fibrinogen.[25] Embp, a very large cell surface protein, binds to fibronectin, facilitating attachment. Attachment to the surface of the catheter thus requires many patient and bacterial factors, presenting many opportunities for intervention and prevention.

Once attached, the biofilm undergoes proliferation and maturation. This is facilitated by complex expression of intercellular adhesive properties.[29] The best studied of these is polysaccharide intercellular adhesion (PIA), which is present in the majority of *S. epidermidis* strains that have been associated with clinical disease.[30] Secretion of PIA aids initial adherence to the catheter surface and interbacterial adhesion as well as facilitates biofilm formation.[25] PIA is also responsible for the formation of the extracellular matrix that surrounds the bacterial cells and renders the biofilm more resistant to phagocytosis by immune cells.[28,31,32] While PIA is the most-studied adherence and proliferation factor, there are strains of *S. epidermidis* that lack PIA. These strains use proteinaceous factors for proliferation. The proteins Aap, Bhp, and Embp have also been shown to facilitate biofilm formation.[30,33] Once an *S. epidermidis* biofilm has formed, it employs quorum sensing to regulate the proliferation and maturation phase. The two major quorum sensing mechanisms in staphylococci are the *agr* and *luxS* systems.[26] These systems provide a mechanism for the bacteria to communicate and control growth and metabolism to optimize survival through changes in gene expression in response to cell density.[26,34] The final stage of a biofilm is detachment, which results in the dispersion of bacteria and colonization of new surfaces.[34] Little is known about the detachment phase of biofilms, but in staphylococci it is thought to be through *agr*-dependent mechanisms and involve the secretion of phenol-soluble modulins (PSMs).[25,34] Each of these phases of biofilm development is under study as potential therapeutic targets for prevention or treatment of biofilm infections. Improving our understanding of biofilm pathophysiology is essential for designing strategies appropriate for prevention, diagnosis, and treatment of shunt infections and other device-associated infections.

4. IMMUNE RESPONSE

Diagnosis and treatment of shunt infections is complicated by the presence of biofilms as well as by the immune privilege of the central nervous system (CNS). The CNS is an immune-privileged site as it features constitutive signals to limit inflammation in the absence of disease and is isolated from the peripheral immune system through the blood-brain barrier (BBB).[35–37] This phenomenon is likely neuroprotective as large inflammatory responses in the CNS would be damaging to the delicate cells there.[35] This is supported by evidence that infections in the perinatal period have detrimental effects on long-term developmental outcomes, suggesting that the infants at highest risk of shunt infection are also at highest risk of poor neurologic outcomes following infection.[38–40] The BBB is the CNS's primary defense against infection, limiting the entry of blood-borne pathogens into the CNS.[35] Unfortunately, at the time a shunt is placed into the lateral ventricle, the BBB is breached by the silicone tubing that runs through the brain parenchyma, which may allow for introduction of bacterial or fungal organisms if the catheter were to be contaminated at the time of insertion.[20]

If organisms are introduced into the brain, the initial CNS response to infection is mediated by the innate immune system through microglial cells that reside in the brain.[35,41–43] Microglia actively probe their environment and initiate the immune cascade to recruit neutrophils and other leukocytes to the CNS.[43] Microglia are able to recognize pathogen-associated molecular patterns on pathogens through the toll-like receptor (TLR) and other pattern recognition receptors.[35,44] They also express several other receptors that participate in immune recognition, including CD14, NOTCH-1 receptors, and complement receptors.[36,37] Once activated, microglia migrate to and accumulate at the site of the infection. There they are able to phagocytose bacteria and become antigen-presenting cells.[42,45] Microglia are also able to release cytokines to further direct the immune response, such as interleukin 1 beta (IL-1β), IL-10, tumor necrosis alpha (TNF-α), and nitric oxide.[37,43] Microglial activation is thought to occur on a continuum from a proinflammatory and potentially neurotoxic (M1) classical activation state to a repair-oriented alternative activation (M2), with shifts along this spectrum as the disease progresses[46,47] (Fig. 2).

While microglia are thought to be primarily responsible for the innate immune response to CNS infection, astrocytes may play a role as well. Their primary function is to provide support for CNS homeostasis.[48,49] Additionally, they function as part of the BBB.[49–51] While participating in the immune response is not their primary function, astrocytes have been noted to express some TLRs and other immune receptors as well as produce several chemokines and cytokines such as IL-6, CXCL1, CCL2, and IL-8.[44,48,52–55] Astrocytes have been shown to produce IL-8 in response to the PIA component of *S. epidermidis* in in vitro studies, reflecting a potential role for astrocytes in the pathophysiology of shunt infection.[56] Importantly, both microglia and astrocytes constitutively signal via CD200 and FGF respectively to maintain their nonreactive, homeostatic states and dampen inflammatory activities in the absence of infection or injury.[46,49,57] The specific roles of microglia and astrocytes in the pathophysiology of shunt infection have not been explicitly studied in vivo, but these glia likely play a significant role in coordinating and balancing the inflammatory response within the CNS.

The further immune response is carried out by the innate immune cells of the periphery that are recruited to the CNS, such as macrophages and neutrophils. Once they receive signals from the microglia and astrocytes in the brain, macrophages and neutrophils make their way to the site of the infection. In an animal model of shunt infection, both *S. aureus* and *S. epidermidis* trigger recruitment of neutrophils, macrophages, and myeloid-derived suppressor cells to the tissue surrounding the infected catheter.[12,58,59] Interestingly, this inflammatory cell recruitment is attenuated in biofilm models of CNS infection in comparison with nonbiofilm models, suggesting that the immune response to biofilm infections is skewed to favor persistence of the biofilm.

4.1 Biofilm Evasion of Immune Response

While hosts can mount a robust immune response to infections within the CNS if needed, the biofilms that form in shunt infections have developed strategies to evade the host immune response. The first of these strategies is the extracellular matrix that encases the biofilm. This matrix prevents the deposition of complement 3b (C3b) and immunoglobulin G (IgG), which protect the bacteria from neutrophil-dependent killing.[60] Additionally, studies have shown that the PIA in the extracellular matrix protects bacteria from being killed by neutrophils.[31] The matrix of candidal and Gram-negative biofilms forms a similar protective barrier against phagocytosis.[61–63] Another strategy by which staphylococcal biofilms avoid the host immune response is through phenol soluble modulins, leukocidins, and α-toxin, which attract and lyse neutrophils and other immune cells.[32,64–67] Biofilms are also capable of skewing the immune response toward an antiinflammatory pathway. For example, it has been shown that myeloid-derived suppressor cells polarize macrophages associated with staphylococcal biofilms in the periphery toward an antiinflammatory M2 phenotype that allows for bacterial persistence.[68] The multiple immune-resistance strategies utilized by biofilms may explain why biofilm bacterial burdens persisted despite an increase in inflammatory cell infiltrates in IL-10 deficient mice in a CNS catheter model of an *S. epidermidis* infection.[59] Importantly, not only did increasing the number and inflammatory state of immune cells responding to infection not reduce the bacterial burdens, but it also increased the adverse neurologic outcomes in response to infection.[59] This highlights the need to carefully evaluate any immunomodulatory strategies to treat biofilm infections in the CNS, as inflammation must be carefully balanced in the brain to avoid neurotoxic injury.

5. SYMPTOMS OF INFECTION

There are some similarities between the symptoms of shunt infections and other CNS infections such as meningitis and brain abscesses, including headaches and fevers. However, the symptoms of shunt infection can be insidious, particularly in the setting of infection with biofilm forming and low-virulence organisms. A high index of suspicion is needed in patients with implanted shunts. Frequently reported symptoms of shunt infections are headache, nausea, lethargy, and mental status changes.[13,14] In addition, other neurologic symptoms such as seizures or an increase in seizure frequency over an established baseline can indicate shunt infection. Tenderness and erythema of the skin overlying the shunt tubing are also indicative of infection, although this is not consistently observed.[4] Depending on the location of the distal end of the shunt, patients may experience a variety of additional symptoms. Those with ventriculoperitoneal shunts may have signs of peritonitis or abdominal tenderness.[69] Those with ventriculopleural shunts may experience signs of pleuritis. In patients with ventriculoatrial shunts, bacteremia or glomerulonephritis can be signs of infection.[14] Finally, shunt infection should be considered in any patient with an indwelling ventricular shunt and fever without a clear source. It is essential to recognize the broad range of symptoms and symptom severity associated with shunt infection to promptly diagnose and treat these infections.

6. DIAGNOSIS OF INFECTION

As with most bacterial infections, culture of a bacteria known or suspected to cause shunt infection from a sterile collection of CSF is considered the gold standard for diagnosing shunt infection.[14,70] While culture is the gold standard for diagnosis of shunt infection, there are many reasons for these cultures to be falsely negative.[71] Due to the biofilm nature of shunt infections, there may be very few planktonic bacteria in the CSF sample; with a low bacterial burden, conventional cultures may not grow the organism.[20,72] Bacterial culture typically requires 24–48 h to isolate organisms for identification and susceptibility.[72] Many common bacteria that cause shunt infections are slow growing or fastidious, which can further delay diagnosis and treatment.[72] *P. acnes*, a common bacterium in shunt infections, is particularly difficult to culture.[73] Further, cultures may never grow in patients who have received antibiotics before CSF was obtained for culture.[73] Evaluation of PCR for a diagnosis of shunt infection has been limited, but suggests that negative PCR could be used to exclude shunt infection.[72] However, this needs to be further studied on a larger scale. Thus, other diagnostic modalities are needed to rapidly and accurately diagnose shunt infections as culture is not always reliable.

When cultures are negative but clinical suspicion of shunt infection is high, clinicians are forced to rely on Gram stain, CSF indices, and other laboratory tests, all of which can be difficult to interpret depending on the age of the patient, recent surgical procedures, and other factors. CSF indices indicative of shunt infection are increased white blood cell count, increased protein, and decreased glucose.[14,70] Unfortunately, there is no combination of CSF indices that have ideal sensitivity and specificity to diagnose shunt infections; CSF WBC is particularly inaccurate in infants.[74,75] Additionally, patients with a CSF shunt infection may have a normal CSF WBC count so CSF WBC should not be used to exclude the possibility of shunt infection.[21]

Other biomarkers for infection include C-reactive protein, procalcitonin, and lactate, all of which may be helpful in distinguishing bacterial infection from viral infection or postsurgical changes.[14,70] Both C-reactive protein and procalcitonin in the serum were elevated in patients with shunt infections in small clinical studies, but these biomarkers can be elevated due to other infections or inflammatory conditions and thus need evaluation in larger studies.[76,77] Lactate has also been evaluated in several studies and has been found to be better than white blood cell count, CSF glucose, or CSF protein levels in distinguishing aseptic from bacterial meningitis. However, these studies included few shunt-infection patients and found that many patients may be missed if lactate alone was utilized.[21,78,79]

Other studies have evaluated the ability of CSF chemokines and cytokines to serve as biomarkers of infection. Chemokines and cytokines are secreted by epithelial, immune, and other cells to recruit and influence the activity of immune cells in the setting of injury or infection.[80] Several proinflammatory cytokines, including interleukin (IL)-8, IL-1β, tumor necrosis factor (TNF)-α, IL-12, and IL-6, are increased in the CSF of patients with bacterial meningitis.[74,80,81] Other chemokines and cytokines of interest include interferon gamma (IFN-γ), monocyte chemoattractant protein 1 (MCP-1), and matrix metallopeptidase-9 (MMP-9), which have been shown to discriminate between meningitis caused by *Streptococcus pneumoniae*, *Neiserria meningitidis*, and *Haemophilus influenzae*.[82] However, the use of chemokines and cytokines to diagnose shunt infection has not been evaluated. As noted above, animal models of shunt infection suggest that the inflammatory response to the biofilm characteristics of shunt infection is attenuated in comparison to planktonic infections; thus, the chemokine and cytokine profiles may be altered or less pronounced in shunt infections in comparison with meningitis.[12]

There are two studies that have looked at a limited number of CSF chemokines specifically in regards to shunt infection. The first study looked at transforming growth factor-β1, TNF-α, vascular endothelial growth factor (VEGF), and total tau as predictors for subsequent shunt infection.[83] This study found that higher levels of VEGF at the time of shunt placement were associated with subsequent shunt infection.[83] However, this study only involved 22 patients and did not examine cytokine levels at the time of presentation of infection but at the time of shunt placement. A different study examined whether soluble membrane attack complex (sMAC) was diagnostic for pyogenic shunt infection.[84] However, sMAC was not able to diagnose shunt infection due to *S. epidermidis*, which accounts for more than 50% of all shunt infections. While many options are under study, currently our ability to diagnose shunt infections is limited by inadequate data to guide recommendations and the complications inherent in a biofilm infection. A better modality to swiftly and accurately diagnose CSF shunt infections would improve our ability to recognize and treat these infections while preventing serious complications.

7. TREATMENT

The optimal management of shunt infections requires both surgical intervention and parenteral antimicrobials. When a shunt infection is suspected, CSF should be collected from the shunt reservoir and the patient then placed on empiric antibiotic therapy while awaiting the results of CSF studies and cultures.[14] Empiric antibiotics should cover *S. epidermidis* and *S. aureus* as well as Gram-negative organisms including *P. aeruginosa*.[14] Typical regimens include vancomycin and cefepime, ceftazidime, or meropenem to provide antipseudomonal coverage.[14] Once a pathogen is identified, antibiotic therapy should be tailored to the pathogen and its susceptibilities. Treatment duration is typically determined by the pathogen, with recommended courses varying from 10 days for *S. epidermidis* or *P. acnes* with minimal symptoms and normal CSF parameters to 21 days for some Gram-negative infections.[14] Direct instillation of antibiotics into the ventricle via the external ventricular drain may be needed if the infection is particularly severe or resistant to parenteral antibiotics.[14] However, no antibiotics have been approved by the Food and Drug Administration for this indication and the efficacy of the approach has not been tested in randomized clinical trials.[13]

After confirmation of shunt infection, most experts would recommend removal of the infected shunt and placement of an external ventricular drain in addition to systemic antibiotic treatment.[13,14] In some situations and some centers, other surgical options have been used, including antimicrobial therapy without surgical removal, distal shunt externalization, or removal of the infected device with immediate replacement of a new shunt.[13,69,85,86] However, treatment with antibiotics alone, without surgical intervention, has been associated with low success and significant mortality, likely due to the biofilm nature of the infection.[13,87,88] Immediate replacement of the shunt, in addition to removal of the infected shunt and parenteral antibiotics, has had success rates of 65%–90% in some studies.[89,90] In the case of peritoneal symptoms and positive cultures of only the distal shunt but not the ventricular component, it may be appropriate to externalize only the distal portion of the shunt.[4] However, in most cases, it is recommended that the entire shunt system be removed in addition to antibiotic treatment. Depending on the clinical status of the patient and the infecting organism, recommendations for delay in replacing a new shunt range from 2 to 10 days.[14] However, there is minimal evidence for the optimal time for external drainage and systemic antibiotics prior to new shunt placement, highlighting the need for ongoing research regarding the management of shunt infections.[4]

8. PREVENTION

Given the difficulty in treating shunt infections, prevention of these infections is essential. As many of the bacteria that cause shunt infections are induced at the time of surgery, perioperative prevention of infection has been an area of significant investigation and progress. Several perioperative risk factors for shunt infection have been identified, including number of staff in the operating room, presence of airborne organisms in the operating room, temperature of the room, surgeon's experience, length of operation, and preparation of the patient.[91–93] To address these risk factors, institutions and working groups have developed operative protocols or bundles to decrease the risk of shunt infection. Several protocols have been published and include both measures for the actual surgical procedure and nonoperative measures.[91–93] Typically, the nonoperative measures include scheduling the procedure as the first of the day, keeping doors closed, limited staff in the room, minimizing movement in the operating room, and preparation of all materials in the operating room.[91,92] Surgical aspects included in the protocol are clipping hair rather than shaving if needed, hand scrub rather than antiseptic cream, careful disinfection of a patient's skin with Chloraprep or betadine, double gloving, minimal touching of shunt components, and keeping procedure time to less than 30 min.[91,92,94,95] Additionally, antibiotic prophylaxis targeted at the skin flora most likely to be introduced at the time of surgery are an important part of the operative infection prevention protocols.[96,97] Antibiotic impregnated shunts (AIS) may also be used to decrease the rate of early shunt infections. Several metaanalyses have shown that AIS are able to reduce the risk of early shunt infection.[98–100] While these analyses are based on small studies, the potential benefit in decreasing the rates of shunt infections has led to the inclusion of AIS in current guidelines and practice at many institutions.[91]

9. CONCLUSIONS AND OUTSTANDING QUESTIONS

Since their introduction in the 1950s, CSF shunts have provided life-saving treatment for hydrocephalus. Unfortunately, these shunts are often complicated by infections. As our understanding of biofilm pathophysiology and immune responses to shunt infections improve, we hope to identify new strategies for diagnosis, treatment, and prevention of these infections. Infants are at highest risk of these infections and deserve further study. Infants may have unique immune and other host factors that alter their responses to infection requiring modifications in our diagnostic and treatment approaches. Additionally, the neurologic complications of shunt infections need to be better characterized so that strategies can be designed to optimize neurologic outcomes with adjunctive immunotherapy, developmental support, and potentially targeted prophylactic strategies for particularly high-risk patients. A combination of basic, translational, and clinical studies is needed to continue to improve our understanding of shunt infections to decrease the significant cost in terms of healthcare expense and, most importantly, decrease morbidity and mortality for these patients.

REFERENCES

1. Del Bigio MR. Ependymal cells: biology and pathology. *Acta Neuropathol* 2010;**119**(1):55–73.
2. Jorgensen J, Williams C, Sarang-Sieminski A. Hydrocephalus and ventriculoperitoneal shunts: modes of failure and opportunities for improvement. *Crit Rev Biomed Eng* 2016;**44**(1–2):91–7.

3. Beuriat PA, Puget S, Cinalli G, Blaublomme T, Beccaria K, Zerah M, et al. Hydrocephalus treatment in children: long-term outcome in 975 consecutive patients. *J Neurosurg Pediatr* 2017;**20**(1):1–9.
4. Duhaime AC. Evaluation and management of shunt infections in children with hydrocephalus. *Clin Pediatr (Phila)* 2006;**45**(8):705–13.
5. Merkler AE, Ch'ang J, Parker WE, Murthy SB, Kamel H. The rate of complications after ventriculo-peritoneal shunt surgery. *World Neurosurg* 2017;**98**:654–8.
6. Simon TD, Hall M, Riva-Cambrin J, Albert JE, Jeffries HE, Lafleur B, et al. Infection rates following initial cerebrospinal fluid shunt placement across pediatric hospitals in the united states clinical article. *J Neurosurg Pediatr* 2009;**4**(2):156–65.
7. Wells DL, Allen JM. Ventriculoperitoneal shunt infections in adult patients. *AACN Adv Crit Care* 2013;**24**(1):6–12 [quiz 13-4].
8. Mounier R, Kapandji N, Birnbaum R, Cook F, Rodriguez C, Nebbad B, et al. Biofilm-associated infection: the hidden face of cerebrospinal fluid shunt malfunction. *Acta Neurochir* 2016;**158**(12):2321–4.
9. Odio C, GH Jr MC, Nelson JD. CSF shunt infections in pediatrics. A seven-year experience. *Am J Dis Child* 1984;**138**(12):1103–8.
10. CDC/NHSN. *CDC/NHSN surveillance definitions for specific types of infections*. https://www.cdc.gov/nhsn/pdfs/pscmanual/17pscnosinfdef_current.pdf; Updated 2017.
11. Gathura E, Poenaru D, Bransford R, Albright AL. Outcomes of ventriculoperitoneal shunt insertion in sub-saharan africa. *J Neurosurg Pediatr* 2010;**6**(4):329–35.
12. Snowden JN, Beaver M, Beenken K, Smeltzer M, Horswill AR, Kielian T. *Staphylococcus aureus* sarA regulates inflammation and colonization during central nervous system biofilm formation. *PLoS One* 2013;**8**(12).
13. Dawod J, Tager A, Darouiche RO, Al Mohajer M. Prevention and management of internal cerebro-spinal fluid shunt infections. *J Hosp Infect* 2016;**93**(4):323–8.
14. Tunkel AR, Hasbun R, Bhimraj A, Byers K, Kaplan SL, Michael Scheld W, et al. 2017 infectious diseases society of america's clinical practice guidelines for healthcare-associated ventriculitis and meningitis. *Clin Infect Dis* 2017;**64**(6). https://doi.org/10.1093/cid/ciw861.
15. Lee JK, Seok JY, Lee JH, Choi EH, Phi JH, Kim SK, et al. Incidence and risk factors of ventriculoperitoneal shunt infections in children: a study of 333 consecutive shunts in 6 years. *J Korean Med Sci* 2012; **27**(12):1563.
16. Raffa G, Marseglia L, Gitto E, Germano A. Antibiotic-impregnated catheters reduce ventriculoperitoneal shunt infection rate in high-risk newborns and infants. *Childs Nerv Syst* 2015;**31**(7):1129–38.
17. Tuan TJ, Thorell EA, Hamblett NM, Kestle JR, Rosenfeld M, Simon TD. Treatment and microbiology of repeated cerebrospinal fluid shunt infections in children. *Pediatr Infect Dis J* 2011;**30**(9):731–5.
18. Kestle JR, Garton HJ, Whitehead WE, Drake JM, Kulkarni AV, Cochrane DD, et al. Management of shunt infections: a multicenter pilot study. *J Neurosurg* 2006;**105(3 Suppl.)**:177–81.
19. Prusseit J, Simon M, dB v, Heep A, Molitor E, Volz S, et al. Epidemiology, prevention and management of ventriculoperitoneal shunt infections in children. *Pediatr Neurosurg* 2009;**45**(5):325–36.
20. Gutierrez-Murgas Y, Snowden JN. Ventricular shunt infections: immunopathogenesis and clinical management. *J Neuroimmunol* 2014.
21. Conen A, Walti LN, Merlo A, Fluckiger U, Battegay M, Trampuz A. Characteristics and treatment outcome of cerebrospinal fluid shunt-associated infections in adults: a retrospective analysis over an 11-year period. *Clin Infect Dis* 2008;**47**(1):73–82.
22. Braxton Jr EE, Ehrlich GD, Hall-Stoodley L, Stoodley P, Veeh R, Fux C, et al. Role of biofilms in neurosurgical device-related infections. *Neurosurg Rev* 2005;**28**(4):249–55.
23. Wang R, Khan BA, Cheung GY, Bach TH, Jameson-Lee M, Kong KF, et al. *Staphylococcus epidermidis* surfactant peptides promote biofilm maturation and dissemination of biofilm-associated infection in mice. *J Clin Invest* 2011;**121**(1):238–48.
24. Stevens NT, Greene CM, O'Gara JP, Bayston R, Sattar MT, Farrell M, et al. Ventriculoperitoneal shunt-related infections caused by *Staphylococcus epidermidis*: pathogenesis and implications for treatment. *Br J Neurosurg* 2012;**26**(6):792–7.
25. Fey PD, Olson ME. Current concepts in biofilm formation of *Staphylococcus epidermidis*. *Future Microbiol* 2010;**5**(6):917–33.
26. Laverty G, Gorman SP, Gilmore BF. Biomolecular mechanisms of staphylococcal biofilm formation. *Future Microbiol* 2013;**8**(4):509–24.

27. Otto M. *Staphylococcus epidermidis* pathogenesis. *Methods Mol Biol* 2014;**1106**:17–31.
28. Rohde H, Frankenberger S, Zähringer U, Mack D. Structure, function and contribution of polysaccharide intercellular adhesin (PIA) to *Staphylococcus epidermidis* biofilm formation and pathogenesis of biomaterial-associated infections. *Eur J Cell Biol* 2010;**89**(1):103–11.
29. Christner M, Franke GC, Schommer NN, Wendt U, Wegert K, Pehle P, et al. The giant extracellular matrix-binding protein of *Staphylococcus epidermidis* mediates biofilm accumulation and attachment to fibronectin. *Mol Microbiol* 2010;**75**(1):187–207.
30. Mack D, Haeder M, Siemssen N, Laufs R. Association of biofilm production of coagulase-negative staphylococci with expression of a specific polysaccharide intercellular adhesin. *J Infect Dis* 1996;**174**(4): 881–4.
31. Vuong C, Kocianova S, Yao Y, Carmody AB, Otto M. Increased colonization of indwelling medical devices by quorum-sensing mutants of *Staphylococcus epidermidis* in vivo. *J Infect Dis* 2004;**190**(8): 1498–505.
32. Cheung GY, Rigby K, Wang R, Queck SY, Braughton KR, Whitney AR, et al. *Staphylococcus epidermidis* strategies to avoid killing by human neutrophils. *PLoS Pathog* 2010;**6**(10).
33. Schaeffer CR, Woods KM, Longo GM, Kiedrowski MR, Paharik AE, Buttner H, et al. Accumulation-associated protein enhances *Staphylococcus epidermidis* biofilm formation under dynamic conditions and is required for infection in a rat catheter model. *Infect Immun* 2015;**83**(1):214–26.
34. Vadyvaloo V, Otto M. Molecular genetics of *Staphylococcus epidermidis* biofilms on indwelling medical devices. *Int J Artif Organs* 2005;**28**(11):1069–78.
35. Ransohoff RM, Perry VH. Microglial physiology: unique stimuli, specialized responses. *Annu Rev Immunol* 2009;**27**:119–45.
36. Aloisi F. Immune function of microglia. *Glia* 2001;**36**(2):165–79.
37. Kettenmann H, Hanisch UK, Noda M, Verkhratsky A. Physiology of microglia. *Physiol Rev* 2011;**91**(2): 461–553.
38. Kosmac K, Bantug GR, Pugel EP, Cekinovic D, Jonjic S, Britt WJ. Glucocorticoid treatment of MCMV infected newborn mice attenuates CNS inflammation and limits deficits in cerebellar development. *PLoS Pathog* 2013;**9**(3).
39. Meyer U, Feldon J, Fatemi SH. In-vivo rodent models for the experimental investigation of prenatal immune activation effects in neurodevelopmental brain disorders. *Neurosci Biobehav Rev* 2009;**33**(7): 1061–79.
40. Bilbo SD, Schwarz JM. Early-life programming of later-life brain and behavior: a critical role for the immune system. *Front Behav Neurosci* 2009;**3**:14.
41. Kielian T, Hickey WF. Proinflammatory cytokine, chemokine, and cellular adhesion molecule expression during the acute phase of experimental brain abscess development. *Am J Pathol* 2000;**157**(2): 647–58.
42. Kielian T, Mayes P, Kielian M. Characterization of microglial responses to *Staphylococcus aureus*: effects on cytokine, costimulatory molecule, and toll-like receptor expression. *J Neuroimmunol* 2002;**130**(1–2): 86–99.
43. Kielian T, Barry B, Hickey WF. CXC chemokine receptor-2 ligands are required for neutrophil-mediated host defense in experimental brain abscesses. *J Immunol* 2001;**166**(7):4634–43.
44. Holm TH, Draeby D, Owens T. Microglia are required for astroglial toll-like receptor 4 response and for optimal TLR2 and TLR3 response. *Glia* 2012;**60**(4):630–8.
45. Fu R, Shen Q, Xu P, Luo JJ, Tang Y. Phagocytosis of microglia in the central nervous system diseases. *Mol Neurobiol* 2014;**49**(3):1422–34.
46. Giunti D, Parodi B, Cordano C, Uccelli A, Kerlero de Rosbo N. Can we switch microglia's phenotype to foster neuroprotection? Focus on multiple sclerosis. *Immunology* 2014;**141**(3):328–39.
47. Graeber MB. Changing face of microglia. *Science* 2010;**330**(6005):783–8.
48. Choi SS, Lee HJ, Lim I, Satoh J, Kim SU. Human astrocytes: secretome profiles of cytokines and chemokines. *PLoS One* 2014;**9**(4).
49. Kang W, Balordi F, Su N, Chen L, Fishell G, Hebert JM. Astrocyte activation is suppressed in both normal and injured brain by FGF signaling. *Proc Natl Acad Sci U S A* 2014;**111**(29):E2987–95.
50. Dong Y, Benveniste EN. Immune function of astrocytes. *Glia* 2001;**36**(2):180–90.
51. Perides G, Tanner-Brown LM, Eskildsen MA, Klempner MS. Borrelia burgdorferi induces matrix metalloproteinases by neural cultures. *J Neurosci Res* 1999;**58**(6):779–90.

52. Liu S, Kielian T. MyD88 is pivotal for immune recognition of citrobacter koseri and astrocyte activation during CNS infection. *J Neuroinflammation* 2011;**8**:35.
53. Farina C, Aloisi F, Meinl E. Astrocytes are active players in cerebral innate immunity. *Trends Immunol* 2007;**28**(3):138–45.
54. Furr SR, Marriott I. Viral CNS infections: role of glial pattern recognition receptors in neuroinflammation. *Front Microbiol* 2012;**3**:201.
55. Hanke ML, Kielian T. Toll-like receptors in health and disease in the brain: mechanisms and therapeutic potential. *Clin Sci (Lond)* 2011;**121**(9):367–87.
56. Stevens NT, Sadovskaya I, Jabbouri S, Sattar T, O'Gara JP, Humphreys H, et al. *Staphylococcus epidermidis* polysaccharide intercellular adhesin induces IL-8 expression in human astrocytes via a mechanism involving TLR2. *Cell Microbiol* 2009;**11**(3):421–32.
57. Quick ED, Leser JS, Clarke P, Tyler KL. Activation of intrinsic immune responses and microglial phagocytosis in an ex vivo spinal cord slice culture model of west nile virus infection. *J Virol* 2014; **88**(22):13005–14.
58. Snowden JN, Beaver M, Smeltzer MS, Kielian T. Biofilm-infected intracerebroventricular shunts elicit inflammation within the central nervous system. *Infect Immun* 2012;**80**(9):3206–14.
59. Gutierrez-Murgas YM, Skar G, Ramirez D, Beaver M, Snowden JN. IL-10 plays an important role in the control of inflammation but not in the bacterial burden in *S. epidermidis* CNS catheter infection. *J Neuroinflammation* 2016;**13**(1):271.
60. Otto M. *Staphylococcus epidermidis*—the 'accidental' pathogen. *Nat Rev Microbiol* 2009;**7**(8):555–67.
61. Gunn JS, Bakaletz LO, Wozniak DJ. What's on the outside matters: the role of the extracellular polymeric substance of gram-negative biofilms in evading host immunity and as a target for therapeutic intervention. *J Biol Chem* 2016;**291**(24):12538–46.
62. Tsui C, Kong EF, Jabra-Rizk MA. Pathogenesis of candida albicans biofilm. *Pathog Dis* 2016;**74**(4).
63. Sandai D, Tabana YM, Ouweini AE, Ayodeji IO. Resistance of candida albicans biofilms to drugs and the host immune system. *Jundishapur J Microbiol* 2016;**9**(11).
64. Dumont AL, Nygaard TK, Watkins RL, Smith A, Kozhaya L, Kreiswirth BN, et al. Characterization of a new cytotoxin that contributes to *Staphylococcus aureus* pathogenesis. *Mol Microbiol* 2011;**79**(3):814–25.
65. DuMont AL, Yoong P, Day CJ, Alonzo 3rd F, WH MD, Jennings MP, et al. *Staphylococcus aureus* LukAB cytotoxin kills human neutrophils by targeting the CD11b subunit of the integrin mac-1. *Proc Natl Acad Sci U S A* 2013;**110**(26):10794–9.
66. Hanamsagar R, Torres V, Kielian T. Inflammasome activation and IL-1beta/IL-18 processing are influenced by distinct pathways in microglia. *J Neurochem* 2011;**119**(4):736–48.
67. Rigby KM, DeLeo FR. Neutrophils in innate host defense against *Staphylococcus aureus* infections. *Semin Immunopathol* 2012;**34**(2):237–59.
68. Heim CE, Vidlak D, Scherr TD, Kozel JA, Holzapfel M, Muirhead DE, et al. Myeloid-derived suppressor cells contribute to *Staphylococcus aureus* orthopedic biofilm infection. *J Immunol* 2014;**192**(8): 3778–92.
69. Conen A, Fux CA, Vajkoczy P, Trampuz A. Management of infections associated with neurosurgical implanted devices. *Expert Rev Anti-Infect Ther* 2017;**15**(3):241–55.
70. Tunkel AR, Hartman BJ, Kaplan SL, Kaufman BA, Roos KL, Scheld WM, et al. Practice guidelines for the management of bacterial meningitis. *Clin Infect Dis* 2004;**39**(9):1267–84.
71. Banks JT, Bharara S, Tubbs RS, Wolff CL, Gillespie GY, Markert JM, et al. Polymerase chain reaction for the rapid detection of cerebrospinal fluid shunt or ventriculostomy infections. *Neurosurgery* 2005;**57**(6): 1237–43 [discussion 1237-43].
72. Simon TD, Pope CE, Browd SR, Ojemann JG, Riva-Cambrin J, Mayer-Hamblett N, et al. Evaluation of microbial bacterial and fungal diversity in cerebrospinal fluid shunt infection. *PLoS One* 2014;**9**(1).
73. van de Beek D, Drake JM, Tunkel AR. Nosocomial bacterial meningitis. *N Engl J Med* 2010;**362**(2): 146–54.
74. Srinivasan L, Kilpatrick L, Shah SS, Abbasi S, Harris MC. Cerebrospinal fluid cytokines in the diagnosis of bacterial meningitis in infants. *Pediatr Res* 2016;**80**(4):566–72.
75. Lenfestey RW, Smith PB, Moody MA, Clark RH, Cotten CM, Seed PC, et al. Predictive value of cerebrospinal fluid parameters in neonates with intraventricular drainage devices. *J Neurosurg* 2007; **107**(3 Suppl):209–12.

76. Berger C, Schwarz S, Schaebitz WR, Aschoff A, Schwab S. Serum procalcitonin in cerebral ventriculitis. *Crit Care Med* 2002;**30**(8):1778–81.
77. Schwarz S, Bertram M, Schwab S, Andrassy K, Hacke W. Serum procalcitonin levels in bacterial and abacterial meningitis. *Crit Care Med* 2000;**28**(6):1828–32.
78. Huy NT, Thao NT, Diep DT, Kikuchi M, Zamora J, Hirayama K. Cerebrospinal fluid lactate concentration to distinguish bacterial from aseptic meningitis: a systemic review and meta-analysis. *Crit Care* 2010;**14**(6):R240.
79. Sakushima K, Hayashino Y, Kawaguchi T, Jackson JL, Fukuhara S. Diagnostic accuracy of cerebrospinal fluid lactate for differentiating bacterial meningitis from aseptic meningitis: a meta-analysis. *J Infect* 2011;**62**(4):255–62.
80. Bociaga-Jasik M, Garlicki A, Ciesla A, Kalinowska-Nowak A, Sobczyk-Krupiarz I, Mach T. The diagnostic value of cytokine and nitric oxide concentrations in cerebrospinal fluid for the differential diagnosis of meningitis. *Adv Med Sci* 2012;**57**(1):142–7.
81. Perdomo-Celis F, Torres MA, Ostos H, Gutierrez-Achury J, Molano V, Duran LF, et al. Patterns of local and systemic cytokines in bacterial meningitis and its relation with severity and long-term sequelae. *Biomark Insights* 2015;**10**:125–31.
82. Grandgirard D, Gaumann R, Coulibaly B, Dangy JP, Sie A, Junghanss T, et al. The causative pathogen determines the inflammatory profile in cerebrospinal fluid and outcome in patients with bacterial meningitis. *Mediat Inflamm* 2013;**2013**.
83. Lee JH, Back DB, Park DH, Cha YH, Kang SH, Suh JK. Increased vascular endothelial growth factor in the ventricular cerebrospinal fluid as a predictive marker for subsequent ventriculoperitoneal shunt infection: a comparison study among hydrocephalic patients. *J Korean Neurosurg Soc* 2012;**51**(6):328–33.
84. Ramos TN, Arynchyna AA, Blackburn TE, Barnum SR, Johnston JM. Soluble membrane attack complex is diagnostic for intraventricular shunt infection in children. *JCI Insight* 2016;**1**(10).
85. Tamber MS, Klimo Jr P, Mazzola CA, Flannery AM. Pediatric hydrocephalus systematic review and evidence-based guidelines task force. Pediatric hydrocephalus: systematic literature review and evidence-based guidelines. Part 8: management of cerebrospinal fluid shunt infection. *J Neurosurg Pediatr* 2014;**14**(Suppl 1):60–71.
86. Simon TD, Kronman MP, Whitlock KB, Gove N, Browd SR, Holubkov R, et al. Variability in management of first cerebrospinal fluid shunt infection: a prospective multi-institutional observational cohort study. *J Pediatr* 2016;**179**:185–191.e2.
87. Schreffler RT, Schreffler AJ, Wittler RR. Treatment of cerebrospinal fluid shunt infections: a decision analysis. *Pediatr Infect Dis J* 2002;**21**(7):632–6.
88. Yogev R. Cerebrospinal fluid shunt infections: a personal view. *Pediatr Infect Dis* 1985;**4**(2):113–8.
89. James HE, Walsh JW, Wilson HD, Connor JD, Bean JR, Tibbs PA. Prospective randomized study of therapy in cerebrospinal fluid shunt infection. *Neurosurgery* 1980;**7**(5):459–63.
90. James HE, Walsh JW, Wilson HD, Connor JD. The management of cerebrospinal fluid shunt infections: a clinical experience. *Acta Neurochir* 1981;**59**(3–4):157–66.
91. Kestle JR, Riva-Cambrin J, Wellons 3rd JC, Kulkarni AV, Whitehead WE, Walker WL, et al. A standardized protocol to reduce cerebrospinal fluid shunt infection: the hydrocephalus clinical research network quality improvement initiative. *J Neurosurg Pediatr* 2011;**8**(1):22–9.
92. Pirotte BJ, Lubansu A, Bruneau M, Loqa C, Van Cutsem N, Brotchi J. Sterile surgical technique for shunt placement reduces the shunt infection rate in children: preliminary analysis of a prospective protocol in 115 consecutive procedures. *Childs Nerv Syst* 2007;**23**(11):1251–61.
93. Schaffzin JK, Harte L, Marquette S, Zieker K, Wooton S, Walsh K, et al. Surgical site infection reduction by the solutions for patient safety hospital engagement network. *Pediatrics* 2015;**136**(5):e1353–60.
94. Gruber TJ, Riemer S, Rozzelle CJ. Pediatric neurosurgical practice patterns designed to prevent cerebrospinal fluid shunt infection. *Pediatr Neurosurg* 2009;**45**(6):456–60.
95. Kestle JR, Holubkov R, Douglas Cochrane D, Kulkarni AV, Limbrick Jr DD, Luerssen TG, et al. A new hydrocephalus clinical research network protocol to reduce cerebrospinal fluid shunt infection. *J Neurosurg Pediatr* 2016;**17**(4):391–6.
96. Klimo Jr P, Van Poppel M, Thompson CJ, Baird LC, Duhaime AC, Flannery AM. Pediatric hydrocephalus: systematic literature review and evidence-based guidelines. Part 6: preoperative antibiotics for shunt surgery in children with hydrocephalus: a systematic review and meta-analysis. *J Neurosurg Pediatr* 2014;**14**(Suppl. 1):44–52.

97. Ratilal B, Costa J, Sampaio C. Antibiotic prophylaxis for surgical introduction of intracranial ventricular shunts: a systematic review. *J Neurosurg Pediatr* 2008;**1**(1):48–56.
98. Klimo Jr P, Thompson CJ, Baird LC, Flannery AM. Pediatric hydrocephalus: Systematic literature review and evidence-based guidelines. Part 7: antibiotic-impregnated shunt systems versus conventional shunts in children: a systematic review and meta-analysis. *J Neurosurg Pediatr* 2014; **14**(Suppl. 1):53–9.
99. Konstantelias AA, Vardakas KZ, Polyzos KA, Tansarli GS, Falagas ME. Antimicrobial-impregnated and -coated shunt catheters for prevention of infections in patients with hydrocephalus: a systematic review and meta-analysis. *J Neurosurg* 2015;**122**(5):1096–112.
100. Parker SL, Anderson WN, Lilienfeld S, Megerian JT, McGirt MJ. Cerebrospinal shunt infection in patients receiving antibiotic-impregnated versus standard shunts. *J Neurosurg Pediatr* 2011;**8**(3):259–65.

CHAPTER 3

Molecular Diagnosis of CNS Viral Infections

Malay K. Das, Tapash Chakraborty
Dibrugarh University, Dibrugarh, India

1. INTRODUCTION

Worldwide, central nervous system (CNS) infections are a matter of serious public health concern because of the high morbidity and mortality associated with many of the cases.[1] The incidence of CNS infection varies widely, with different causes and diseases in different areas of the world over all age groups.[2] CNS infections can be caused by many infectious agents including bacteria, viruses, parasites, fungi, spirochetes, and even prions.[3] But in most CNS infections, the causative organism is not recognized. However, epidemiological studies on CNS infections originating from the 1960s and 1980s[4–6] showed that viruses are the most common cause of both aseptic meningitis and encephalitis. Though the CNS is one of the best-protected organs of the body, it is not impervious to infections and in some cases pathogens may gain access to areas of the CNS.[7] CNS infection is extremely serious and life threatening with an unpredictable neurological outcome.[8]

Viruses have evolved mechanisms that enable them to easily infiltrate the nervous system. Two main methods of viral entry have been identified as the transneuronal spread and hematogenous spread.[9] The mechanism behind transneuronal spread is not entirely known yet, but it involves the virus escaping the immune system by traveling up the axons of the nerves.[10] There are two main ways that a virus is thought to enter the brain via hematogenous spread. The first is by infecting an immune cell, which then carries the virus to the nervous tissue. Viral examples of this include the John Cunningham (JC) virus that infects B cells and HIV, which infects CD4 T cells and macrophages to infiltrate the brain. The second is by crossing the blood capillaries as a free virus or in leukocytes.[11,12]

Different types of viruses have attained the ability to get entry to the brain. Table 1 represents a list of very common well-known viruses that have been reported to infect the CNS.[13–15]

CNS infections can be broadly divided into meningitis, encephalitis, and myelitis. Any of these infections can coexist and the patient may develop meningoencephalitis, encephalomyelitis, or meningoencephalomyelitis.[16]

The Microbiology of Central Nervous System Infections
https://doi.org/10.1016/B978-0-12-813806-9.00003-2

Table 1 Major CNS viruses causing neuroviral infections

DNA virus family	Herpesviruses	Herpes simplex virus type 1 (HSV-1)
		Herpes simplex virus type 2 (HSV-2)
		Human cytomegalovirus (CMV)
		Varicella-zoster virus (VZV)
		Epstein-Barr virus (EBV)
		Human herpes virus type 6 (HHV-6)
		Human herpes virus type 7 (HHV-7)
	Polyomaviruses	John Cunningham virus (JCV)
RNA virus family	Rhabdoviruses	Rabies virus
	Paramyxoviruses	Measles virus
		Mumps virus
	Retroviruses	Human immunodeficiency virus (HIV)

Identifying the causative agent responsible for infecting the CNS poses tremendous diagnostic challenges, and only in 30% of the cases of suspected viral encephalitis is a specific organism identified. Traditionally, the definitive diagnosis depends on the culture of the virus from cerebrospinal fluid (CSF) or brain tissue; identification of viral particles, inclusions, antigens, or nucleic acid in brain tissue; or demonstration of virus-specific intrathecal antibody synthesis.[17] This chapter discusses the techniques used for the diagnosis of CNS viral infections with a special emphasis on the molecular diagnosis techniques.

2. TECHNIQUES FOR THE DIAGNOSIS OF NEUROVIRAL INFECTIONS

For effective treatment of any infection, identifying the causative organism is the most crucial part of the treatment regime. Doctors prescribe different laboratory tests just to identify the microorganism responsible for the infection. Once the organism is identified, the treatment procedure becomes easier. Diagnosing the CNS infection becomes more important as the brain is the vital organ for the survival of a patient. A delay in diagnosis or a wrong test result may lead to disastrous consequences. Physicians have tried different diagnostic techniques to determine the exact causative organism for CNS infections. Physicians are still trying to develop new techniques to determine the causative organisms in a fast and more patient-friendly way. In the past, invasive methods were used to obtain samples for diagnosis, such as brain biopsy. Today, with the advancement of technology, less-invasive means are frequently being used, such as neuroimaging and analysis of the CSF.

2.1 Neuroimaging

Neuroimaging or brain imaging involves the use of different techniques such as a CT scan or MRI to image the structure, function, and pharmacology of the CNS, either directly or indirectly. This technique is relatively newer in medicine, neuroscience, and psychology. Neuroimaging can be divided into two main categories. The first is structural imaging. This deals with the structure of the nervous system and the diagnosis of large-scale intracranial diseases such as a tumor or brain injury. The second is functional imaging, which is used to diagnose metabolic diseases and lesions on a smaller scale, such as Alzheimer's disease. It is also used for neurological and cognitive psychology research and building brain-computer interfaces.[18–20]

CT and MRI images are utilized to visualize inflammations and lesions created by the viral infections of CNS. An MRI is used to visualize deep white matter and temporal lobe lesions, which are not well defined by a CT scan.[21]

2.2 Lumbar Puncture and CSF Analysis

This is a valuable method for diagnosing viral infections of the CNS. Analysis of the CSF typically involves determining the patient's total white cell count, glucose level, and protein level in the CSF. Viral infection of the CNS tends to increase the total white cell count while increasing the level of protein. The levels of glucose tend to be decreased by a viral infection due to increased glucose consumption. Table 2 represents an overview of the results generally found in different types of CNS infections.[22–24]

2.3 Serology

The scientific study of serum and other bodily fluids is known as serology. Practically this term means identification of antibodies in the serum for diagnostic purposes. Our body develops antibodies in response to an infection, against other foreign proteins, or to one's own proteins like in cases of autoimmune diseases.[25]

Serological tests are prescribed for diagnostic purposes when an infection is suspected, such as in rheumatic illnesses, and in one's blood group determination. Serology is useful in diagnosing viral infections of the CNS when PCR analysis returns negative results.[26]

Serological tests can be used for the identification of antibodies to many viruses. Diagnosis usually depends on a fourfold rise in antibody titer between the acute and convalescent phase of the infection. Detection of specific immunoglobulin M (IgM) antibodies in CSF can strongly suggest CNS invasion, which in turn accelerates the treatment of many viral infections in the early stages of disease.[27] Though the identification of specific antibodies in both CSF and serum samples strongly suggests the presence of its corresponding antigen, it is very important to differentiate between intrathecal antibodies produced within the brain and the presence of antibodies in CSF through transudation from serum.[28,29]

Table 2 Overview of the CSF analysis indicating various types of CNS infections

	Appearance	Opening pressure	WBC count	Glucose level	Protein level
Normal results in adults	Clear	10–20 cm H_2O	0–5 cells/μL (<2 polymorphonucleocytes [PMN])	>60% of serum glucose	<45 mg/dL
Bacterial meningitis	Clear, cloudy, or purulent	Elevated (>25 cm H_2O)	>100 cells/μL (>90% PMN)	Low (<40% of serum glucose)	Elevated (>50 mg/dL)
Aseptic (viral) meningitis	Clear	Normal or elevated	10–1000 cells/μL (lymph but PMN early)	>60% serum glucose (may be low in HSV infection)	Elevated (>50 mg/dL)
Fungal meningitis	Clear or cloudy	Elevated	10–500 cells/μL	Low	Elevated
Tuberculosis	Clear or opaque	Elevated	50–500 cells/μL (early PMN then lymph)	Low	Elevated
Subarachnoid hemorrhage	Xanthochromia, bloody, or clear	Elevated	(1 additional WBC per 1000 RBCs is considered normal correction)	Normal	Elevated
Multiple sclerosis	Clear	Normal	0–20 cells/μL (lymph)	Normal	Mildly elevated (45–75 mg/dL)
Guillain-Barré syndrome	Clear or xanthochromia	Normal or elevated	Normal or elevated	Normal	Elevated

However, accurate antibody measurements are difficult in the diagnosis of viral encephalitis because some viruses lack a specific antigen within their serotypes against which IgM is raised, which makes it too difficult to develop a routine IgM assay. Moreover, it may take two or more days before intrathecal antibodies develop.[30,31] Further, the persistence of viral antibody levels from past infections may complicate the interpretation of serological results. Therefore, serological tests are less useful in the diagnosis of acute encephalitis. In addition, many immunological methods used in the detection of viral antigens from various body specimens are usually of low diagnostic value when applied to CSF samples, making them of limited use for the routine diagnosis of viral encephalitis.[32]

2.4 Brain Biopsy

Brain biopsy involves the removal of a small piece of brain tissue for the diagnosis of abnormalities of the brain. It has been used to diagnose tumors, Alzheimer's disease, different types of infections, inflammation, and other brain diseases. Examined under a microscope, the biopsy sample provides the information required for treatment.[33]

Due to the development of less-invasive, patient-friendly diagnosis techniques, brain biopsies are rarely used for the diagnosis of CNS viral infections these days. But some viral infections of the CNS are tough to diagnose without histological and electron microscopic evidence. In these cases, brain biopsies are only performed when the information gained by the brain biopsy will outweigh the risks.[34]

2.5 Molecular Methodologies

Since the introduction of molecular techniques in diagnostic laboratories, viral detection has been entirely transformed and changed. Molecular methods have numerous advantages over conventional methods, including high sensitivity, reliability, and rapidity as the results are available within a very short period (maximum of 72 h from sample collection). Therefore they are of significant value for patient management.

Recently, a variety of molecular techniques have been developed and adapted for use in many clinical as well as research laboratories for the detection, identification, and characterization of many viral agents in CSF samples. However, the most widely used techniques are nucleic acid amplification (NAA) with different postamplification analyses.

2.5.1 CSF Nucleic Acid Amplification Using Polymerase Chain Reaction

The polymerase chain reaction (PCR), developed by Kary Mullis in 1983, is a technique used in molecular biology to amplify a single copy or a few copies of a piece of DNA across several orders of magnitude, generating thousands to millions of copies of a particular DNA sequence. In 1993, Mullis won the Nobel Prize in Chemistry along with Michael Smith for his work on PCR. This technique is utilized to amplify small amounts of viral nucleic acid from CSF for the identification of the virus responsible for the CNS infection. PCR is an easy and cheap tool to amplify a focused segment of DNA. It is useful in various fields like in the diagnosis and monitoring of genetic diseases, identification of criminals, studying the function of a targeted segment, etc.[35,36] Other applications of PCR include DNA cloning for sequencing, DNA-based phylogeny, or functional analysis of genes; the diagnosis of hereditary diseases; the identification of genetic fingerprints; and the detection of pathogens in nucleic acid tests for the diagnosis of infectious diseases.

PCR technique has revolutionized the diagnosis of viral CNS infections. CSF PCR is rapid, inexpensive, and only minimally invasive. Unfortunately, validation of the sensitivity and specificity of CSF PCR by comparison to a "gold standard" such as brain biopsy is only rarely available.

PCR is frequently used to for rapid identification of specific DNA viruses from the CSF, while reverse transcriptase PCR is commonly used to identify RNA viruses in the CSF. The accuracy of this diagnostic tool is limited by the amount of virus present in the CSF. Viral replication tends to peak early and then decline to undetectable levels in CNS infections. Within the first 5 days of symptom onset, before the decline of viral replication, PCR assays have a higher incidence of detecting CNS infection.

PCR is well established in routine clinical laboratories for the diagnosis of many viruses in biological specimens, including CSF samples for the evaluation of CNS infections. The diagnosis of HSV encephalitis was one of its early applications in clinical laboratories.[37] Currently, it is finding increasing application for the diagnosis of viral CNS infections, particularly after the development of variants of the basic method.[38–41]

Currently, a large variety of NAA techniques that form the starting point of sophisticated postamplification analyses is available.[41–43] In diagnostic virology laboratories, either a single method or a combination of these techniques is used for the diagnosis of a particular virus. In addition, highly automated PCR machines using real-time PCR procedures that carry out the entire process from sample extraction through amplification to nucleic acid detection and quantitation are available in many diagnostic laboratories. These have reduced the time required to obtain results while minimizing contamination. The accuracy, sensitivity, and specificity of many assays have also improved.[38,41] Consequently in recent times, PCR has become a powerful diagnostic tool, especially in neurovirology.[26]

Following NAA, several molecular techniques may be used to analyze amplified nucleic acids, some of which are only applied in research laboratories. The most common postamplification molecular techniques are microarrays, sequencing, and Luminex technology.

2.5.2 DNA Microarray

DNA microarray technology is a powerful postamplification analysis used after PCR, multiplex PCR, RT-PCR, or any other NAA technique. It consists of a single solid phase "chip," usually a glass or silicone slide, carrying immobilized short oligonucleotide probes that are used to cross-hybridize and subsequently detect or analyze hundreds to thousands of genes and target molecules amplified from single specimens in the same assay.[44,45]

Many microarray platforms are available. The most commonly used for virus identification and discovery are the ViroChip, which was developed in 2003 by Wang et al., and the GreeneChip, which was developed in 2007 by Palacios et al.[46,47] In both, a maximum of 70-mer oligonucleotide probes representing different regions of DNA from fully sequenced reference viral genomes in Genbank is utilized. They have been updated regularly to meet the continuous submission of new viral sequences in the genomic databases. The two microarray chips differ from each other in platform fabrication, probe design, reaction conditions, and bioinformatics. However, both have been used for the identification of viruses to species and, recently, to subspecies level along with the discovery of new variants of already-known viruses.[41,48,49]

Over recent years, microarray technology has been used to provide superior sensitivity, specificity, and reproducibility of results in many fields, including discovery, identification and molecular typing, measurement of multiple gene expression, and the determination of antiviral drug resistance. Most if not all of these applications have been reviewed extensively.[41,44,48]

Although this technology has so far been used only in research laboratories, its remarkable improvement in terms of sensitivity, specificity, efficiency, discriminatory power, and reproducibility in many applications makes it highly attractive to researchers. This is recognized by the dramatic increase in the number of published articles that used microarrays in all fields, including microbiology, since 2000.[45]

This molecular technique promises to fit into routine virology clinical laboratories soon and may be particularly useful in the diagnosis of CNS viral infections. In 2004, Boriskin et al. demonstrated the ability of microarray to detect different viruses from CSF samples by constructing a microarray with 38 genes targeting 13 viral causes of meningitis and encephalitis.[50] Subsequently, researchers improved and modified this technique for the detection of viruses in CSF samples.[51–53] To date, this technology has not been applied successfully for the discovery of new viruses in CSF samples. However, its implementation in diagnostic laboratories may have a great impact on patient management, epidemiological studies, and more.[44]

Even though this molecular technology has contributed to the characterization of viruses in the CNS, as with other molecular diagnostic techniques, it has some limitations. Apart from the cost of the technology, there are difficulties in management and interpretation of data that require a highly experienced worker. The sensitivity of this technology in virus discovery is low (106–107 copies) compared to amplification methods that target either a specific pathogen or a combination of known pathogens.[48] This could be due to many DNA sequences generated by the procedure, which has the same efficiency in competing for hybridization within the same reaction. Another drawback of the technology is that it targets either a specific virus or the likely viral group and therefore some suggestion of the putative pathogen is always required.

2.5.3 Luminex Technology

Luminex is a microsphere-based suspension array technology. It uses different sets (based on the ratio of fluorochrome used in each bead) of 5.6 μm polystyrene microspheres as a solid phase in suspension. These microspheres are coated with capture reagents such as oligonucleotides, antibodies, or receptors, allowing the hybridization and detection of specific analytes using a laser along with a flow cytometer.[45] In a single reaction, up to 100 microsphere sets can carry different reactants, such as different nucleic acid sequences, on its surface. This can allow the detection and simultaneous measurement of up to 100 different analytes within a single sample with excellent sensitivity and specificity.[54] Different formats of this technology with a variety of applications are now available as reviewed by Dunbar (2006) along with Miller and Tang (2009).[45,54] Some of these applications are detection and typing of respiratory viruses, antibody detection, human leukocyte antigen typing, molecular analysis of infectious organisms, and detection of multiple cytokines and chemokines. However, this technology has not yet been applied for the diagnosis of CNS infections. Furthermore, because it depends on a hybridization reaction for the detection or identification of any pathogen, some suggestion of the causative pathogen is required.

2.5.4 DNA Sequencing

DNA sequencing is the most accurate post-NAA analysis. With sequencing, the nucleotide sequences of the entire amplified DNA can be determined.

Several methods have been developed for DNA sequencing. The earliest, most efficient, reliable, and hence most popular is the chain termination method developed by Sanger and Coulson (1975). It was modified to semiautomated dye-terminator sequencing in the early 1980s.[55,56] Full automation of this method with computer analysis is now available in many molecular diagnostic and research laboratories. Alternative technologies exist, including pyrosequencing, which was developed in 1996.[57,58] Pyrosequencing of DNA is carried out by detecting the release of pyrophosphate when a nucleotide is incorporated in the synthesis of a complementary DNA strand.

Recently, automated next-generation platforms with robust high-throughput sequencing technologies have been developed. These platforms use dye-terminator, pyrosequencing, or other techniques. The most common next-generation sequencing platforms are the pyrosequencing system of 454 life sciences, dye terminator Solexa/Illumina sequencing technology, SOLiD DNA sequencing by ligation from Applied Biosystems, and Helicos-True single-molecule sequencing.[59,60] Updated versions of these technologies with enhanced performance are also now available. These sequencing technologies can sequence variable lengths of base pairs and produce millions to billions of sequence reads per run. Although they are currently only available in research laboratories, they are proven to be highly sensitive and simultaneously very efficient through avoiding subcloning of amplified nucleic acids. The data generated from these methods require the availability of highly skilled workers in computational processing and bioinformatics. To date, the maximum read length obtained from 454 pyrosequencing is 400–500 bp compared to the Sanger method, in which 1 kb length can be read. It can also take a long time (4–5 days) to finish a reading of 1 Gb in a single run.[41,49,60–62]

Next-generation sequencing technologies have potential application in many fields, from human genome projects to pathogen discovery. Their application in virology following unbiased DNA amplification demonstrated their ability in identification and characterization of new viruses in biological samples other than CSF samples. Some of the new human viruses identified are a new arenavirus in transplant recipients,[47] a new polyomavirus in human Merkel cell carcinoma,[63] Lujo hemorrhagic fever virus,[64] and new picornaviruses (Klassevirus and Saliviruses) (Table 3).[65]

Table 3 Comparison of the different diagnostic techniques utilized in the diagnosis of CNS viral infections

Technique	Method	Advantage	Disadvantage
Brain biopsy[33]	A small piece of brain tissue is collected and manually observed for the virus.	Direct method. Viruses can be identified directly.	Invasive method; poor patient compliance; high risk for brain damage during sample collection.
Serology[25]	It involves identification of specific antibodies in the patient's serum to know about the virus.	The test is simple, fast, and noninvasive yet gives great results.	Nonantibody producing viruses cannot be identified. Antibodies from previous infections may lead to wrong results.
CSF analysis[22]	It involves determining the total white cell count, glucose level and protein level in the patient's CSF.	The test is simple, fast, less-invasive and gives an idea of the virus involved in the infection.	Gives idea about the type of causative organism, but no specific data can be obtained.

Continued

Table 3 Comparison of the different diagnostic techniques utilized in the diagnosis of CNS viral infections—cont'd

Technique	Method	Advantage	Disadvantage
Neuroimaging[18]	It involves the use of techniques like CT scan and MRI to image the structure, function, pharmacology of the CNS.	Noninvasive; fast; more appropriate when the infection produces any lesions or tumors in the brain.	Requires highly experienced personnel to identify the type of virus involved.
PCR[36]	It involves amplifying the small amounts of viral nucleic acid present in CSF of the infected patient and identifying the virus from that DNA/RNA responsible for the CNS infection.	Less invasive; fast; highly accurate.	Samples need to be collected within a specific period after the infection; otherwise nucleic acid in CSF may be absent leading to false results.
DNA microarray[45]	It involves cross-hybridization of viral genes with immobilized short oligonucleotides and subsequent detection of the genes.	Less invasive; fast; can detect >100 genes in a single run.	Less sensitive than PCR. It targets either a specific virus or the likely viral group, hence identifying a new virus is difficult.
Luminex technology[54]	It contains polystyrene microspheres suspension coated with capture reagents such as oligonucleotides, antibodies or receptors, allowing the hybridization and detection of a specific gene using a laser along with flow cytometer.	Less invasive; fast; can detect >100 genes in a single run.	Because it depends on a hybridization reaction for the detection or identification of any pathogen, some suggestion of the causative pathogen is required.
DNA sequencing[56]	The sequence of the entire DNA is determined which leads to an error free virus identification.	Less invasive; fast; fully automated.	Costly, yet to be used in full commercial scale.

3. PATENTS ON MOLECULAR DIAGNOSIS OF CNS VIRAL INFECTIONS

A few patents based on PCR technology have been reported elsewhere. Rao et al. in 2013 obtained a patent for the development of a PCR-based method for simultaneous detection and discrimination of bacterial, fungal, parasitic, and viral infections of the ocular and CNS.[66] Chen et al. developed another PCR-based method and gene composition for the detection of a CNS virus.[67] A few patents for the treatment of CNS viral infection have also been reported elsewhere.[68–71]

4. FUTURE SCOPE FOR DIAGNOSIS OF CNS VIRAL INFECTIONS

The recognition of viruses infecting humans is increasing steadily, but it is still incomplete. >95% of CNS infections remain undiagnosed internationally. These cases with suspected viral etiologies make it difficult to interpret negative results. Consequently, physicians can only use negative results with moderate confidence to rule out CNS infection. Conversely, with an increasing number of emerging infectious syndromes with a suspected viral origin, there is an increased demand for the development of advanced assays for virus discovery. The same modern advanced assays will also be important in understanding the pathogenic role of any newly discovered pathogen. They will also help to improve the management of infected people, especially immune-compromised patients; facilitate the development of specific antiviral compounds; enhance epidemiological studies; and aid the development of methods to control diseases.[72,73]

Over the last 15–20 years, many molecular techniques have been developed for novel virus discovery, each with advantages and drawbacks. Degenerate PCR, which amplifies a conserved region of a viral family, followed by low-density DNA microarray and sequencing (to detect new family members), is among these methods. Degenerate PCR has been proven sensitive and effective in the detection of new viruses; nevertheless, it can only be used to detect related members of a relatively select range of known viral families such as the recently discovered Parechovirus HPeV10 isolated from stool samples of children with gastroenteritis in Sri Lanka.[74]

All the above-discussed molecular methodologies for the detection of a CNS viral infection has the same unsolved problem: they require some existing information on the sequences of the infecting viral genes. Based on the gene information available, these techniques compare the results and finally give a conclusion. But, if there is a new virus in the patient's samples or if the existing viruses have undergone mutation, then all the techniques fail to identify the infecting virus as there is no information available on their gene sequences. To solve this issue, a new technique termed "viral metagenomics" has been successfully developed and utilized for the discovery and identification of many previously uncharacterized human, animal, insect, and plant viruses in environmental and biological samples.[49,75–78] Viral metagenomics for virus discovery involves sequence-independent amplification (SIA) followed by subcloning, large-scale sequencing, and bioinformatics. The methodology usually starts with nucleic acid extraction. If it is not known whether the expected viral genome is DNA or RNA, then a purified nucleic acid from each sample should be separated into two different reaction tubes (for expected RNA and DNA viruses) and processed separately. Later, all expected nucleic acids; *ss*RNA, *ds*RNA, *ss*DNA, and *ds*DNA should undergo conversion to *ds*DNA, after which SIA can be carried out followed by postamplification analysis and bioinformatics.[79,80] Though viral metagenomics has the potential to revolutionize the diagnosis process of viral infections, until now it is in the development stage and will take some time for use in a full commercial scale.

REFERENCES

1. Ludlow M, Kortekaas J, Herden C, Hoffmann B, Tappe D, Trebst C, et al. Neurotropic virus infections as the cause of immediate and delayed neuropathology. *Acta Neuropathol* 2016;**131**(2):159–84.
2. Sejvar J. Neuroepidemiology and the epidemiology of viral infections of the nervous system. *Handb Clin Neurol* 2014;**123**:67–87.
3. Walker MD, Zunt JR. Neuroparasitic infections: cestodes, trematodes, and protozoans. *Semin Neurol* 2005;**25**(3):262–77.
4. Meyer H, Johnson R, Crawford I, Dascomb H, Rogers N. Central nervous system syndromes of 'viral' etiology. A study of 713 cases. *Am J Med* 1960;**29**:334–47.
5. Lennette EH, Magoffin RL, Knouf EG. Viral central nervous system disease an etiologic study conducted at the Los Angeles County General Hospital. *J Am Med Assoc* 1962;**179**(9):687–95.
6. Beghi E, Nicolosi A, Kurland LT, Mulder DW, Hauser WA, Shuster L. Encephalitis and aseptic meningitis, Olmsted County, Minnesota, 1950–1981: I. Epidemiology. *Ann Neurol* 1984;**16**(3):283–94.
7. Koyuncu OO, Hogue IB, Enquist LW. Virus infections in the nervous system. *Cell Host Microbe* 2013;**13**(4):379–93.
8. Johnson RT. Microbiology of the nervous system. In: Baron S, editor. *Medical microbiology*. 4th ed. Galveston, TX: University of Texas Medical Branch; 1996. p. 1–15.
9. Young VA, Rall GF. Making it to the synapse: measles virus spread in and among neurons. *Curr Top Microbiol Immunol* 2009;**330**:3–30.
10. Rhoades RE, Tabor-Godwin JM, Tsueng G, Feuer R. Enterovirus infections of the central nervous system. *Virology* 2011;**411**(2):288–305.
11. Tavazzi E, Ferrante P, Khalili K. Progressive multifocal leukoencephalopathy: an unexpected complication of modern therapeutic monoclonal antibody therapies. *Clin Microbiol Infect* 2011;**17**(12):1776–80.
12. McGavern DB, Kang SS. Illuminating viral infections in the nervous system. *Nat Rev Immunol* 2011;**11**(5):318–29.
13. Handique SK. Viral infections of the central nervous system. *Neuroimaging Clin N Am* 2011;**21**(4):777–94.
14. Bulakbasi N, Kocaoglu M. Central nervous system infections of herpesvirus family. *Neuroimaging Clin N Am* 2008;**18**(1):53–84.
15. Somand D, Meurer W. Central nervous system infections. *Emerg Med Clin North Am* 2009;**27**(1):89–100.
16. Solomon T, Hart IJ, Beeching NJ. Viral encephalitis: a clinician's guide. *Pract Neurol* 2007;**7**(5):288–305.
17. Davis LE, Tyler KL. Molecular diagnosis of CNS viral infections. *J Neurol Neurosurg Psychiatry* 2005;**76**(1):10.
18. Petersen SE, Fiez JA, Corbetta M. Neuroimaging. *Curr Opin Neurobiol* 1992;**2**(2):217–22.
19. Mueller SG, Weiner MW, Thal LJ, Petersen RC, Jack C, Jagust W, et al. The Alzheimer's disease neuroimaging initiative. *Neuroimaging Clin N Am* 2005;**15**(4):869–77.

20. Shah DH, Patel KM, Patel JB, Patel JS, Garg CS, Jyoti SD. Interface between neuroimaging and brain mapping in cognitive psychology. *Int J Drug Dev Res* 2011;**3**(2):51–63.
21. Lelyveld VS, Atanasijevic T, Jasanoff A. Challenges for molecular neuroimaging with MRI. *Int J Imaging Syst Technol* 2010;**20**(1):71–9.
22. Reiber H, Peter JB. Cerebrospinal fluid analysis: disease-related data patterns and evaluation programs. *J Neurol Sci* 2001;**184**(2):101–22.
23. Prajapati BS, Prajapati RB, Vora HD. Analysis of cerebrospinal fluid (CSF) in children. *Pediatr Infect Dis* 2015;**7**(1):22–6.
24. Reiber H. Proteins in cerebrospinal fluid and blood: barriers, CSF flow rate and source-related dynamics. *Restor Neurol Neurosci* 2003;**21**(3–4):79–96.
25. Mahfoud F, Grtner B, Kindermann M, Ukena C, Gadomski K, Klingel K, et al. Virus serology in patients with suspected myocarditis: utility or futility? *Eur Heart J* 2011;**32**(7):897–903.
26. DeBiasi RL, Tyler KL. Molecular methods for diagnosis of viral encephalitis. *Clin Microbiol Rev* 2004;**17**(4):903–25.
27. Kumar R. Aseptic meningitis: diagnosis and management. *Indian J Pediatr* 2005;**72**:57–63.
28. Steiner I, Budka H, Chaudhuri A, Koskiniemi M, Sainio K, Salonen O, et al. Viral encephalitis: a review of diagnostic methods and guidelines for management. *Eur J Neurol* 2005;**12**(5):331–43.
29. Chaudhuri A, Kennedy PGE. Diagnosis and treatment of viral encephalitis. *Postgrad Med J* 2002;**78**(924): 575–83.
30. Lu H-J, Bloch KC, Tang Y-W. Molecular techniques in the diagnosis of central nervous system infections. *Curr Infect Dis Rep* 2002;**4**(4):339–50.
31. Davies NWS, Brown LJ, Gonde J, Irish D, Robinson RO, Swan AV, et al. Factors influencing PCR detection of viruses in cerebrospinal fluid of patients with suspected CNS infections. *J Neurol Neurosurg Psychiatry* 2005;**76**(1):82–7.
32. Kennedy PGE. Neurovirological methods and their applications. *J Neurol Neurosurg Psychiatry* 2003;**74**(8):1016–22.
33. Warren JD, Schott JM, Fox NC, Thom M, Revesz T, Holton JL, et al. Brain biopsy in dementia. *Brain* 2005;**128**(9):2016–25.
34. Gilkes CE, Love S, Hardie RJ, Edwards RJ, Scolding NJ, Rice CM. Brain biopsy in benign neurological disease. *J Neurol* 2012;**259**(5):995–1000.
35. Gibbs RA. Polymerase chain reaction techniques. *Curr Opin Biotechnol* 1991;**2**(1):69–75.
36. Powledge TM. The polymerase chain reaction. *Adv Physiol Educ* 2004;**28**(1–4):44–50.
37. Klapper PE, Cleator GM, Dennett C, Lewis AG. Diagnosis of herpes encephalitis via southern blotting of cerebrospinal fluid DNA amplified by polymerase chain reaction. *J Med Virol* 1990;**32**(4):261–4.
38. Read SJ, Burnett D, Fink CG. Molecular techniques for clinical diagnostic virology. *J Clin Pathol* 2000;**53**(7):502–6.
39. Cinque P, Bossolasco S, Lundkvist A. Molecular analysis of cerebrospinal fluid in viral diseases of the central nervous system. *J Clin Virol* 2003;**26**(1):1–28.
40. Gill P, Ghaemi A. Nucleic acid isothermal amplification technologies: a review. *Nucleosides Nucleotides Nucleic Acids* 2008;**27**(3):224–43.
41. Liu YT. A technological update of molecular diagnostics for infectious diseases. *Infect Disord Drug Targets* 2008;**8**:183–8.
42. Bej AK, Mahbubani MH, Atlas RM, Salkl RK. Amplification of nucleic acids by polymerase chain reaction (PCR) and other methods and their applications. *Crit Rev Biochem Mol Biol* 1991;**26**(3–4): 301–34.
43. Wolcott MJ. Advances in nucleic acid-based detection methods. *Clin Microbiol Rev* 1992;**5**(4):370–86.
44. Clewley JP. A role for arrays in clinical virology: fact or fiction? *J Clin Virol* 2004;**29**(1):2–12.
45. Miller MB, Tang YW. Basic concepts of microarrays and potential applications in clinical microbiology. *Clin Microbiol Rev* 2009;**22**(4):611–33.
46. Wang D, Urisman A, Liu YT, Springer M, Ksiazek TG, Erdman DD, et al. Viral discovery and sequence recovery using DNA microarrays. *PLoS Biol* 2003;**1**(2):257–60.
47. Palacios G, Druce J, Du L, Tran T, Birch C, Briese T, et al. A new arenavirus in a cluster of fatal transplant-associated diseases. *New Engl J Med* 2008;**358**(10):991–8.

48. Quan PL, Briese T, Palacios G, Ian Lipkin W. Rapid sequence-based diagnosis of viral infection. *Antivir Res* 2008;**79**(1):1–5.
49. Tang P, Chiu C. Metagenomics for the discovery of novel human viruses. *Future Microbiol* 2010;**5**(2): 177–89.
50. Boriskin YS, Rice PS, Stabler RA, Hinds J, Al-Ghusein K, Vass K, et al. DNA microarrays for virus detection in cases of central nervous system infection. *J Clin Microbiol* 2004;**42**(12):5811–8.
51. Conejero-Goldberg C, Wang E, Yi C, Goldberg TE, Jones-Brando L, Marincola FM, et al. Infectious pathogen detection arrays: viral detection in cell lines and postmortem brain tissue. *BioTechniques* 2005;**39**(5):741–9.
52. Zheng ZB, Wu YD, Yu XL, Shang SQ. DNA microarray technology for simultaneous detection and species identification of seven human herpes viruses. *J Med Virol* 2008;**80**(6):1042–50.
53. Wada K, Mizoguchi S, Ito Y, Kawada J-I, Yamauchi Y, Morishima T, et al. Multiplex real-time PCR for the simultaneous detection of herpes simplex virus, human herpesvirus 6, and human herpesvirus 7. *Microbiol Immunol* 2009;**53**(1):22–9.
54. Dunbar SA. Applications of Luminex(R) xMAPtrade mark technology for rapid, high-throughput multiplexed nucleic acid detection. *Clin Chim Acta* 2006;**363**(1–2):71–82.
55. Smith LM, Sanders JZ, Kaiser RJ, Hughes P, Dodd C, Connell CR, et al. Fluorescence detection in automated DNA sequence analysis. *Nature* 1986;**321**(6071):675–9.
56. Smith LM, Fung S, Hunkapiller MW, Hunkapiller TJ, Hood LE. The synthesis of oligonucleotides containing an aliphatic amino group at the 5′ terminus: synthesis of fluorescent DNA primers for use in DNA sequence analysis. *Nucleic Acids Res* 1985;**13**(7):2399–412.
57. Ronaghi M, Karamohamed S, Pettersson B, Uhlén M, Nyrén PL. Real-time DNA sequencing using detection of pyrophosphate release. *Anal Biochem* 1996;**242**:84–9.
58. Ronaghi M, Uhlén M, Nyrén P. PyroSequencing: a {DNA} sequencing method based on real-time pyrophosphate detection. *Science* 1998;**281**:363–5.
59. Ansorge WJ. Next-generation DNA sequencing techniques. *New Biotechnol* 2009;**25**(4):195–203.
60. Kircher M, Kelso J. High-throughput DNA sequencing—concepts and limitations. *BioEssays* 2010; **32**(6):524–36.
61. Pettersson E, Lundeberg J, Ahmadian A. Generations of sequencing technologies. *Genomics* 2009; **93**(2):105–11.
62. Schoenfeld T, Liles M, Wommack KE, Polson SW, Godiska R, Mead D. Functional viral metagenomics and the next generation of molecular tools. *Trends Microbiol* 2010;**18**(1):20–9.
63. Feng H, Shuda M, Chang Y, Moore PS. Clonal integration of a polyomavirus in human Merkel cell carcinoma. *Science* 2008;**319**(5866):1096–100.
64. Briese T, Paweska JT, McMullan LK, Hutchison SK, Street C, Palacios G, et al. Genetic detection and characterization of Lujo virus, a new hemorrhagic fever-associated arenavirus from southern Africa. *PLoS Pathog* 2009;**5**(5).
65. Greninger AL, Runckel C, Chiu CY, Haggerty T, Parsonnet Ganem JD, Derisi JL. The complete genome of klassevirus—a novel picornavirus in pediatric stool. *Virol J* 2009;**6**(1):82.
66. Rao CM, Rao IS, Ramchander PV, Madhavan HN, Sharma S, Satpathy G, et al. *Method for simultaneous detection and discrimination of bacterial, fungal, parasitic and viral infections of eye and central nervous system*; 2013 [US 8465951 B2].
67. Chen F, Kong LI, Lee MC, Jules C, Tabb MM, Aye M. *Methods and compositions for detecting CNS viruses*; 2009 [US 2009/0226889 A1].
68. Chaturvedi PR. *Treatment of the cns effects of hiv with vx478, alone or in combination with azt or 3tc*; 1997 [CA 2238471 A1].
69. Mckenzie BS, Curwen PF, Maraskovsky E. *Treatment of neurological conditions*; 2011 [WO 2011032204 A1].
70. Saikh KU. *A small molecule inhibitor of myd88 for therapeutic treatment against alphavirus and staphylococcal enterotoxin infections and toxin exposure*; 2015 [WO 2015191111 A2].
71. Cantin EM, Chandran R, Lundberg P. *Compositions and methods for treating herpes simplex virus infections and related diseases*; 2010 [US 2010/0040601 A1].
72. Meyding-Lamadé U, Strank C. Herpesvirus infections of the central nervous system in immunocompromised patients. *Ther Adv Neurol Disord* 2012;**5**(5):279–96.

73. Gupta RK, Kumar S. Central nervous system tuberculosis. *Neuroimaging Clin N Am* 2011;**21**(4): 795–814.
74. Pham NTK, Trinh QD, Takanashi S, Abeysekera C, Abeygunawardene A, Shimizu H, et al. Novel human parechovirus, Sri Lanka. *Emerg Infect Dis* 2010;**16**(1):130–2.
75. Breitbart M, Hewson I, Felts B, Mahaffy JM, Nulton J, Salamon P, et al. Metagenomic analyses of an uncultured viral community from human feces. *J Bacteriol* 2003;**185**(20):6220–3.
76. Breitbart M, Salamon P, Andresen B, Mahaffy JM, Segall AM, Mead D, et al. Genomic analysis of uncultured marine viral communities. *Proc Natl Acad Sci U S A* 2002;**99**(22):14250–5.
77. Ambrose HE, Clewley JP. Virus discovery by sequence-independent genome amplification. *Rev Med Virol* 2006;**16**(6):365–83.
78. Delwart EL. Viral metagenomics. *Rev Med Virol* 2007;**17**(2):115–31.
79. Alavandi SV, Poornima M. Viral metagenomics: a tool for virus discovery and diversity in aquaculture. *Indian J Virol* 2012;**23**(2):88–98.
80. Agindotan BO, Ahonsi MO, Domier LL, Gray ME, Bradley CA. Application of sequence-independent amplification (SIA) for the identification of RNA viruses in bioenergy crops. *J Virol Methods* 2010;**169**(1):119–28.

FURTHER READING

1. Read SJ, Mitchell JL, Fink CG. LightCycler multiplex PCR for the laboratory diagnosis of common viral infections of the central nervous system. *J Clin Microbiol* 2001;**39**(9):3056–9.

CHAPTER 4

Vaccines Against Central Nervous System Infections: Past Achievements and Future Challenges

Manish Sadarangani[*,†]
[*]University of British Columbia, Vancouver, BC, Canada
[†]University of Oxford, Oxford, United Kingdom

1. INTRODUCTION

Meningitis and encephalitis are devastating infections of the central nervous system (CNS) with high mortality and morbidity rates. Although most individuals with meningitis (inflammation of the meninges) or encephalitis (inflammation of the brain parenchyma) have elements of both to a lesser or greater extent, they are often considered as separate entities with different but overlapping causes. Bacterial meningitis has a case-fatality rate of approximately 5%–10% in high-income countries, and this has remained relatively unchanged since the 1950s.[1–11] In addition, many survivors suffer significant long-term neurological sequelae, including epilepsy, hearing and vision loss, developmental delay, and cognitive difficulties.[12–19] The coexistence of septicemia in a large number of cases also contributes to these complications, and can lead to long-term non-neurological sequelae such as limb loss and damage to bony growth plates. Rates of these sequelae depend to a large extent on the pathogen and age at time of infection. There are less data on outcomes following viral meningitis, with a small number of conflicting studies. Most affected individuals are thought to experience a full recovery, but some studies have suggested transient complications such as fatigue, irritability, reduced concentration, and muscle pain, weakness, or spasm.[20] Others have reported subtler, longer-term problems, including memory difficulties and mild cognitive defects.[21–24] In contrast, viral encephalitis has a very poor outcome, resulting in a 7%–20% case-fatality rate[25–27] with up to 50% of survivors reporting deficits such as memory loss, epilepsy, learning disability, and functional impairment with prolonged followup.[28–32]

During the initial pathogenesis of meningitis, organisms infect a mucosal surface (most commonly the nasopharynx, but also the gastrointestinal tract) before invading the blood stream and subsequently crossing the blood-brain barrier. The time from initial invasion to involvement of the CNS can be extremely short (often hours). A similar process is also thought to occur for most encephalitis pathogens, although some may travel

The Microbiology of Central Nervous System Infections
https://doi.org/10.1016/B978-0-12-813806-9.00004-4

from the initial site of infection into the CNS via a neuronal pathway, such as the herpes simplex virus (HSV). This is important when considering protection against such infections via vaccination. At an individual level, a high level of ongoing immunoprotective factors (usually specific antibodies) is required due to the rapidity of the onset of infection, because there may not be time for the host to mount an anamnestic immune response. At a population level, prevention of initial infection at the mucosal surface will reduce transmission of the organism(s) in a highly immunized population and thereby indirectly protect unimmunized individuals as well as directly protect the immunized person. This herd protection has been critical for the success of protein-polysaccharide conjugate vaccines against bacterial meningitis pathogens over the last 25 years.

2. EPIDEMIOLOGY OF MENINGITIS

Historically, most attention has been focused on bacterial meningitis due to its poor outcomes and its potential treatment with antibiotics. In countries that include conjugate vaccines against the major bacterial meningitis pathogens (*Haemophilus influenzae*, *Streptococcus pneumoniae*, and *Neisseria meningitidis*) in their routine infant immunization schedules, viral causes predominate. Bacterial meningitis is most common in young infants aged <3 months, and is predominantly due to Group B Streptococcus, *Escherichia coli*, and other Gram-negative organisms, as well as *Listeria monocytogenes* in this age group; there are currently no licensed vaccines against any of these pathogens. Enteroviruses and human parechoviruses are the most common viruses found in meningitis. Enteroviruses affect individuals of all ages, but are most common in infants; parechoviruses usually occur in younger infants <3 months of age, and infections are predominantly due to type 3.[33] Measles and mumps were major causes of both meningitis and encephalitis prior to the routine use of measles and mumps vaccines, but are now rare causes of CNS infections. Other viral causes include HSV (which is usually considered as an encephalitis cause), varicella zoster virus, Epstein-Barr virus, cytomegalovirus, adenoviruses, human immunodeficiency virus, rubella, influenza, parainfluenza, and rotavirus. *Mycobacterium tuberculosis* is a major cause of meningitis worldwide, despite widespread use of the Bacille de Calmette et Guérin vaccine. A large number of other infectious agents, including nonpyogenic and atypical bacteria, fungi, protozoa, and helminthes, also cause meningitis.

3. EPIDEMIOLOGY OF ENCEPHALITIS

Most cases of acute infectious encephalitis are caused by viruses, with HSV being the most common virus identified in cases of severe encephalitis. Almost all the viruses mentioned above as causes of meningitis are also considered as significant encephalitis pathogens, with enterovirus being the most common. There are also a large number of arthropod-transmitted agents that cause encephalitis and are restricted geographically based on the distribution of the relevant vector. These include West Nile virus, La Crosse virus, St Louis encephalitis virus, dengue, rabies, Eastern equine encephalitis virus, and Western equine encephalitis virus, among others. Bacteria can cause encephalitis and can follow infection with any of the agents more usually considered as meningitis pathogens listed above. Other bacterial causes include *Bartonella henselae, Coxiella burnetii, Borrelia burgdorferi, Mycoplasma pneumoniae, Rickettsia,* and *M. tuberculosis*. Less common causes are *Toxoplasma gondii, Naegleria fowleri, Baylisascaris procyonis,* and Creutzfeldt-Jakob Disease.

This chapter will focus on pathogens for which vaccines have been developed and had a major impact on meningitis and encephalitis epidemiology.

4. VACCINES AGAINST *H. INFLUENZAE*

H. influenzae is categorized into different serotypes (a–f) on the basis of its polysaccharide capsule, with some organisms being nonencapsulated and designated "nontypable." Prior to the use of *H. influenzae* vaccines, the majority of invasive disease was caused by *H. influenzae* type b (Hib); the incidence of invasive disease due to Hib was 12–54 per 100,000 per year in children under 5 years old in Europe[34–36] and 20–88 per 100,000 per year in the United States.[37–39] Approximately 60% of these children had meningitis. Hib vaccines have specifically targeted the type b capsule, which consists of polyribosylribitol phosphate (PRP).

Initial Hib vaccines were based on purified PRP alone, and >90% of children aged 4–5 years had an antibody response (≥2-fold rise in anti-PRP IgG) to this vaccine.[40] In contrast, younger children and infants responded poorly.[36,41–43] This poor response in younger children, particularly those <2 years of age, has subsequently been shown as a consistent feature of purified polysaccharide vaccines (see later sections on vaccines against *S. pneumoniae* and *N. meningitidis*) and is because the purified polysaccharide fails to recruit T helper cells as part of the immune response. In younger children, the T-independent B cell response is poor and does not result in good antibody responses. In later studies, Hib polysaccharide vaccine was reported to have an efficacy of between 55% and 92% against invasive Hib disease in children aged over 18–24 months,[39,44–46] and was licensed in the United States in 1985 for children aged 24–72 months.[47]

These polysaccharide vaccines were partially effective in reducing the overall rates of Hib disease, but were unable to protect those at highest risk—infants and children <2 years of age. Conjugate vaccines were subsequently developed whereby the PRP polysaccharide is chemically linked to a protein carrier. The protein component elicits T cell help following vaccination, resulting in a robust immune response in all age groups. Additional benefits of conjugate vaccines include induction of a memory B cell response, resulting in anamnestic responses following booster doses. Multiple Hib conjugate vaccines have now been developed, differing in their carrier protein and precise conjugation chemistry. The most common carrier proteins used for conjugate vaccines, including Hib, are CRM_{197} (a detoxified form of diphtheria toxin) and tetanus toxoid. Studies of these conjugate vaccines demonstrated vaccine efficacy of 87%–100% in the United States, Finland, the United Kingdom, Chile, and Gambia after two or three doses in infants and young children.[48–54] Early postlicensure studies in the United States, Australia, and Israel confirmed the findings of the efficacy trials.[55–58]

Hib meningitis and other invasive Hib diseases have almost completely disappeared in populations where Hib vaccines are used. For example, in the United Kingdom there were 2.9–6.7 admissions per 100,000 children per year for Hib meningitis between 1968 and 1992, with a peak of 6.72 per 100,000 per year in 1992.[59] After the introduction of routine Hib vaccination in 1992, there was a 94% decrease to 0.4 admissions per 100,000 children per year within 2 years. This trend in hospital admissions was mirrored by data from laboratory reports of Hib in the United Kingdom, which revealed a 98% reduction in invasive Hib disease in children younger than 5 years by 6 years after Hib vaccine introduction.[60,61] A subsequent small rise in admissions in the early 2000s occurred, peaking in 2003; following the introduction of a Hib booster vaccine dose in toddlers this decreased by 2008.[59,61,62] Interestingly, the increase in total Hib disease reported in laboratory surveillance data in this later resurgence reached almost a third of the peak number of cases in the early 1990s, whereas the increase in *H. influenzae* meningitis admissions based on hospital data was only about a sixth of the previous peak.[59,63,64] This suggests that cases seen in the early 2000s may have been less severe than previous cases, possibly indicating partial immunity from infant immunization or a lower susceptibility of older children to meningitis during Hib infection compared with infants. This resurgence was thought to be caused by multiple factors: (i) children aged 1–4 years who had been vaccinated as infants had a greater than expected waning of anti-PRP antibody concentrations postvaccination, such that antibody levels fell below that required for protection against invasive disease; (ii) there was a decrease in herd immunity after the cessation of the catch-up campaign (implemented when the vaccine was introduced); and (iii) a change in the combination Hib vaccine used.[62,63,65] A similar increase in invasive Hib disease occurred in the Netherlands in 2002,[66] but not in Finland.[67] Country-specific differences might be the result of varied vaccine schedules.

Hib conjugate vaccines have the additional advantage of preventing pharyngeal colonization of the organism, thereby reducing transmission of the organism in the population. This results in indirect protection of individuals who are not vaccinated, or herd protection,[68] as long as there is high coverage of the vaccine in a population. This has been confirmed in multiple populations following Hib vaccine introduction and has further contributed to the dramatic reductions in Hib meningitis observed since the early 1990s.[69–71]

The disappearance of Hib meningitis and other Hib invasive diseases over the last 25 years has made the occurrence of other *H. influenzae* serotypes more apparent.

H. influenzae type a (Hia) has now emerged as a significant cause of bacterial meningitis and other diseases that had been similarly caused by Hib, such as pneumonia, bone and joint infections, and bacteremia. This serotype appears to be most predominant in indigenous populations of Australia, the United States, and Canada. Studies of indigenous and/or rural communities in Northern Canada and Alaska have reported Hia incidence rates up to 419 per 100,000 per year.[72] Development of conjugate Hia vaccines is ongoing although a number of key issues remain unresolved, including the precise epidemiology of Hia disease, the determination of high risk groups which would be targeted for any immunization program, better understanding of Hia transmission, and establishment of the protective level of antibody required. Once a vaccine is developed it would also have to go through the usual preclinical studies and phase 1, 2, and 3 clinical trials, so prevention of Hia meningitis remains several years away.

5. VACCINES AGAINST *S. PNEUMONIAE*

S. pneumoniae is classified into different serotypes based on its polysaccharide capsule, but unlike *H. influenzae*, where there are only six serotypes, there are >90 serotypes of *S. pneumoniae*. Similar to *H. influenzae*, initial infection with *S. pneumoniae* occurs in the nasopharynx and results in asymptomatic carriage, during which time transmission between individuals can occur. Bacteria are then eliminated from the infected individual by the immune system, or go on to cause disease, including meningitis. Pneumococcal disease is a major cause of morbidity and mortality worldwide. In 2008, there were ~476,000 deaths globally due to pneumococcal disease and ~5% of these deaths were in children aged <5 years.[73] The peak incidence of invasive pneumococcal disease is in children <2 years, but the case-fatality rate is higher in older adults. In Europe the incidence of pneumococcal meningitis was 1–8 cases per 100,000 per year in children <5 years prior to widespread use of the 7 valent pneumococcal conjugate vaccine (PCV7). In the prevaccine era (1998–99) in the United States, the incidence of pneumococcal meningitis was 1.1 per 100,000 per year, with the highest rate of 10 per 100,000 per year observed in children aged <2 years.[74] Following introduction of Hib vaccines, *S. pneumoniae* was the leading cause of bacterial meningitis.

PCV7 contains polysaccharides from serotypes 4, 6B, 9V, 14, 18C, 19F, and 23F, with each polysaccharide conjugated to a carrier protein. The 10-valent vaccine PCV10 also covers serotypes 1, 5, and 7F. PCV13 includes these serotypes plus 3, 6A, and 19A. Higher valency conjugate vaccines are currently undergoing clinical trials. PCVs have markedly reduced the disease burden wherever they have been used throughout the world. PCV7 was first licensed in 2000 in the United States, using a schedule of three doses at ages 2, 4, and 6 months, followed by a booster at age 12–15 months (3 + 1). PCV7 has largely been succeeded by PCV10 and/or PCV13 because of their broader serotype coverage. In clinical trials of different PCVs, vaccine efficacy of the 3 + 1 regimen was 83%–100% against vaccine-type pneumococcal disease.[75] Subsequent trials reported efficacy of 71%–83% for a 3+0 schedule and 92% for a 2+1 schedule (two doses at ages <7 months and a booster at age >11 months).

Following introduction of PCV7 in the United States in 2000, the incidence of pneumococcal meningitis declined from 1.1 per 100,000 per year to 0.8 per 100,000 per year between 1998–99 and 2004–05.[74] The decrease in the vaccinated age group of children under 2 years was 64%, but there was also a marked and significant decrease of 54% in older adults aged 65 years and above, who were not in the vaccinated group. Rates of PCV7-serotype meningitis declined by 73% from 0.7 per 100,000 per year to 0.2 per 100,000 per year overall, with a 93% reduction in children aged <2 years. In the United Kingdom the incidence of invasive pneumococcal disease overall decreased by 56% in children under 2 years of age after introduction of PCV7 and prior to use of PCV13, from 54 per 100,000 per year to 24 per 100,000 per year.[76] The decrease in PCV7-serotype IPD was 98%. Two years after introduction of PCV7 for vaccination of "high-risk" children in France, there was a 39% reduction in the incidence of pneumococcal meningitis in all children <2 years.[77] In the United States, introduction of PCV13 has further reduced the proportion of pneumococcal meningitis caused by PCV13-serotypes, from 54% in 2007–09 to 27% in 2011–13.[78] The reduction in the older, unvaccinated population is due to PCVs reducing nasopharyngeal carriage and therefore pathogen transmission, resulting in herd protection and indirect protection of those who have not been vaccinated, similar to Hib conjugate vaccines. A carriage study in Alberta, Canada, found that in 2003, pre-PCV introduction, 77% of *S. pneumoniae* isolated from young children were PCV13-serotypes, but in 2012, following use of PCV7 and PCV13, only 6% were PCV13-serotypes.[79]

The impact of PCV was also observed in a reduction in hospital admissions due to pneumococcal meningitis in a study from the United Kingdom. Between 1968 and 1985, the admission rate for children with pneumococcal meningitis was 1.1–2.3 per 100,000 per year in England; this increased during the 1990s and 2000s to a peak of 4.5 per 100,000 per year in 2006. PCV7 was introduced in the United Kingdom in 2006, and a decrease in hospital admissions due to pneumococcal meningitis was observed after 2006, with the overall hospital admission rate in 2011 being similar to the rate in the early 1990s of 2.0 per 100,000 per year.[59] The effect of vaccination is less pronounced than that observed following Hib vaccine introduction because PCVs cover only 7, 10, or 13 of the commonest >90 serotypes of *S. pneumoniae*. While decreases in vaccine serotypes have been marked, this has been partially offset by smaller increases in nonvaccine serotypes.

Ongoing challenges remain in prevention of pneumococcal meningitis. The large number of serotypes is a challenge for vaccine developers. While the current conjugate vaccines target the most common serotypes observed in North America, a significant proportion of disease is caused by nonvaccine serotypes. These nonvaccine serotypes have increased since PCVs have been used, although overall rates of pneumococcal disease remain much lower than in the prevaccine era. Serotype distribution is also different in other parts of the world, most notably resource-poor regions, where disease rates remain high.[80] Higher valency vaccines are in development, but the number of serotypes that can be included in a vaccine is limited by cost and the difficulties in formulating large numbers of polysaccharides into the small volume required for a vaccine. A 23-valent plain polysaccharide vaccine is used in older adults in many countries, but does not have a major impact on pneumococcal meningitis because this disease manifestation occurs primarily in younger children in whom the polysaccharide vaccine is ineffective. Alternative approaches for "universal" pneumococcal vaccines are being utilized, including whole cell, protein, and DNA vaccines, which target nonpolysaccharide components of the bacteria and are therefore not serotype-restricted. One whole cell vaccine has been effective in a mouse model and is now in a phase 1 clinical trial in healthy adult volunteers.[81]

6. VACCINES AGAINST *N. MENINGITIDIS*

N. meningitidis is categorized into capsular groups on the basis of the polysaccharide capsule, similarly to *H. influenzae* and *S. pneumoniae*. Six of the 12 groups (A, B, C, W, X, and Y) are responsible for almost all cases of invasive disease worldwide.[82] It is a common nasopharyngeal commensal, found in approximately 10% of the population, but bacteria occasionally enter the bloodstream to cause devastating invasive diseases such as meningitis and septicaemia.[83] In Europe, disease incidence ranges between 0.1 and 1.8 cases per 100,000 per year with highest rates occurring in infants under 1 year of age (>10 per 100,000 per year). The majority of cases in Europe are now caused by group B organisms (68% in 2012), followed by group C (17%), Y (8%), and W (4%). Between 2008 and 2012 there was a decrease in the number of cases caused by group B and C organisms. In part this can be attributed to the introduction of capsular group C (MenC) conjugate vaccines in a number of countries; however, until 2015 there were no licensed vaccines targeting MenB and this decrease is probably part of the natural fluctuations in disease incidence.[84] In the United States, the disease rate was 1.1 cases per 100,000 population in 1999, but had fallen to 0.14 cases per 100,000 by 2014. In the United States, capsular group B organisms cause most disease in infants. Beyond 1 year of age, 85% of disease cases are approximately equally distributed between capsular groups B, C, and Y strains. The arrival and eventual decline, through development of natural immunity, of hyperinvasive lineages is now recognized as a major driver of changes in disease rates over time. For example, disease caused by capsular group W strains has increased in the United Kingdom in recent years as a result of a hyperinvasive clone that appears to have originated in Latin America.[85] The highest incidence of capsular group A disease has occurred in a band across sub-Saharan Africa, the "meningitis belt," with annual endemic rates of 10–25 per 100,000 population and rates in epidemic years as high as 1000 per 100,000 population.

The first meningococcal vaccines in regular use were plain polysaccharide vaccines. These purified polysaccharide group-specific vaccines target disease caused by groups A, C, W, and Y. They are effective in protecting older children and adults against disease, but as with other polysaccharide vaccines, the resulting immune response does not involve recruitment of T helper cells. Therefore these vaccines do not induce immunological memory, induce short-term protection only, are associated with immunological hyporesponsiveness (reduced responses after administration of booster doses), and do not elicit a response in children under 2 years of age.[86–91] In the 1990s meningococcal conjugate vaccines were successfully developed.[92] The first monovalent MenC conjugate vaccines were licensed and introduced into the UK routine immunization program in 1999, and between 1999 and 2001 there was a reduction in MenC cases of 87% among the vaccinated groups.[93] Following the success of monovalent MenC conjugate vaccines, the same technology was used to develop conjugate vaccines targeting other capsular groups. In 2005, the first quadrivalent ACWY conjugate vaccine was licensed in the United States, and multiple quadrivalent conjugate vaccines are now available with different protein carriers (diphtheria toxoid, CRM_{197}, and tetanus toxoid). The rate of meningococcal disease caused by groups C, W, and Y in the United States decreased in 11–19 year olds from 1.1 per 100,000 per year in 1998–99 to 0.05 per 100,000 per year in 2010–11 after introduction of the MenACWY conjugate vaccine, although this rate was already decreasing prior to vaccine introduction in 2005. An affordable monovalent MenA-tetanus toxoid conjugate vaccine has been produced for use in the meningitis belt of sub-Saharan Africa.[94] It was rolled out in 2010 and has resulted in a 94% reduction in disease incidence in Chad,[95,96] with no reports of group A meningococcal meningitis among vaccinated individuals. However, other capsular groups have since caused significant outbreaks in the meningitis belt, demonstrating the need for multivalent meningococcal conjugate vaccines for this population.

The polysaccharide capsule of group B meningococci is not immunogenic in humans due to structural homology between the polysaccharide and human tissue, leading to immunological tolerance. The key component of the serogroup B capsule is an α2-8-linked sialic acid homopolymer, which contains epitopes that are cross-reactive with the polysialylated form of the neural cell adhesion molecule.[97] Alternative strategies for the development of MenB vaccines have therefore been required, and these have focussed on subcapsular components, particularly outer membrane proteins (OMPs). OMPs, however, are antigenically diverse so a successful vaccine must contain either different OMPs and/or multiple variants of the same OMP to ensure broad coverage of protection against the diverse range of strains observed in countries with endemic disease. Two new vaccines designed to induce protection against a number of different MenB strains have recently been licensed in Europe, North America, and several other areas.

rLP2086 is a bivalent vaccine containing two variants of the immunogenic outer membrane lipoprotein factor H binding protein (fHBP).[98] In a phase 3 clinical trial, three doses of rLP2086 in 11–18 year olds resulted in seroconversion (≥4-fold rise in serum bactericidal antibody (SBA) titer) in 87%–93% of participants (https://clinicaltrials.gov, identifier NCT00808028). In one study, sera from 75% to 100% of adolescents and young adults immunized with rLP2086 had bactericidal activity against a panel of representative strains.[99] The four-component vaccine 4CMenB contains three recombinant OMPs (neisserial heparin binding antigen, neisserial adhesin A, and fHBP) formulated with the outer membrane vesicle derived from a New Zealand outbreak strain.[100] Following three doses of 4CMenB given at 2, 4, and 6 months of age, 85%–95% of infants had an SBA titer of ≥1:4 (the presumed protective threshold when using human complement in the assay) against target strains that expressed the same antigen variants as those present in the vaccine. This increased to 93%–100% after a booster dose at 12 months.[101] In more recent studies, 84%–100% of infants vaccinated with the same schedule had an SBA titer of ≥1:5 against strains expressing identical variants of the vaccine antigens.[102] At 40–44 months of age, 41%–76% of children had SBA titers ≥1:4 following four doses of vaccine at 2, 4, 6, and 12 months, compared with 0%–68% of unvaccinated children when strains matched for vaccine antigen variants were used in the SBA assay. The magnitude, duration, and breadth of the antibody response was greater among adolescents and adults than in infants.[103] 4CMenB has recently been used to control outbreaks of meningococcal disease at two United States universities[104] and in 2014 was used in a vaccination campaign for hyperendemic disease in the Saguenay-Lac-Saint-Jean region of Quebec, Canada.[105] No cases of MenB disease have been identified in vaccinated individuals in this region up to May 2017 (P. de Wals, personal communication). In September 2015, 4CMenB was introduced into the routine infant immunization schedule in the United Kingdom with a schedule of three doses at 2, 4, and 12 months of age. Early data indicate a vaccine effectiveness of 83% against all capsular group B meningococcal diseases following two doses at 2 and 4 months of age,[106] but effectiveness is anticipated to be higher against strains targeted by the vaccine antigens.

The reduction in nasopharyngeal meningococcal carriage by all conjugate vaccines has been an important contribution to their extraordinary success at the population level. The MenC conjugate vaccine in the United Kingdom reduced transmission of group C *N. meningitidis*, thereby providing herd protection—indirectly protecting susceptible unvaccinated individuals.[107] Following the introduction of the vaccine, the number of cases among unvaccinated age groups fell by 67%, corresponding to a similar reduction in MenC carriage rates seen in vaccinated young adults.[108] Given that the highest rates of meningococcal carriage occur in adolescents and young adults,[109] many countries include an adolescent booster dose of MenC or MenACWY conjugate vaccine to maintain herd protection. This provides a highly effective way of maintaining protection of the population,[110] as long as high vaccine uptake rates can be maintained. Data from the mass vaccination campaigns in the meningitis belt of sub-Saharan Africa suggest that the MenA conjugate vaccine is also able to successfully protect against carriage and generate herd protection.[95]

The recent exciting development of new meningococcal vaccines has led to the prospect of protection against all disease-causing capsular groups of *N. meningitidis*, which are predominant across Europe and North America, including group B for the first time. However, despite data demonstrating that these new vaccines are highly immunogenic, a number of questions remain unanswered and the impact these vaccines will have is difficult to predict. Strain coverage and vaccine efficacy have been estimated from in vitro assays, but only postimplementation surveillance that includes detailed typing of strains causing disease will be able to determine vaccine effectiveness. The meningococcus has a remarkable capacity to evade the human immune system through variation of antigens decorating the surface of the organism. While current vaccines have attempted to circumvent this by including multiple components of the same or different antigens, it is possible that there will be sufficient variation in the future to evade vaccine-induced protection directed at specific protein variants, highlighting the importance of ongoing typing of circulating strains. The vaccines have been shown to induce anamnestic responses in clinical trials, and 4CMenB had a modest effect on carriage, suggesting that it may be possible to induce a degree of herd protection though no direct evidence yet exists. Further studies are needed to identify any impact on carriage of disease-causing isolates, which would have a significant influence on clinical and cost-effectiveness of vaccine programs.

7. VACCINES AGAINST MEASLES AND MUMPS

Measles and mumps vaccines, most commonly administered as the measles-mumps-rubella (MMR) vaccine, with or without varicella, are not typically considered to be vaccines against CNS infections because the most common disease manifestations are outside the CNS, and their effect on these diseases has been demonstrated.[111] However, recent longitudinal studies from the United Kingdom have highlighted the importance of these vaccines in reducing meningitis and encephalitis, and it is appropriate for them to be thought of as vaccines against neuropathogens.

Measles is endemic worldwide, and is ubiquitous in the absence of vaccination. Measles vaccines have had a dramatic effect on disease incidence, most notably in resource-poor countries where disease burden is high and outcomes are poor. Deaths due to measles have decreased by 78% worldwide, from approximately 733,000 in 2000 to 164,000 in 2008 because of immunization.[112] Measles is most commonly recognized as a febrile viral exanthema and neurological manifestations are uncommon. Rates of acute infectious/postinfectious encephalitis are 1 per 1000–2000 cases, with the less common late complication of subacute sclerosing panencephalitis occurring in

1 per 100,000 cases. This invariably fatal manifestation occurs several years after measles infection and children infected in the first 2 years of life are at higher risk.[112]

Prior to the introduction of the mumps vaccine, mumps was most commonly reported among school-aged children, causing parotitis in almost all cases, clinical meningitis in up to 17%, and clinical encephalitis in up to 2%.[113] Overall, involvement of the CNS is reported as occurring in up to 62% of cases.[114–120] The wide range of rates is because the spectrum of CNS involvement ranges from asymptomatic CSF pleocytosis to frank meningoencephalitis. If lumbar punctures are routinely done, >50% of infections may show evidence of meningitis.[115,119–121] Mumps encephalitis is a more serious complication than mumps meningitis and occurs less frequently.[118,122,123] Prior to mumps vaccine use, mumps was the leading cause of viral encephalitis in the United States, accounting for approximately one third of the cases, and was responsible for most of the encephalitis deaths.[124,125] Other serious CNS manifestations of mumps include cerebellar ataxia, transverse myelitis, a poliomyelitis-like disease, and hydrocephalus.[126–133]

In the United Kingdom, the hospital admission rate for children with mumps meningitis from 1979 to 1985 was 0.6–2.1 per 100,000 per year. Hospital admissions for mumps meningitis almost disappeared from 1990 onward after the introduction of the MMR vaccination in 1988. Of note, a small peak in admissions occurred in 2005 (0.05 per 100,000 per year). This was consistent with a documented mumps outbreak in England at that time. This was partly caused by the presence of an unimmunized cohort in the population and incomplete immunization coverage, demonstrating the importance of maintaining high two-dose vaccine coverage even when the prevalence of background disease is low.[134] In the same analysis, only one case of measles meningitis was recorded during 1995–2011, with no specific data before 1995 due to the absence of a specific ICD code.

Before the introduction of the single-dose MMR in the United Kingdom in 1988, the hospital admission rate due to measles encephalitis was 0.32 per 100,000 per year. From 1990 to 1995, the admission rate fell by 88% to 0.04 per 100,000 per year. After introduction of the two-dose MMR schedule, the measles encephalitis hospital admission rate (in 1997–2011) was 0.01 per 100,000 per year, corresponding to a 97% decrease when compared with presingle dose MMR rates, and a 78% decrease when compared with postsingle dose MMR rates. Hospital admissions for mumps encephalitis fell from a pre-MMR high of 0.91 per 100,000 in 1983 to a postsingle dose MMR low of 0.01 per 100,000 in 1991. The hospital admission rate for mumps encephalitis differed between the vaccination periods: 0.60 per 100,000 per year in 1979–85, 0.02 in 1990–95 (post-single dose), and 0.01 in 1997–2011 (post-two dose MMR schedule). Compared with the presingle dose MMR period, these figures correspond to a 97% decrease in hospital admissions after introduction of the single dose MMR vaccine and a 98% decrease following use of the two-dose MMR vaccine. In addition to this demonstrated specific effect on mumps and measles encephalitis, the same study also reported a reduction in cases of encephalitis without known cause in the post-MMR compared with pre-MMR era, and it is likely that many of these cases were therefore also due to measles or mumps. Similar effects of MMR on measles and mumps encephalitis have been reported in smaller studies from Finland and the Czech Republic.[122,135] What is also clear from this study is that a measles-only vaccine used with low coverage (<60%) was insufficient to fully interrupt transmission.[136] After introduction of the single-dose MMR vaccine in 1988 and an increase in vaccine coverage to >90%,[136] possibly due to an increase in public awareness, a sustained decline in measles encephalitis occurred. In addition, there was an increase in measles encephalitis admissions in 1994, and while laboratory notifications were most increased in children aged over 10 years,[137,138] hospital admissions for encephalitis were highest among children aged 1–4 years, possibly reflecting increased susceptibility of young children to complications such as CNS manifestations.[139]

Current vaccines against measles and mumps remain highly effective and are able to target all viral strains, in contrast to vaccines against the bacterial pathogens described above. Measles remains a target for elimination for the World Health Organization (WHO), with a target to eliminate measles (and rubella) from at least five WHO regions by 2020.[140] As highlighted by the United Kingdom studies, the biggest challenge remains in maintaining high coverage for these vaccines to ensure ongoing population protection. Whenever vaccine coverage has fallen, measles and mumps epidemics have occurred, highlighting the fragile nature of the current balance between humans and these viruses. The importance of these and other vaccines needs to be emphasized to the public and healthcare workers on a regular and frequent basis, especially in an era when the diseases they prevent are relatively rare and media concerns about vaccine safety are common.

8. CONCLUSION

Meningitis and encephalitis are now much less common than at any time in the last few decades, largely due to the successful introduction of bacterial conjugate vaccines against Hib, *S. pneumoniae,* and *N. meningitidis*, and widespread use of vaccines against measles and mumps. Viral causes of meningitis and encephalitis now predominate, in particular enteroviruses and human parechoviruses. At present there are no broadly protective vaccines against these pathogens, and no plans to develop or introduce such vaccines in the near future as the infections they cause are largely thought to be mild and self-limiting in comparison to the devastating diseases caused by what are now vaccine-preventable pathogens. However, the natural history of these infections has been poorly studied and further data are required to help inform vaccine development strategies. The other major shortcoming in consideration of future vaccines to prevent meningitis and encephalitis is the absence of an aetiological diagnosis in a large number of cases—up to 40%–60%

of meningitis and encephalitis cases in children and adults do not have a pathogen identified. Novel laboratory techniques based on better pathogen diagnosis through molecular methods and better classification of host immune responses to aid pathogen group identification (i.e., bacterial vs. viral based on host gene expression) are currently used in the research setting. Some of these pathogen diagnostic tools are starting to be used more routinely in clinical laboratories. Increased use of these methods will improve identification of current causative organisms of meningitis and encephalitis and inform future prevention strategies, including vaccination.

DECLARATION OF INTERESTS

MS has been an investigator on investigator-initiated research grants awarded to his institution from Pfizer, Merck, and VBI Vaccines but has not received any personal payments.

REFERENCES

1. Pelton S, Sadarangani M, Glennie L, Levin M. Clinical aspects of meningococcal disease. In: Feavers I, Pollard AJ, Sadarangani M, editors. *Handbook of meningococcal disease management*. Switzerland: Springer; 2016.
2. Havens PL, Garland JS, Brook MM, Dewitz BA, Stremski ES, Troshynski TJ. Trends in mortality in children hospitalized with meningococcal infections, 1957 to 1987. *Pediatr Infect Dis J* 1989;**8**(1):8–11.
3. Edwards MS, Baker CJ. Complications and sequelae of meningococcal infections in children. *J Pediatr* 1981;**99**(4):540–5.
4. Gedde-Dahl TW, Lettenstrom GS, Bovre K. Coverage for meningococcal disease in the Norwegian morbidity and mortality statistics. With reference to incidence and fatality trends. *NIPH Ann* 1980;**3**(2):31–5.
5. Andersen BM. Mortality in meningococcal infections. *Scand J Infect Dis* 1978;**10**(4):277–82.
6. Farries JS, Dickson W, Greenwood E, Malhotra TR, Abbott JD, Jones DM. Meningococcal infections in Bolton, 1971–74. *Lancet* 1975;**2**(7925):118–20.
7. Stiehm ER, Damrosch DS. Factors in the prognosis of meningococcal infection. Review of 63 cases with emphasis on recognition and management of the severely ill patient. *J Pediatr* 1966;**68**(3):457–67.
8. Banks HS. Meningococcosis; a protean disease. *Lancet* 1948;**2**(6530):635 [passim].
9. Glasscock EL. Meningococcus infections. *J Pediatr* 1935;**6**:763–9.
10. Dunn CH. The serum treatment of epidemic cerebrospinal meningitis. *JAMA* 1908;**4**:15–21.
11. Flexner S, Jobling JW. An analysis of four hundred cases of epidemic meningitis treated with the anti-meningitis serum. *J Exp Med* 1908;**10**(5):690–733.
12. Pace D, Pollard AJ. Meningococcal disease: clinical presentation and sequelae. *Vaccine* 2012;**30**(Suppl. 2): B3–9.
13. Buysse CM, Oranje AP, Zuidema E, et al. Long-term skin scarring and orthopaedic sequelae in survivors of meningococcal septic shock. *Arch Dis Child* 2009;**94**(5):381–6.
14. Buysse CM, Raat H, Hazelzet JA, Hop WC, Maliepaard M, Joosten KF. Surviving meningococcal septic shock: health consequences and quality of life in children and their parents up to 2 years after pediatric intensive care unit discharge. *Crit Care Med* 2008;**36**(2):596–602.
15. Buysse CM, Vermunt LC, Raat H, et al. Surviving meningococcal septic shock in childhood: long-term overall outcome and the effect on health-related quality of life. *Crit Care* 2010;**14**(3):R124.
16. Stein-Zamir C, Shoob H, Sokolov I, Kunbar A, Abramson N, Zimmerman D. The clinical features and long term sequelae of invasive meningococcal disease in children. *Pediatr Infect Dis J* 2014;**33**(7):777–9.

17. Borg J, Christie D, Coen PG, Booy R, Viner RM. Outcomes of meningococcal disease in adolescence: prospective, matched-cohort study. *Pediatrics* 2009;**123**(3):e502–9.
18. Judge D, Nadel S, Vergnaud S, Garralda ME. Psychiatric adjustment following meningococcal disease treated on a PICU. *Intensive Care Med* 2002;**28**(5):648–50.
19. Sander J, Bay D, Gedde-Dahl TW, et al. Late sequelae after meningococcal disease. A controlled study in young men. *NIPH Ann* 1984;**7**(1):3–11.
20. Cherry JD, Nielsen KA. Aseptic meningitis and viral meningitis. In: Feigin RD, Cherry JD, Demmler-Harrison GJ, Kaplan SL, editors. *Textbook of pediatric infectious diseases*. 6th ed. Philadelphia, USA: Saunders; 2009.
21. Baker RC, Kummer AW, Schultz JR, Ho M, Gonzalez del Rey J. Neurodevelopmental outcome of infants with viral meningitis in the first three months of life. *Clin Pediatr* 1996;**35**(6):295–301.
22. Sells CJ, Carpenter RL, Ray CG. Sequelae of central-nervous-system enterovirus infections. *N Engl J Med* 1975;**293**(1):1–4.
23. Farmer K, MacArthur BA, Clay MM. A follow-up study of 15 cases of neonatal meningoencephalitis due to Coxsackie virus B5. *J Pediatr* 1975;**87**(4):568–71.
24. Wilfert CM, Thompson Jr. RJ, Sunder TR, O'Quinn A, Zeller J, Blacharsh J. Longitudinal assessment of children with enteroviral meningitis during the first three months of life. *Pediatrics* 1981;**67**(6):811–5.
25. Davison KL, Crowcroft NS, Ramsay ME, Brown DW, Andrews NJ. Viral encephalitis in England, 1989–1998: what did we miss? *Emerg Infect Dis* 2003;**9**(2):234–40.
26. Raschilas F, Wolff M, Delatour F, et al. Outcome of and prognostic factors for herpes simplex encephalitis in adult patients: results of a multicenter study. *Clin Infect Dis* 2002;**35**(3):254–60.
27. Easton A. *Life after encephalitis: a narrative approach*. Abingdon: Routledge; 2016.
28. Dagsdóttir HM, Sigurðardóttir B, Gottfreðsson M, et al. Herpes simplex encephalitis in Iceland 1987–2011. *Springerplus* 2014;**3**:524.
29. Michaeli O, Kassis I, Shachor-Meyouhas Y, Shahar E, Ravid S. Long-term motor and cognitive outcome of acute encephalitis. *Pediatrics* 2014;**133**(3):e546–52.
30. Dowell E, Easton A, Solomon T. *The consequences of encephalitis, report of a postal survey*. Malton, UK: The Encephalitis Society; 2000.
31. Mailles A, De Broucker T, Costanzo P, Martinez-Almoyna L, Vaillant V, Stahl JP. Long-term outcome of patients presenting with acute infectious encephalitis of various causes in France. *Clin Infect Dis* 2012;**54**(10):1455–64.
32. Aygün AD, Kabakuş N, Celik I, et al. Long-term neurological outcome of acute encephalitis. *J Trop Pediatr* 2001;**47**(4):243–7.
33. Harvala H, Robertson I, Chieochansin T, McWilliam Leitch EC, Templeton K, Simmonds P. Specific association of human parechovirus type 3 with sepsis and fever in young infants, as identified by direct typing of cerebrospinal fluid samples. *J Infect Dis* 2009;**199**(12):1753–60.
34. Watt JP, Levine OS, Santosham M. Global reduction of Hib disease: what are the next steps? Proceedings of the meeting Scottsdale, Arizona, September 22–25, 2002. *J Pediatr* 2003;**143(6 Suppl.)**:S163–87.
35. Reinert P, Liwartowski A, Dabernat H, Guyot C, Boucher J, Carrere C. Epidemiology of *Haemophilus influenzae* type b disease in France. *Vaccine* 1993;**11**(Suppl. 1):S38–42.
36. Peltola H, Kayhty H, Virtanen M, Makela PH. Prevention of *Hemophilus influenzae* type b bacteremic infections with the capsular polysaccharide vaccine. *N Engl J Med* 1984;**310**(24):1561–6.
37. Adams WG, Deaver KA, Cochi SL, et al. Decline of childhood *Haemophilus influenzae* type b (Hib) disease in the Hib vaccine era. *JAMA* 1993;**269**(2):221–6.
38. Bisgard KM, Kao A, Leake J, Strebel PM, Perkins BA, Wharton M. *Haemophilus influenzae* invasive disease in the United States, 1994–1995: near disappearance of a vaccine-preventable childhood disease. *Emerg Infect Dis* 1998;**4**(2):229–37.
39. Murphy TV, Osterholm MT, Pierson LM, et al. Prospective surveillance of *Haemophilus influenzae* type b disease in Dallas County, Texas, and in Minnesota. *Pediatrics* 1987;**79**(2):173–80.
40. Makela PH, Peltola H, Kayhty H, et al. Polysaccharide vaccines of group A *Neisseria meningtitidis* and *Haemophilus influenzae* type b: a field trial in Finland. *J Infect Dis* 1977;**136(Suppl.)**:S43–50.

41. Kayhty H, Karanko V, Peltola H, Makela PH. Serum antibodies after vaccination with *Haemophilus influenzae* type b capsular polysaccharide and responses to reimmunization: no evidence of immunologic tolerance or memory. *Pediatrics* 1984;**74**(5):857–65.
42. Smith DH, Peter G, Ingram DL, Harding AL, Anderson P. Responses of children immunized with the capsular polysaccharide of *Hemophilus influenzae*, type b. *Pediatrics* 1973;**52**(5):637–44.
43. Kayhty H, Peltola H, Karanko V, Makela PH. The protective level of serum antibodies to the capsular polysaccharide of *Haemophilus influenzae* type b. *J Infect Dis* 1983;**147**(6):1100.
44. Peltola H, Kayhty H, Sivonen A, Makela H. *Haemophilus influenzae* type b capsular polysaccharide vaccine in children: a double-blind field study of 100,000 vaccinees 3 months to 5 years of age in Finland. *Pediatrics* 1977;**60**(5):730–7.
45. Osterholm MT, Rambeck JH, White KE, et al. Lack of efficacy of Haemophilus b polysaccharide vaccine in Minnesota. *JAMA* 1988;**260**(10):1423–8.
46. Shapiro ED, Murphy TV, Wald ER, Brady CA. The protective efficacy of Haemophilus b polysaccharide vaccine. *JAMA* 1988;**260**(10):1419–22.
47. Centers for Disease. Polysaccharide vaccine for prevention of *Haemophilus influenzae* type b disease. *MMWR Morb Mortal Wkly Rep* 1985;**34**(15):201–5.
48. Black SB, Shinefield HR, Fireman B, Hiatt R, Polen M, Vittinghoff E. Efficacy in infancy of oligosaccharide conjugate Haemophilus influenzae type b (HbOC) vaccine in a United States population of 61,080 children. The Northern California Kaiser Permanente Vaccine Study Center Pediatrics Group. *Pediatr Infect Dis J* 1991;**10**(2):97–104.
49. Peltola H, Eskola J, Kayhty H, Takala AK, Makela PH. Clinical comparison of the *Haemophilus influenzae* type B polysaccharide-diphtheria toxoid and the oligosaccharide-CRM197 protein vaccines in infancy. *Arch Pediatr Adolesc Med* 1994;**148**(6):620–5.
50. Santosham M, Wolff M, Reid R, et al. The efficacy in Navajo infants of a conjugate vaccine consisting of *Haemophilus influenzae* type b polysaccharide and *Neisseria meningitidis* outer-membrane protein complex. *N Engl J Med* 1991;**324**(25):1767–72.
51. Mulholland K, Hilton S, Adegbola R, et al. Randomized trial of *Haemophilus influenzae* type-b tetanus protein conjugate vaccine [corrected] for prevention of pneumonia and meningitis in Gambian infants. *Lancet* 1997;**349**(9060):1191–7.
52. Booy R, Moxon ER, MacFarlane JA, Mayon-White RT, Slack MP. Efficacy of *Haemophilus influenzae* type B conjugate vaccine in Oxford region. *Lancet* 1992;**340**(8823):847.
53. Lagos R, Horwitz I, Toro J, et al. Large scale, postlicensure, selective vaccination of Chilean infants with PRP-T conjugate vaccine: practicality and effectiveness in preventing invasive *Haemophilus influenzae* type b infections. *Pediatr Infect Dis J* 1996;**15**(3):216–22.
54. Fritzell B, Plotkin S. Efficacy and safety of a *Haemophilus influenzae* type b capsular polysaccharide-tetanus protein conjugate vaccine. *J Pediatr* 1992;**121**(3):355–62.
55. Vadheim CM, Greenberg DP, Eriksen E, et al. Eradication of *Haemophilus influenzae* type b disease in southern California. Kaiser-UCLA Vaccine Study Group. *Arch Pediatr Adolesc Med* 1994;**148**(1):51–6.
56. Moulton LH, Chung S, Croll J, Reid R, Weatherholtz RC, Santosham M. Estimation of the indirect effect of *Haemophilus influenzae* type b conjugate vaccine in an American Indian population. *Int J Epidemiol* 2000;**29**(4):753–6.
57. Horby P, Gilmour R, Wang H, McIntyre P. Progress towards eliminating Hib in Australia: an evaluation of *Haemophilus influenzae* type b prevention in Australia, 1 July 1993 to 30 June 2000. *Commun Dis Intell Q Rep* 2003;**27**(3):324–41.
58. Gortzak-Uzan L, Fraser D, Dagan R. Epidemiology of invasive Hemophilus influenzae B infections in Bedouins and Jews; conjugate Hib vaccines. *Harefuah* 1998;**135**(5–6):175–80. 256.
59. Martin NG, Sadarangani M, Pollard AJ, Goldacre MJ. Hospital admission rates for meningitis and septicaemia caused by *Haemophilus influenzae, Neisseria meningitidis*, and *Streptococcus pneumoniae* in children in England over five decades: a population-based observational study. *Lancet Infect Dis* 2014;**14**(5):397–405.
60. Heath PT, McVernon J. The UK Hib vaccine experience. *Arch Dis Child* 2002;**86**(6):396–9.
61. Lee YC, Kelly DF, LM Y, et al. *Haemophilus influenzae* type b vaccine failure in children is associated with inadequate production of high-quality antibody. *Clin Infect Dis* 2008;**46**(2):186–92.

62. Ladhani S, Slack MP, Heys M, White J, Ramsay ME. Fall in *Haemophilus influenzae* serotype b (Hib) disease following implementation of a booster campaign. *Arch Dis Child* 2008;**93**(8):665–9.
63. Ladhani SN, Ramsay M, Slack MP. The impact of *Haemophilus influenzae* serotype B resurgence on the epidemiology of childhood invasive *Haemophilus influenzae* disease in England and Wales. *Pediatr Infect Dis J* 2011;**30**(10):893–5.
64. Ramsay ME, McVernon J, Andrews NJ, Heath PT, Slack MP. Estimating *Haemophilus influenzae* type b vaccine effectiveness in England and Wales by use of the screening method. *J Infect Dis* 2003;**188**(4):481–5.
65. McVernon J, Andrews N, Slack MP, Ramsay ME. Risk of vaccine failure after *Haemophilus influenzae* type b (Hib) combination vaccines with acellular pertussis. *Lancet* 2003;**361**(9368):1521–3.
66. Rijkers GT, Vermeer-de Bondt PE, Spanjaard L, Breukels MA, Sanders EA. Return of Haemophilus influenzae type b infections. *Lancet* 2003;**361**(9368):1563–4.
67. Peltola H, Salo E, Saxen H. Incidence of *Haemophilus influenzae* type b meningitis during 18 years of vaccine use: observational study using routine hospital data. *BMJ* 2005;**330**(7481):18–9.
68. Takala AK, Santosham M, Almeido-Hill J, et al. Vaccination with *Haemophilus influenzae* type b meningococcal protein conjugate vaccine reduces oropharyngeal carriage of *Haemophilus influenzae* type b among American Indian children. *Pediatr Infect Dis J* 1993;**12**(7):593–9.
69. Takala AK, Eskola J, Leinonen M, et al. Reduction of oropharyngeal carriage of Haemophilus influenzae type b (Hib) in children immunized with an Hib conjugate vaccine. *J Infect Dis* 1991;**164**(5):982–6.
70. Gallo G, Ciofi degli Atti ML, Cerquetti M, Piovesan C, Tozzi AE, Salmaso S. Impact of a regional Hib vaccination programme in Italy. *Vaccine* 2002;**20**(7–8):993–5.
71. Perdue DG, Bulkow LR, Gellin BG, et al. Invasive *Haemophilus influenzae* disease in Alaskan residents aged 10 years and older before and after infant vaccination programs. *JAMA* 2000;**283**(23):3089–94.
72. Tsang RS, Bruce MG, Lem M, Barreto L, Ulanova M. A review of invasive *Haemophilus influenzae* disease in the Indigenous populations of North America. *Epidemiol Infect* 2014;**142**(7):1344–54.
73. World Health Organization. *Pneumococcal vaccines WHO position paper;* 2012.
74. Hsu HE, Shutt KA, Moore MR, et al. Effect of pneumococcal conjugate vaccine on pneumococcal meningitis. *N Engl J Med* 2009;**360**(3):244–56.
75. Conklin L, Loo JD, Kirk J, et al. Systematic review of the effect of pneumococcal conjugate vaccine dosing schedules on vaccine-type invasive pneumococcal disease among young children. *Pediatr Infect Dis J* 2014;**33**(Suppl. 2):S109–18.
76. Miller E, Andrews NJ, Waight PA, Slack MP, George RC. Herd immunity and serotype replacement 4 years after seven-valent pneumococcal conjugate vaccination in England and Wales: an observational cohort study. *Lancet Infect Dis* 2011;**11**(10):760–8.
77. Ruckinger S, van der Linden M, Reinert RR, von Kries R, Burckhardt F, Siedler A. Reduction in the incidence of invasive pneumococcal disease after general vaccination with 7-valent pneumococcal conjugate vaccine in Germany. *Vaccine* 2009;**27**(31):4136–41.
78. Olarte L, Barson WJ, Barson RM, et al. Impact of the 13-valent pneumococcal conjugate vaccine on pneumococcal meningitis in US children. *Clin Infect Dis* 2015;**61**(5):767–75.
79. Ricketson LJ, Wood ML, Vanderkooi OG, et al. Trends in asymptomatic nasopharyngeal colonization with streptococcus pneumoniae after introduction of the 13-valent pneumococcal conjugate vaccine in Calgary, Canada. *Pediatr Infect Dis J* 2014;**33**(7):724–30.
80. O'Brien KL, Wolfson LJ, Watt JP, et al. Burden of disease caused by *Streptococcus pneumoniae* in children younger than 5 years: global estimates. *Lancet* 2009;**374**(9693):893–902.
81. Oliveira ML, Areas AP, Campos IB, et al. Induction of systemic and mucosal immune response and decrease in *Streptococcus pneumoniae* colonization by nasal inoculation of mice with recombinant lactic acid bacteria expressing pneumococcal surface antigen A. *Microbes Infect* 2006;**8**(4):1016–24.
82. Jafri RZ, Ali A, Messonnier NE, et al. Global epidemiology of invasive meningococcal disease. *Popul Health Metrics* 2013;**11**(1):17.
83. Cartwright KA, Stuart JM, Jones DM, Noah ND. The Stonehouse survey: nasopharyngeal carriage of meningococci and Neisseria lactamica. *Epidemiol Infect* 1987;**99**(3):591–601.

84. Christensen H, Trotter CL, Hickman M, Edmunds WJ. *Re*-evaluating cost effectiveness of universal meningitis vaccination (Bexsero) in England: modelling study. *BMJ* 2014;**349**.
85. Ladhani SN, Beebeejaun K, Lucidarme J, et al. Increase in endemic *Neisseria meningitidis* capsular group W sequence type 11 complex associated with severe invasive disease in England and Wales. *Clin Infect Dis* 2015;**60**(4):578–85.
86. Reingold AL, Broome CV, Hightower AW, et al. Age-specific differences in duration of clinical protection after vaccination with meningococcal polysaccharide A vaccine. *Lancet* 1985;**2**(8447):114–8.
87. Jokhdar H, Borrow R, Sultan A, et al. Immunologic hyporesponsiveness to serogroup C but not serogroup A following repeated meningococcal A/C polysaccharide vaccination in Saudi Arabia. *Clin Diagn Lab Immunol* 2004;**11**(1):83–8.
88. Granoff DM, Gupta RK, Belshe RB, Anderson EL. Induction of immunologic refractoriness in adults by meningococcal C polysaccharide vaccination. *J Infect Dis* 1998;**178**(3):870–4.
89. Peltola H, Makela H, Kayhty H, et al. Clinical efficacy of meningococcus group A capsular polysaccharide vaccine in children three months to five years of age. *N Engl J Med* 1977;**297**(13):686–91.
90. Gold R, Lepow ML, Goldschneider I, Draper TL, Gotschlich EC. Clinical evaluation of group A and group C meningococcal polysaccharide vaccines in infants. *J Clin Invest* 1975;**56**(6):1536–47.
91. Artenstein MS, Gold R, Zimmerly JG, Wyle FA, Schneider H, Harkins C. Prevention of meningococcal disease by group C polysaccharide vaccine. *N Engl J Med* 1970;**282**(8):417–20.
92. Ravenscoft R, Feavers I. Conjugate vaccines. In: Frosch M, MCJ M, editors. *Handbook of meningococcal disease: infection biology, vaccination, clinical management.* Weinheim, Germany: Wiley-VCH; 2006.
93. Trotter CL, Andrews NJ, Kaczmarski EB, Miller E, Ramsay ME. Effectiveness of meningococcal serogroup C conjugate vaccine 4 years after introduction. *Lancet* 2004;**364**(9431):365–7.
94. LaForce FM, Konde K, Viviani S, Preziosi MP. The Meningitis Vaccine Project. *Vaccine* 2007;**25**(Suppl. 1):A97–100.
95. Daugla DM, Gami JP, Gamougam K, et al. Effect of a serogroup A meningococcal conjugate vaccine (PsA-TT) on serogroup A meningococcal meningitis and carriage in Chad: a community study [corrected]. *Lancet* 2014;**383**(9911):40–7.
96. Kristiansen PA, Ba AK, Sanou I, et al. Phenotypic and genotypic characterization of meningococcal carriage and disease isolates in Burkina Faso after mass vaccination with a serogroup a conjugate vaccine. *BMC Infect Dis* 2013;**13**:363.
97. Finne J, Leinonen M, Makela PH. Antigenic similarities between brain components and bacteria causing meningitis. Implications for vaccine development and pathogenesis. *Lancet* 1983;**2**(8346):355–7.
98. Nissen MD, Marshall HS, Richmond P, et al. In: *A randomized, placebo-controlled, double-blind, phase 1 trial of ascending doses of meningococcal group B rLP2086 vaccine. European Society for Pediatric Infectious Diseases annual meeting, Graz, Austria*; 2008.
99. Anderson AS, Hao L, Jiang Q, et al. Potential impact of the bivalent rLP2806 vaccine on *Neisseria meningitidis* carriage and invasive serogroup B disease. *Hum Vaccin Immunother* 2013;**9**(3):471–9.
100. O'Ryan M, Stoddard J, Toneatto D, Wassil J, Dull PM. A multi-component meningococcal serogroup B vaccine (4CMenB): the clinical development program. *Drugs* 2014;**74**(1):15–30.
101. Findlow J, Borrow R, Snape MD, et al. Multicenter, open-label, randomized phase II controlled trial of an investigational recombinant meningococcal serogroup B vaccine with and without outer membrane vesicles, administered in infancy. *Clin Infect Dis* 2010;**51**(10):1127–37.
102. Vesikari T, Esposito S, Prymula R, et al. Immunogenicity and safety of an investigational multicomponent, recombinant, meningococcal serogroup B vaccine (4CMenB) administered concomitantly with routine infant and child vaccinations: results of two randomised trials. *Lancet* 2013;**381**(9869):825–35.
103. Nolan T, O'Ryan M, Wassil J, Abitbol V, Dull P. Vaccination with a multicomponent meningococcal B vaccine in prevention of disease in adolescents and young adults. *Vaccine* 2015;**33**(36):4437–45.
104. McNamara LA, Shumate AM, Johnsen P, et al. First use of a serogroup B meningococcal vaccine in the US in response to a university outbreak. *Pediatrics* 2015;**135**(5):798–804.
105. Turner D. In: *Clinical experience with the meningococcal B vaccine, Bexsero.* 33rd annual meeting of the European Society for Paediatric Infectious Diseases, Leipzig, Germany; 2015 [p. ESPID-0546].
106. Parikh SR, Andrews NJ, Beebeejaun K, et al. Effectiveness and impact of a reduced infant schedule of 4CMenB vaccine against group B meningococcal disease in England: a national observational cohort study. *Lancet* 2016;**388**(10061):2775–82.

107. Ramsay ME, Andrews NJ, Trotter CL, Kaczmarski EB, Miller E. Herd immunity from meningococcal serogroup C conjugate vaccination in England: database analysis. *BMJ* 2003;**326**(7385):365–6.
108. Maiden MC, Stuart JM. Carriage of serogroup C meningococci 1 year after meningococcal C conjugate polysaccharide vaccination. *Lancet* 2002;**359**(9320):1829–31.
109. Christensen H, May M, Bowen L, Hickman M, Trotter CL. Meningococcal carriage by age: a systematic review and meta-analysis. *Lancet Infect Dis* 2010;**10**(12):853–61.
110. Sadarangani M, Scheifele DW, Halperin SA, et al. The impact of the meningococcal serogroup C conjugate vaccine in Canada between 2002 and 2012. *Clin Infect Dis* 2014;**59**(9):1208–15.
111. Goldacre MJ, Maisonneuve JJ. Hospital admission rates for measles and mumps in England: historical perspective. *Lancet* 2013;**382**(9889):308–9.
112. Strebel PM, Papania MJ, Fiebelkorn AP, Halsey NA. Measles vaccine. In: Plotkin SA, Orenstein WA, Offit PA, editors. *Vaccines*. Philadelphia, USA: Elsevier; 2013.
113. Rubin SA, Plotkin SA. Mumps vaccine. In: Plotkin SA, Orenstein WA, Offit PA, editors. *Vaccines*. Elsevier; 2013.
114. Reed D, Brown G, Merrick R, Sever J, Feltz E. A mumps epidemic on St. George Island, Alaska. *JAMA* 1967;**199**(13):113–7.
115. Falk WA, Buchan K, Dow M, et al. The epidemiology of mumps in southern Alberta 1980–1982. *Am J Epidemiol* 1989;**130**(4):736–49.
116. McGuiness AC, Gall EA. Mumps at army camps in 1943. *War Med* 1944;**5**:95–104.
117. Dayan GH, Quinlisk MP, Parker AA, et al. Recent resurgence of mumps in the United States. *N Engl J Med* 2008;**358**(15):1580–9.
118. Russell RR, Donald JC. The neurological complications of mumps. *Br Med J* 1958;**2**(5087):27–30.
119. Bang HO, Bang J. Involvement of the central nervous system in mumps. *Acta Med Scand* 1943;**113**:487–505.
120. Bruyn HB, Sexton HM, Brainerd HD. Mumps meningoencephalitis; a clinical review of 119 cases with one death. *Calif Med* 1957;**86**(3):153–60.
121. Brown JW, Kirkland HB, Hein GE. Central nervous system involvement during mumps. *Am J Med Sci* 1948;**215**(4):434–41.
122. Koskiniemi M, Vaheri A. Effect of measles, mumps, rubella vaccination on pattern of encephalitis in children. *Lancet* 1989;**1**(8628):31–4.
123. Koskiniemi M, Donner M, Pettay O. Clinical appearance and outcome in mumps encephalitis in children. *Acta Paediatr Scand* 1983;**72**(4):603–9.
124. Bennett AD, Modlin J, Orenstein WA. Brandling. Current status of mumps in the United States. *J Infect Dis* 1975;**132**(1):106–9.
125. Wagenvoort JH, Harmsen M, Boutahar-Trouw BJ, Kraaijeveld CA, Winkler KC. Epidemiology of mumps in the Netherlands. *J Hyg (Lond)* 1980;**85**(3):313–26.
126. Cohen HA, Ashkenazi A, Nussinovitch M, Amir J, Hart J, Frydman M. Mumps-associated acute cerebellar ataxia. *Am J Dis Child* 1992;**146**(8):930–1.
127. Nussinovitch M, Prais D, Volovitz B, Shapiro R, Amir J. Post-infectious acute cerebellar ataxia in children. *Clin Pediatr (Phila)* 2003;**42**(7):581–4.
128. Bansal R, Kalita J, Misra UK, Kishore J. Myelitis: a rare presentation of mumps. *Pediatr Neurosurg* 1998;**28**(4):204–6.
129. Gibbons JL, Miller HG, Stanton JB. Para-infectious encephalomyelitis and related syndromes; a critical review of the neurological complications of certain specific fevers. *Q J Med* 1956;**25**(100):427–505.
130. Kilham L, Levens J, Enders JF. Nonparalytic poliomyelitis and mumps meningoencephalitis; differential diagnosis. *J Am Med Assoc* 1949;**140**(11):934–6.
131. Johnstone JA, Ross CA, Dunn M. Meningitis and encephalitis associated with mumps infection. A 10-year survey. *Arch Dis Child* 1972;**47**(254):647–51.
132. Ogata H, Oka K, Mitsudome A. Hydrocephalus due to acute aqueductal stenosis following mumps infection: report of a case and review of the literature. *Brain Dev* 1992;**14**(6):417–9.
133. Timmons GD, Johnson KP. Aqueductal stenosis and hydrocephalus after mumps encephalitis. *N Engl J Med* 1970;**283**(27):1505–7.

134. Yung CF, Andrews N, Bukasa A, Brown KE, Ramsay M. Mumps complications and effects of mumps vaccination, England and Wales, 2002–2006. *Emerg Infect Dis* 2011;**17**(4):661–7 [quiz 766].
135. Orlikova H, Maly M, Lexova P, et al. Protective effect of vaccination against mumps complications, Czech Republic, 2007–2012. *BMC Public Health* 2016;**16**:293.
136. Department of Health. Japanese encephalitis. In: Salisbury DM, Begg NT, editors. *Immunisation against infectious disease*. London, UK: Her Majesty's Stationery Office; 1996.
137. Ramsay M, Gay N, Miller E, et al. The epidemiology of measles in England and Wales: rationale for the 1994 national vaccination campaign. *Commun Dis Rep CDR Rev* 1994;**4**(12):R141–6.
138. Gay N, Miller E, Hesketh L, et al. Mumps surveillance in England and Wales supports introduction of two dose vaccination schedule. *Commun Dis Rep CDR Rev* 1997;**7**(2):R21–6.
139. Centers for Disease Control and Prevention. *Epidemiology and prevention of vaccine-preventable diseases*. Washington, DC, USA: Public Health Foundation; 2015.
140. World Health Organization. http://www.who.int/mediacentre/factsheets/fs286/en/.

CHAPTER 5

Metal Nanoparticles in Management of Diseases of the Central Nervous System

Avinash P. Ingle*, Sudhir Shende*, Indarchand Gupta*,†, Dipali Nagaonkar*, Raksha Pandit*, Priti Paralikar*, Mahendra Rai*
*SGB Amravati University, Amravati, India
†Government Institute of Science, Aurangabad, India

1. INTRODUCTION

The central nervous system (CNS) is of major concern to human health as all bodily activities are controlled by the brain. In the past two decades, an increasing infections are major cause for substantial CNS morbidity. The causes of infections associated with the CNS comprise a large number of agents such as viruses, bacteria, fungi, and even parasites.[1] The incidence and etiology of CNS infections differ in time with age, vaccination policies, geographical region, comorbidities, etc.[1] Hence, it is very important to document the causative agents for CNS infections to advance public health measures as well as clinical practices. The burden has been predominantly high in low- and middle-income countries like India where the infections are associated with variable pathogens.[2] It has been reported that the most common bacterial CNS pathogens are *Streptococcus pneumoniae* and *Neisseria meningitidis*, occurring in both children and adults.[3] *Haemophilus influenzae* type b is a causative agent for bacterial meningitis in children, but its incidence has decreased in areas where an effectual vaccine has been included into the national vaccination program.[4] The zoonotic pathogen from pigs (i.e., *Streptococcus suis*) is one of the most important causes of meningitis in the adult population of Southeast Asia.[5] The fungal infections of CNS were considered negligible until the 1970s, but in recent years the number has been increased due to extensive use of antibiotics, cytotoxic drugs, and corticosteroids. The immunocompromised patients with underlying malignancy, AIDS, and organ transplantation are all candidates for acquiring fungal infections in the brain. *Cryptococcus neoformans*, *Coccidioides immitis*, *Candida albicans,* and *Histoplasma capsulatum* are widespread causes of fungal meningitis. *Candida* spp., *Aspergillus* spp., Zygomycetes, and a few of the melanized fungi are identified as the origin for mass lesions in the brain. However, fungi such as *C. neoformans*, *Cladophialophora bantiana*, *Exophiala dermatitidis*,

The Microbiology of Central Nervous System Infections
https://doi.org/10.1016/B978-0-12-813806-9.00005-6

Ramichloridium mackenzie, and *Ochroconis gallopava* are considered as true neurotropic fungi. Most fungi causing CNS infections are saprobes with worldwide distribution.[6]

Globally and particularly for India, the actual burden of CNS diseases has not been significantly estimated due to the lack of data as it is not related with a single aspect of damage, injury, or diseased condition. Sometimes major injuries to the CNS leads to internal infections in the brain. The prevention of these infections that lead to brain damage and affect the CNS is a matter of great concern with research priority as the entire anticipation of infection removes all menace of CNS sequelae. Therefore, there is a need to treat the CNS sequelae and the rehabilitation of individuals with nervous system morbidity, which is important for the billions who live with the CNS infections. The prevention and treatment of infections that affect the CNS requires the identification of responsible pathogens, the pathogen reservoirs, and the potential points at which the pathogen life cycle could be interrupted. There are two types of CNS diseases—infectious and noninfectious—which are described in the following section.

2. TYPES OF CNS DISEASES

2.1 Infectious CNS Diseases

In infectious CNS diseases, the most common are cerebral malaria, viral encephalitis, alterations of cognition/behavior (AIDS), sleep perturbations (African trypanosomiasis), meningitis, etc. Cerebral malaria has the most severe neurological complication of infection with *Plasmodium falciparum*.[7] The mortality rate of cerebral malaria is very high (about 20%–50%); some surviving patients sustain long-term brain injury that manifests as long-term neurocognitive impairments.[7,8] The characteristic features of a postmalaria neurological syndrome are seizure, acute onset of confusion, myoclonus, ataxia, aphasia, and tremors, which are described in patients effectively treated for cerebral malaria where the parasitemia is absent. This syndrome is accredited to immunologic mechanisms.[8] According to WHO, human immunodeficiency virus (HIV) and other associated opportunistic infections have infected around 2.3 million people in 2012, and 1.6 million died due to AIDS-related illnesses worldwide.[9] HIV-related neurological syndromes are distinguished as primary HIV infections, secondary or opportunistic infections, and treatment-related neurological diseases. The primary HIV infection leads to acute aseptic viral meningitis or meningoencephalitis.[10] The most common HIV-associated neurocognitive disorder is a neurodegenerative situation characterized by cognitive, motor, and behavioral abnormalities and it has been reported increasing in people aged 50 and above.[11] The opportunistic infections such as tuberculous meningitis caused by *Mycobacterium tuberculosis*,[12,13] cryptococcal meningitis caused due to *C. neoformans*, and *Cryptococcus gattii* are most common in immuno-compromised patients suffering from AIDS and, cerebral toxoplasmosis.[14] Rabies and Japanese encephalitis virus (JEV) globally affected 60,000 people and are responsible for an estimated yearly mortality of 17,000 people.[15–17] The Japanese encephalitis cases are limited to Asia while rabies is a bane throughout Southeast Asia, Latin America, and Africa; it occurs less frequently in other global areas. Herpes simplex virus (HSV) encephalitis and Rabies, which is present worldwide, lead to great mortality without treatment.[18] Among these, it has been found that HSV was the principle cause of acute encephalitis in studies from the western world.[19–21] From the reports of Southeast Asia and the Western Pacific regions, the leading cause of acute encephalitis is due to JEV infections, especially in children and young adults.[22] On the other hand, in other countries of Southeast Asia (Bangladesh and India), the Nipah virus and enterovirus are the major concern with regard to public health.[23–25] In addition to these, there are several other CNS infections such as the Congenital cytomegalovirus (CMV) infection in the United States[26]; African trypanosomiasis, also called sleeping sickness or sleep perturbations, caused by *Trypanosoma brucei gambiense* (TbG); and *Trypanosoma brucei rhodesiense,*[27] Dengue and Chikangunya are leading cause of arboviral encephalitis worldwide.[15,28]

2.2 Noninfectious CNS Diseases

Besides the infectious CNS diseases, the noninfectious diseases are comparatively less. For instance, this includes the chronic inflammatory disorders such as multiple sclerosis (MS), neuromyelitis optica (NMO), or CNS vasculitis, etc.[29,30] MS is the most common autoimmune disorder and demyelinating disease of the CNS in which the insulating cover of nerve cells in the brain and spinal cord are damaged.[31,32] Since 2013, the number of people with MS is between 2 million and 2.5 million worldwide, with rates varying widely among regions and populations.[33,34] Jean-Martin Charcot was the first person to describe MS. The development of numerous scars on the white matter of the brain and spinal cord is referred to as MS.[35] The number of new cases that develop each year is about 2.5 per 100,000,[33] and the disease usually begins between the ages 20 and 50; it is comparatively twice as common in women than in men.[34] Like MS, NMO is also a noninfectious autoimmune CNS disease, commonly known as Devic's disease, is a rare neurological condition.[36] The damage to the brain, spinal cord, or nerves leads to this neurological condition. Though inflammation may also affect the brain, the lesions are variable from those that are observed in MS.[37] Spinal cord lesions lead to varying degrees of weakness or paralysis in the arms or legs, blindness, loss of sensation, and bladder and bowel dysfunction.[38] The CNS vasculites are characterized by means of necrosis and an inflammation of the blood vessel wall. The primary angiitis of the CNS is a rare disorder affecting both medium- and small-sized vessels. The major symptoms of cerebral vasculitis are headache, stroke, and encephalopathy. Diagnosis of this disease is based on laboratory findings such as magnetic resonance imaging (MRI) and 18-fluorodeoxyglucose positron emission tomography, etc.[39] The noninfectious CNS disease frequency is comparatively lower than that of infectious CNS diseases.

As discussed above, the prevalence of infectious CNS diseases is higher than that of noninfectious, which is related to the pathogens. There are several reports on the evaluation of antimicrobial activity of the metal nanoparticles against human pathogens, which demonstrated the noteworthy antimicrobial effects against the tested human pathogens such as bacteria,[40–42] fungi,[43–47] viruses,[48–50] and some other parasites. Therefore, the metal nanoparticles that showed antimicrobial activity could be used in the treatment of these CNS pathogenic conditions. Apart from this, nanotechnology is also used in the diagnosis and drug delivery in CNS diseases. This chapter mainly focuses on various aspects related to CNS diseases involving current treatment strategies and their limitations, nanoneuromedicines and their necessity, the role of nanotechnology in the diagnosis and treatment of CNS diseases, and toxicological issues.

3. TREATMENT STRATEGIES IN THE MANAGEMENT OF CNS DISEASES AND THEIR LIMITATIONS

Inadequate knowledge of the pathogenesis of pathogens causing CNS diseases contributes to mortality. Treatment strategies for the management of CNS diseases totally depend upon identification of the causal organism and clinical features associated with infectious CNS diseases.[51,52] For example, antibiotics were recommended in the initial treatment of bacterial meningitis, based on the patient's history and antibiotic-resistance pattern.[53,54] Rifampin, ceftriaxone, azithromycin, and ciprofloxacin were recommended for CNS infections caused by *Neisseria meningitidis*; however, its dose depends on the age of the patient.[54] Antibiotics require enough time to kill and to prevent the recurrence of bacterial infections in the CNS. But, this antibiotic therapy totally depends on causative bacteria, the antibacterial agent, and sensitivity. For noncomplicated cases, the requirements are 5–7 days for *N. meningitidis*, 10–14 days for *S. pneumoniae*, 14–21 days for GBS and Gram-negative bacilli, and 7–10 days for *H. influenzae*.[51,52,55] The main limitation of antibiotics therapy is that the drug should have the ability to reach the infection site and maintain effective concentration to treat bacterial meningitis. It is important to determine a systematic dosing plan of effective antibiotics because of different pharmacodynamics and pharmacokinetics profiles of antibiotics. Concentration and half-life in the cerebrospinal fluid (CSF) of antibiotics depend upon the penetration and elimination abilities of the blood-brain barriers (BBB).[54]

The formulation of an appropriate meningococcal conjugate vaccine is recommended for the increased risk of meningococcal disease. In the United States, four vaccines were licensed for use against *N. meningitidis*. Regular vaccination is not recommended for infants aged 2 months to 10 years against meningococcal disease. Due to differences in serogroup composition, the same vaccine products should be used at increased risk of meningococcal disease for all doses in an infant. The use of meningococcal conjugate vaccine and a two-dose primary series is recommended in children aged 2 through 10 years.[56,57] Similarly, the treatment of cerebral malaria includes management associated with complications and specific antimalarial therapy. The primary treatment for cerebral malaria involves the use of parenteral artemisinin.[8] Artemisinin includes agents such as artesunate and artemether. Artesunate has been reported to reduce mortality by 34.7% in Asian adults. It also reduces the occurrence of convulsion, coma, and hypoglycemia.[8] Children with cerebral malaria were treated with intravenous quinine, intramuscular artemisinin, and quinidine or derivatives. In the United States, cerebral malaria was treated with quinidine because of a lack of artemisinin derivatives and quinine.[58] Cerebral malaria management with quinine and artimisinin derivatives has been associated with a high-reoccurrence rate. This problem is decreased with the addition of different antibiotics such as sulfadoxine, tetracycline, and clindamycin to the treatment therapy.[59,60]

Rituximab was used for the treatment of NMO, which is an autoimmune inflammatory disease associated with the CNS. Beneficiary effects of rituximab were reported by Jacob.[61] The study reported that NMO successfully treated 25 patients with rituximab. Similarly, fruitful outcomes of rituximab treatment were reported in two different studies with 23 and 10 patients, respectively, with NMO.[62,63] Successful therapeutic strategies have been difficult to achieve due to complexity in CNS disorders; limited diffusion of drugs or biological agents across the BBB further limits the utility of delivery methods.

4. NANONEUROMEDICINES AND ITS NECESSITY

It is well known and also discussed above that many potential drugs have been discovered to treat several disorders of the CNS, but certain factors such as the BBB and blood-cerebrospinal fluid barrier (BCSFB) always restrict the therapeutic success of these drugs.[64] The ineffectiveness and failure of treatment strategies recently available for the management of CNS disorders necessitates the development of newer and effective treatment strategies to overcome these limitations. Considering the recent advancement in the field of nanotechnology, particularly in biomedical applications, it is believed that nanoscience will be a key enabling technology for delivering a translational bridge between basic and clinical research.[65]

Currently, many researchers around the globe are working on multidisciplinary aspects of nanotechnology to develop effective nanoneuromedicines, which can easily overcome the major obstacles in CNS therapeutics with no or negligible side effects. It was evident that nanoparticles and their combined use with therapeutic agents can be an effective tool in brain drug targeting for safer therapies in the future.[66] Apart from this, nanotechnology can play a crucial role in diagnosis, imaging, and drug delivery for CNS diseases. It is also believed that nanotechnology can possibly help in the development of innovative methods of regenerating CNS tissues.[67] Moreover, the important advantages of using nanomaterial for drug delivery in CNS include specific transport of the drug to the target tissue over the BBB with a controlled rate. Use of biodegradable carrier nanomaterial can help in lessening the toxicity to nontarget tissues and organs. The issues related to poorly soluble or total insolubility of therapeutic compounds in aqueous solutions can be addressed by using nanotechnology. In addition, the application of nanomaterials in the formulation of CNS therapeutic compounds can solve the difficulties in their oral or parental transport.[68,69]

Generally, in the past, polymer-based nanoparticles (PNPs), solid-lipid nanoparticles (SLNPs), liposomes, micelles, etc., have been commonly used as nanocarriers in drug delivery. However, recent advances in nanotechnology have developed new and efficient carrier materials such as dendrimers, nanoemulsions, nano gels, and nanosuspensions.[64,70] Although, different nanomaterials can be potentially used as carriers in CNS drug delivery, the effective results should have promising features like (i) nanomaterials used as carriers should be biodegradable, nontoxic, and biocompatible[71,72]; (ii) nanomaterials should be flexible so that their physical properties can easily be manipulated according to the mode of delivery[73]; (iii) nanomaterials with amended chemical properties should have the ability to achieve organ- or cell-specific drug delivery; and (iv) most importantly, the formulation used should be cost-effective. All these features of nanomaterials enhance the efficacy and specificity of CNS drug delivery. Overall, nanotechnology will be helpful for the safe and efficient management of CNS diseases. However, the role of nanotechnology in diagnosis, imaging, and drug delivery for the management of CNS diseases is separately discussed in this chapter.

5. NANOTECHNOLOGY AND NANOMATERIALS IN THE MANAGEMENT OF CNS DISEASES

The management of CNS diseases can be achieved by diagnosis and treatment of these diseases. In this context, nanotechnology together with the use of various nanomaterials offer applications in both diagnosis and treatment of CNS diseases. Diagnostic application of nanotechnology involves nanomaterials-mediated neuroimaging; and treatment applications involve the use of nanomaterials as the carrier in drug delivery.

5.1 Nanoparticles-Aided Neuroimaging

Neuroimaging is the prime step in the diagnosis of CNS disorders and can bring new possibilities of improved and early treatment of the disorders. However, current neuroimaging techniques viz., MRI, optical imaging (OI), computed tomography (CT), etc., have certain limitations such as less BBB penetration, insufficient sensitivity to specific diagnostic biomarkers, toxicity of contrast agents, etc.[74] To overcome these shortcomings of the prevalent techniques of neuroimaging, researchers have focused on nanoparticles for the proliferation of novel imaging modalities. Nanoparticles as probes with advanced optical properties can offer better means of visualization, characterization, and quantification of bioprocesses occurring within the brain.[75] For the purpose of neuroimaging, several classes of nanomaterials such as magnetic and paramagnetic nanoparticles and polymeric nanomaterials have gained immense attraction as novel contrast agents.[65] However, care should be taken while employing nanomaterials for imaging purposes in terms of size-driven toxicity as the large-sized nanoparticles tend to remain for a longer time in blood circulation.[76] Mostly magnetic nanoparticles and organic nanomaterials discussed below are used in neuroimaging.

5.1.1 Magnetic Nanoparticles

The use of iodinated contrast agents in the circulatory system for creation of contrast between blood vessels and parenchyma has proved a revolutionary step in visualizing the CNS. Magnetic nanoparticles such as iron oxide nanoparticles have been intensively investigated as contrast agent enhancers. Iron oxide nanoparticles play an indispensable role as front-runners over their high cost and less-sensitive conventional counterparts. Varallyay et al.[77] have tested the feasibility of steady state-cerebral blood volume measurements (SS-CBV, complementary to the conventional MRI) by using ferumoxytol, the superparamagnetic iron oxide nanoparticles (SPIO) in patients with CNS neoplasms. At increasing ferumoxytol doses, the SS-CBV maps were achieved with improved contrast to noise ratios and higher resolution images.[77] Similarly, Millward et al.[78] showed that very small SPIO can be useful in detecting CNS lesions in patients suffering from MS as well as experimental animal models of autoimmune encephalomyelitis. Prior to this, the lesions were not detectable by conventional MRI contrast agents. In another study, Farrell et al.[79] demonstrated the superiority of ferumoxytol, ultra-small SPIO nanoparticles over gadolinium-based contrast agents (GBCAs) for diagnosis of neuroinflammatory lesions. The study indicated the potential use of ferumoxytol for improved correlation between cerebral blood volume and CNS inflammation in patients with CNS disorders.[79] Ferromagnetic nanoparticles are being applied in MRI-based CNS pathology for the diagnosis of Alzheimer's disease,[80] epilepsy,[81] strokes,[82] autoimmune neuroinflammation,[83] etc.

5.1.2 Organic Nanomaterials

Several classes of organic nanomaterials, including polymeric nanoparticles, quantum dots (QDs), dendrimers, carbon nanotubes, etc., have been investigated for the molecular imaging of the brain and the diagnosis of CNS diseases. Tosi et al.[84] synthesized polymeric polylactide-*co*-glycolide (PLGA) nanoparticles conjugated with an opioid glycopeptide (g7) and a near-infrared (NIR) probe. The in vivo time-domain OI of the nanoconjugate showed that nanoparticles can be easily visualized by OI while g7 glycopeptide and can enhance the permeability across the BBB. Nano-scaled semiconductor fluorescent QDs have been proven as potential tools to visualize specific brain processes as well as the diagnosis of brain diseases. For example, Feng et al.[85] designed QDs conjugated with an antiAβ antibody (QDs Aβ-Ab) to track the Aβ accumulation in an Alzheimer's disease-induced mouse model in vivo, suggesting the potential of QDs for early molecular imaging. In another study, the hydrophilic QDs injected in a wild-type mouse brain were taken up by microglia. The study concluded that QDs can provide better means of visualization of interaction of microglia and neurons independently with improved image resolution.[86] Dendrimers, another class of nanomaterials, have also been found to be applicable in molecular imaging as well as drug delivery to treat brain cancer. For example, the dendrimer-grafted, gadolinium-functionalized nanogrphene oxide NPs having anticancer drug epirubicin (EPI) and Let-7 g miRNA (Gd-NGO-EPI-Let-7 g) were prepared and found to be potent cancer cell growth inhibitors. Moreover, the complex can be used as a contrast agent in an MRI to detect tumor loci and a BBB opening as well as to measure the quantity of the delivered drug to the tumor cells.[87] Recently, David-Pur et al.[88] designed a carbon nanotube technology-based neuroprosthetic device for electrical recording and stimulation. They further demonstrated the applications of the device for neuronal imaging purposes.

5.2 Nanomaterials-Mediated Drug Delivery

The nanoparticles-mediated drug delivery approach plays a significant role in the treatment of CNS disorders. It enables the controlled release of drugs while also helping in targeted drug delivery. Various nanomaterials utilized for the brain drug delivery include liposomes, polymeric nanoparticles, organic and inorganic nanoparticles, and dendrimers.[66]

Polymeric nanoparticles are made up of polymers that interact with the drugs; it can pass through the BBB and be distributed in the proper manner. Polymeric nanoparticles enhance the stability of drugs. Drugs may be adsorbed on the surface of the nanoparticles; conjugated or functionalized nanoparticles are widely used in drug delivery.[89] Recently, researchers revealed that polymeric nanoparticles conjugated with transferrin and fluorescein-loaded magnetic nanoparticles can be used in drug delivery and managing the CNS disorder.[90] Polymeric nanoparticles like tri-block *co*polymer nanoparticles composed of poly lactic-*co*-glycolic acid (PLGA), poly histidine, and poly ethylene glycol (PEG) can be used as effective vehicles for drug delivery to the CNS patient. PLGA nanoparticles conjugated with magnetic nanoparticles, polymeric nanoparticles conjugated with biotin, ligand-conjugated PLGA nanoparticles, and sialic acid functionalized nanoparticles were used for drug delivery.[91] Use of nanoparticles encapsulated or conjugated peptide is one of the novel approaches for the management of CNS disorders. Cell-penetrating peptides generally utilized in the treatment of CNS diseases mainly include the trans activating transcriptional activator (TAT), angiopep, penertratin, TP, Prion peptide, Syn B, etc.[92] It was found that polymeric nanoparticles such as PLGA-coated polysorbate[80] can be effectively used for the targeted delivery of the drug rivastigmine, which can be used in the treatment of the CNS disorder.[72] Polymeric nanoparticles-coated transferrin and polymeric nanoparticles loaded with lactoferrin are both used in drug delivery for the CNS disorder.[93] Today, researchers have investigated whether dual-conjugated nanoparticles can be used in the efficient delivery of the drug. Out of two peptides, one peptide is used for crossing the BBB and the other for targeted drug delivery. The efficacy of PEGylated PLGA nanoparticles conjugated with two antimicrobial peptides (TGN and QSH) was demonstrated. In this conjugated complex, TGN helps in the crossing of the BBB whereas QSH was utilized for treating and targeted drug delivery.[94] PLGA nanoparticles conjugated with lactide polyglycoside basic fibroblast growth factor (bFGF) and *Solanum tuberosum* can be used in the management of the CNS disorder.[95] Wen et al.[96] demonstrated the preparation of PLGA-conjugated odorranalectin nanoparticles and investigated whether conjugated nanoparticles were effective in CNS treatment. Mathew et al.[97] reported PLGA-loaded curcumin nanoparticles, which were further conjugated with the peptide Tet-1. It is the peptide that has a strong affinity toward the neuron cells. The conjugated nanoparticles were found to be an efficient drug in the treatment of Alzheimer's.

Chitosan (CS) nanoparticle-loaded drugs such as bromocriptine, rasagline, and ropindrole are also used in drug delivery for a CNS patient.[98] It was reported that the distribution and stability of the rivastigmine drug is enhanced after being loaded with CS nanoparticles. A study reported that CS nanoparticles-based drug delivery can act as a suitable candidate for drug delivery to the CNS, as CS nanoparticles can enter the neuronal cells by the process of endocytosis and accumulate around the nucleus and passes via the capillaries to the targeted site.[99] CS-nanoconjugated hormones, CS-loaded SmRho nanoparticles, and CS-sodium lauryl sulfate nanoparticles are the other suitable carriers for drug delivery in CNS patients.[99] It was found that the conjugated nanoparticles-loaded drug doxorubicin can be used in drug delivery for the brain tumor patient. Amoxicillin-loaded chitosan nanoparticles, chitosan nanoparticles-formulated cyclodextrin, and chitosan-encapsulated beta cyclodextrin can also be used in the delivery for CNS patients.[100] Dolichyl P and Brain-Derived Neutroponic Factor (BDNF) play important roles in the protection of CNS patients. Therefore, researchers have started exploiting nanomaterials, which can be used in the delivery of these biomolecules and thereby help in the management of CNS drug delivery.[101]

Lipid-based nanoparticles can also act as the carrier for drug delivery in a CNS patient. Researchers investigated whether stearylamine and esterquat-coated, cholesterol-mediated cationic SLN can be used in the controlled release of the drug in the CNS patients. Chen et al.[102] developed curcumin-loaded SLN and donepezed-loaded SLN, reporting that it can be used in the treatment of CNS patients. Zhuang et al.[103] reported that SLNPs can be used in the drug delivery of vinpocetine; it can be used in the treatment of a CNS disorder. Ferulic acid-loaded SLN was also used in drug delivery, which helps in the formation of reactive oxygen species and provides suitable atmosphere for the delivery of cholinergic drugs.[104] Huperzine-loaded SLN was prepared and found to be effective in the therapy of CNS patients.[105] Kakkar et al.[106] reported the preparation of curcumin-loaded SLN, and investigated whether curcumin inhibits the Aβ formation and aggregation of Aβ pleated can be used in the treatment of Alzheimer's disease.

A liposome-mediated drug delivery system is widely used for the treatment of CNS disorders. It enables the biodistribution of the drugs and helps in targeted drug delivery. Different types of liposomes involved in the drug delivery for CNS diseases are nanoemulsion, nanoparticles, phytosomes, cochleates, niosomes, liposphers, ethosomes, transformers, nanomers, virosomes, and iscorns.[89] Recently, it was reviewed whether liposome decorated or loaded nanoparticles are more effective in drug delivery as compared to the nondecorated liposome.[107] Yang et al.[108] prepared rivastigmine-loaded liposome and cell-penetrating, peptide-modified liposome for the proper distribution of the drug and the stability of the drugs. Ligand-functionalized liposome was used for delivery of the drug galantamine. It was found that cadmium telluride (CdTe) nanoparticles inhibit the formation of amyloid beta fibrils, which is responsible for Alzheimer's disease. CdTe nanoparticles get attached with Aβ plaque and help in the prevention of fibril growth. Nanoparticles arrest the growth of Aβ fibril and can prevent the progression of Alzheimer's disease.[109] Fig. 1 represents the schematic illustration of

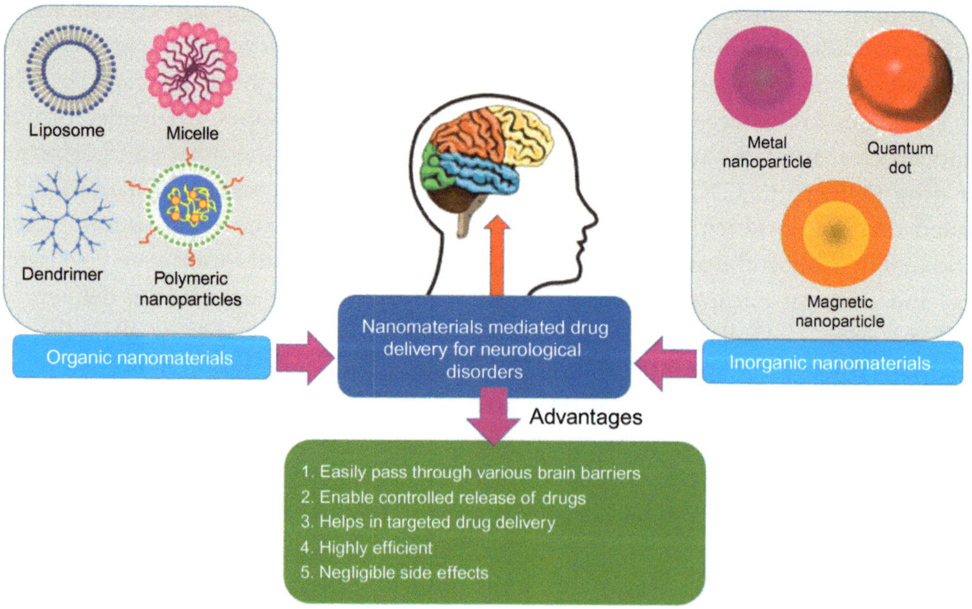

Fig. 1 Schematic illustration of various nanomaterial-mediated drug delivery for the treatment of neurological disorders and their advantages.

various nanomaterials-mediated drug delivery for the treatment of neurological disorders and their advantages.

6. NANOPARTICLES-BASED NEUROTOXICITY

Although application of nanoparticles has been increasing enormously in recent times, the concerns regarding their adverse effects have been realized by scientists and the public at large. Nanoparticles exerts strong antimicrobial properties and hence control the growth of microbes including bacteria, fungi, and viruses; but at the same time, it has ability to interact with human cells and interfere its normal functions. Hence, there is a need to have a close look at such interactions. As far as the nervous system is concerned, nanoparticles may be hazardous. Neurons are very delicate cells among the cells of other organs of animals and humans. Therefore, any interaction of nanoparticles with them may show more damage.

Nanoparticles have shown toxicity to brain cells as they can pass through the various membranes.[110] The biological half-life of silver is more in the brain compared to any other organ; hence, that can show a greater damage to brain cells as compared to others.[111] It is reported that AgNPs can enter into the brain through the BBB and the blood-cerebrospinal fluid[112] and can be translocated into various regions of the brain.[113] Therefore, it may cause a risk to patients undergoing treatment with drug-loaded AgNPs. Further, they can induce the impairment of normal neurons[114] and microglia cells[115,116] and may ultimately intensify brain pathology.[117] Aminosilane-coated iron oxide nanoparticles (AmS-IONPs) have promising applications in multifunctional drug delivery systems. At a low concentration, they were reported to show low toxicity to neurons, endothelial cells, and astrocytes, whereas at higher concentration they demonstrated toxicity to astrocytes.[118] Depending on size and time, Fe_2O_3 nanoparticles were also reported to translocate in the mouse olfactory bulb where they were found to be accumulated in the axons of olfactory neurons, mitochondria, and lysosomes of hippocampus cells.[119] Al_2O_3 nanoparticles, if inhaled, can be translocated to the brain via the bloodstream by damaging the BBB. They can also enter into the CNS via the olfactory nerve and eventually trigger the apoptosis and necrosis in neurons through inducing oxidative stress and alteration of cell signaling.[120]

The excitability of neurons is mainly dependent on the voltage-gated sodium currents. They define the properties of neurons at large viz. the action potential of neurons and their transmittance across the synaptic terminals. Nanoparticles can cause the local depolarization of neurons. They can further lead to the degradation of the neural cytoskeleton, perturbations of pre- and postsynaptic proteins, and mitochondrial dysfunction, which may result in cell death.[114,116,121,122] Nanoparticles can enter into the brain though various routes. The higher the retention of nanoparticles in the blood, the higher the possibility that nanoparticles will be in the brain-blood capillaries.[123] In such conditions, they can cross the BBB. High-nanoparticle concentration can cause them to cross

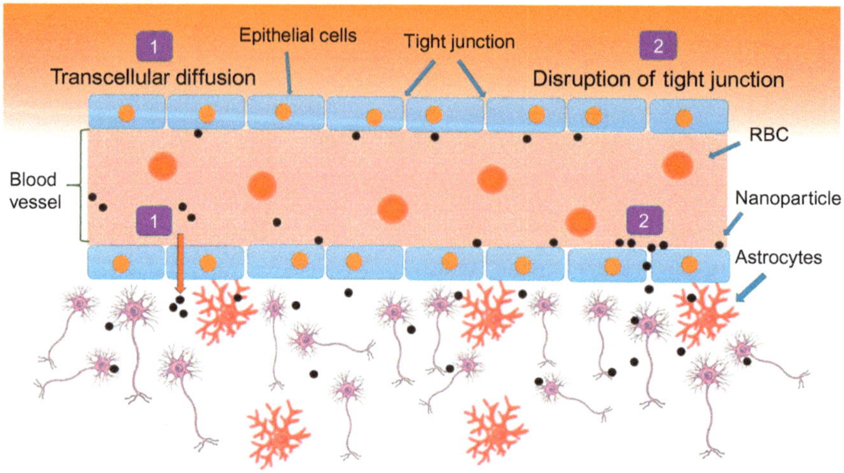

Fig. 2 Schematic illustration of possible mechanisms of nanoparticle transport across the blood brain barrier. (1) Transcellular diffusion: nanoparticles entered the brain cell by endocytosis and/or transcytosis. (2) Disruption of tight junction: damage to the tight junction facilitates entry of nanoparticles in the brain. In both cases, once nanoparticles entered the brain, they may damage astrocytes and normal neurons.

the endothelial layer so as to reach the brain. An increase in the fluidity of the endothelial cell membrane will help to increase the permeability of the nanoparticles to cross the BBB. Moreover, the disruption of tight junctions present between the endothelial cells will also create a gap so as to increase the nanoparticle permeability.[124] This mechanism is illustrated in Fig. 2. Nanoparticles in a combination of drugs can be transported across the BBB through transcellular diffusion endocytosed or transcytosed, leading to their transfer in the brain.[123] The smaller nanoparticles can easily enter into the brain whereby they can induce toxicity, if accumulated. If they are used beyond certain concentrations, then they may show a toxic effect on human cells. Therefore, there is a need to further explore the possible harmful effects of nanoparticles in vitro followed by in vivo studies.

7. CONCLUSIONS

Applications of nanotechnology in neurosciences are of utmost medical need to generate hope for patients suffering from CNS disorders. The development of a new generation of nanoneuromedicines with reduced side effects can help in the control of prolonged treatment and in efficient targeted drug delivery. Various nanomaterials such as magnetic nanoparticles and organic nanomaterials help in the diagnosis of CNS disorders. In addition, polymeric nanoparticles, liposomes, SLNPs, etc., loaded with various drugs can be effectively used in targeted drug delivery, which ultimately helps in the management of CNS disorders. Although, it is confirmed that nanoparticle-mediated drug delivery has several advantages, many aspects related to its toxicity are still a matter of great concern because there is no specific method to identify the toxicity level and targeted drug release in the CNS. The benefits offered by nanotechnology in CNS can't be denied but widespread applications of nanomaterials pose the risk of toxicity. Hence, extensive studies need to be performed to evaluate the exact toxicity of nanomaterials in the CNS. Moreover, specific guidelines should be followed to avoid the harmful effects of nanotechnology.

REFERENCES

1. Giri A, Arjyal A, Koirala S, Karkey A, Dongol S, Thapa SD, et al. Aetiologies of central nervous system infections in adults in Kathmandu, Nepal: a prospective hospital-based study. *Sci Rep* 2013;**3**:2382. https://doi.org/10.1038/srep02382.
2. John CC, Carabin H, Montano SM, Bangirana P, Zunt JR, Peterson PK. Global research priorities for infections that affect the nervous system. *Nature* 2015;**527**:S178–86.
3. Bhimraj A. Acute community-acquired bacterial meningitis in adults: an evidence-based review. *Cleve Clin J Med* 2012;**79**:393–400.
4. Peltola H. Worldwide *Haemophilus influenzae* type b disease at the beginning of the 21st century: global analysis of the disease burden 25 years after the use of the polysaccharide vaccine and a decade after the advent of conjugates. *Clin Microbiol Rev* 2000;**13**:302–17.
5. Mai NTH, Hoa NT, Nga TVT, Linh LD, Chau TTH, Sinh DX, et al. *Streptococcus suis* meningitis in adults in Vietnam. *Clin Infect Dis* 2008;**46**:659–67.
6. Chakrabarti A. Epidemiology of central nervous system mycoses. *Neurol India* 2007;**55**(3):191–7.
7. Idro R, Marsh K, John CC, Newton CRJ. Cerebral malaria; mechanisms of brain injury and strategies for improved neuro-cognitive outcome. *Pediatr Res* 2010;**68**(4):267–74.
8. Misra UK, Kalita J, Prabhakar S, Chakravarty A, Kochar D, Nair PP. Cerebral malaria and bacterial meningitis. *Ann Indian Acad Neurol* 2011;**14**(Suppl. 1):S35–9.
9. WHO, World Health Organization. *Number of people (all ages) living with HIV*. http://www.who.int/gho/hiv/epidemic_status/cases_all/en/; 2015.
10. Kennedy PGE. Viral encephalitis: causes, differential diagnosis, and management. *J Neurol Neurosurg Psychiatry* 2004;**75**(Suppl. I):i10–5.
11. Ances BM, Ellis RJ. Dementia and neurocognitive disorders due to HIV-1 infection. *Semin Neurol* 2007;**27**:86–92.
12. Nair N, Wares F, Sahu S. Tuberculosis in the WHO South-East Asia region. *Bull World Health Organ* 2010;**88**:164.
13. Changal KH, Raina AH. Central nervous system manifestations of tuberculosis: a review article. *J Mycobact Dis* 2014;**4**:2. https://doi.org/10.4172/2161-1068.1000146.
14. Tan IL, Smith BR, von Geldern G, Mateen FJ, McArthur JC. HIV-associated opportunistic infections of the CNS. *Lancet Neurol* 2012;**11**:605–17.
15. Labeaud AD, Bashir F, King CH. Measuring the burden of arboviral diseases: the spectrum of morbidity and mortality from four prevalent infections. *Popul Health Metrics* 2011;**9**(1):1–11.
16. Dutta KBA. Japanese encephalitis: a tale of inflammation and degeneration in the central nervous system. In: Peterson PK, Toborek M, editors. *Neuroinflammation and neurodegeneration*. New York, NY: Springer Verlag; 2014. p. 309–35.
17. Fooks AR, Banyard AC, Horton DL, Johnson N, McElhinney LM, Jackson AC. Current status of rabies and prospects for elimination. *Lancet* 2014;**384**:1389–99.
18. Cassady KA, Whitley RJ. Pathogenesis and pathophysiology of viral infections of the central nervous systems. In: Scheld WM, Whitley RJ, Marra CM, editors. *Infections of the central nervous system*. Baltimore, MD: Lippincott Williams & Wilkins; 2004. p. 137–57.

19. Khetsuriani N, Holman RC, Anderson LJ. Burden of encephalitis associated hospitalizations in the United States, 1988–1997. *Clin Infect Dis* 2002;**35**:175–82.
20. Davison KL, Crowcroft NS, Ramsay ME, Brown DW, Andrews NJ. Viral encephalitis in England, 1989–1998: what did we miss? *Emerg Infect Dis* 2003;**9**:234–40.
21. Huppatz C, Durrheim DN, Levi C, Dalton C, Williams D, Clements MS, et al. Etiology of encephalitis in Australia, 1990–2007. *Emerg Infect Dis* 2009;**15**:1359–65.
22. WHO, World Health Organization. *Fourth biregional meeting on the control of Japanese Encephalitis (JE)*. vol. 2012. World Health Organization, Regional Office for South-East Asia; 2009.
23. Rotbart HA. Viral meningitis. *Semin Neurol* 2000;**20**:277–92.
24. Lo MK, Rota PA. The emergence of Nipah virus, a highly pathogenic paramyxovirus. *J Clin Virol* 2008;**43**:396–400.
25. Potharaju NR. Incidence rate of acute encephalitis syndrome without specific treatment in India and Nepal. *Indian J Community Med* 2012;**37**:240–51.
26. Swanson EC, Schleiss MR. Congenital cytomegalovirus infection: new prospects for prevention and therapy. *Pediatr Clin N Am* 2013;**60**:335–49.
27. WHO, World Health Organization. *Human African trypanosomiasis. Lowest caseload recorded as the world prepares to defeat sleeping sickness*. http://www.who.int/trypanosomiasis_african/en/; 2016.
28. Bhatt S, Gething PW, Brady OJ, Messina JP, Farlow AW, Moyes CL, et al. The global distribution and burden of dengue. *Nature* 2013;**496**:504–7.
29. Makhani N, Banwell B. Non-infectious central nervous system inflammatory disorders. *Semin Pediatr Neurol* 2011;**18**(2):91–4.
30. Makhani N, Bigi S, Banwell B, Shroff M. Diagnosing neuromyelitis optica. *Neuroimaging Clin N Am* 2013;**23**(2):279–91.
31. NINDS, National Institute of Neurological Disorders and Stroke. *Multiple sclerosis information page*; 2015 [Retrieved 1 April 2017].
32. Berer K, Krishnamoorthy G. Microbial view of central nervous system autoimmunity. *FEBS Lett* 2014;**588**(22):4207–13.
33. WHO, World Health Organization. *Atlas: multiple sclerosis resources in the world 2008 (PDF)*. Geneva: World Health Organization; 2008. p. 15–6.
34. Milo R, Kahana E. Multiple sclerosis: geoepidemiology, genetics and the environment. *Autoimmun Rev* 2010;**9**(5):A387–94.
35. Clanet M. Jean-Martin Charcot. 1825 to 1893. *Int MS J* 2008;**15**(2):59–61.
36. Lucchinetti CF, Yong G, Bogdan FGP, Kazuo F, Yasuto I, et al. The pathology of an autoimmune astrocytopathy: lessons learned from neuromyelitis optica. *Brain Pathol* 2014;**24**(1):83–97.
37. Pittock SJ, Weinshenker BG, Lucchinetti CF, Wingerchuk DM, Corboy JR, Lennon VA. Neuromyelitis optica brain lesions localized at sites of high aquaporin 4 expression. *Arch Neurol* 2006;**63**(7): 964–8.
38. Wingerchuk DM. Neuromyelitis optica. *Int MS J* 2006;**13**(2):42–50.
39. Berlit P. Diagnosis and treatment of cerebral vasculitis. *Ther Adv Neurol Disord* 2010;**3**(1):29–42.
40. Soo-Hwan K, Hyeong-Seon L, Deok-Seon R, Soo-Jae C, Dong-Seok L. Antibacterial activity of silver-nanoparticles against *Staphylococcus aureus* and *Escherichia coli*. *Korean J Microbiol Biotechnol* 2011;**39**(1):77–85.
41. Thomas R, Janardhanan A, Varghese RT, Soniya EV, Mathew J, Radhakrishnan EK. Antibacterial properties of silver nanoparticles synthesized by marine *Ochrobactrum* sp. *Braz J Microbiol* 2014;**45**(4): 1221–7.
42. Perito B, Giorgetti E, Marsili P, Muniz-Miranda M. Antibacterial activity of silver nanoparticles obtained by pulsed laser ablation in pure water and in chloride solution. *Beilstein J Nanotechnol* 2016;**7**:465–73.
43. Gajbhiye M, Kesharwani J, Ingle A, Gade A, Rai M. Fungus mediated synthesis of silver nanoparticles and their activity against pathogenic fungi in combination with fluconazole. *Nanomedicine* 2009;**5**(4): 382–6.
44. Nasrollahi A, Pourshamsian K, Mansourkiaee P. Antifungal activity of silver nanoparticles on some of fungi. *Int J Nano Dimens* 2011;**1**(3):233–9.
45. Pulit J, Banach M, Szczygłowska R, Bryk M. Nanosilver against fungi. Silver nanoparticles as an effective biocidal factor. *Acta Biochim Pol* 2013;**60**(4):795–8.

46. Mallmann EJJ, Cunha FA, Castro BNMF, Maciel AM, Menezes EA, et al. Antifungal activity of silver nanoparticles obtained by green synthesis. *Rev Inst Med Trop Sao Paulo* 2015;**57**(2):165–7.
47. Elgorban AM, El-Samawaty ARM, Yassin MA, Sayed SR, Syed FA, Elhindi KM, et al. Antifungal silver nanoparticles: synthesis, characterization and biological evaluation. *Biotechnol Biotechnol Equip* 2016;**30**(1):56–62.
48. Galdiero S, Falanga A, Vitiello M, Cantisani M, Marra V, Galdiero M. Silver nanoparticles as potential antiviral agents. *Molecules* 2011;**16**:8894–918.
49. Gaikwad S, Ingle A, Gade A, Rai M, Falanga A, Incoronato N, et al. Antiviral activity of mycosynthesized silver nanoparticles against herpes simplex virus and human para influenza virus type 3. *Int J Nanomedicine* 2013;**8**:4303–14.
50. Mori Y, Ono T, Miyahira Y, Nguyen VQ, Matsui T, Ishihara M. Antiviral activity of silver nanoparticle/chitosan composites against H1N1 influenza A virus. *Nanoscale Res Lett* 2013;**8**(1):93. https://doi.org/10.1186/1556-276X-8-93.
51. Kim KS. Acute bacterial meningitis in infants and children. *Lancet Infect Dis* 2010;**10**(1):32–42.
52. Tan YC, Gill AK, Kim KS. Treatment strategies for central nervous system infections: an update. *Expert Opin Pharmacother* 2015;**16**(2):187–203.
53. Brouwer MC, Tunkel AR, van de Beek D. Epidemiology, diagnosis, and antimicrobial treatment of acute bacterial meningitis. *Clin Microbiol Rev* 2010;**23**(3):467–92.
54. Shin SH, Kim KS. Treatment of bacterial meningitis: an update. *Expert Opin Pharmacother* 2012;**13**(15): 2189–206.
55. Prasad K, Karlupia N, Kumar A. Treatment of bacterial meningitis: an overview of cochrane systematic reviews. *Respir Med* 2009;**103**(7):945–50.
56. CDC. Prevention and control of meningococcal disease: recommendations of the Advisory Committee on Immunization Practices (ACIP). *MMWR* 2005;**54**(RR-7):1–28.
57. CDC. Report from the Advisory Committee on Immunization Practices (ACIP): decision not to recommend routine vaccination of all children aged 2–10 years with quadrivalent meningococcal conjugate vaccine (MCV4). *MMWR* 2008;**57**:462–5.
58. John CC, Idro RI. Cerebral malaria in children. *Infect Med* 2003;**20**(1):53–8.
59. van Hensbroek MB, Onyiorah E, Jaffar S, Schneider G, Palmer A, Frenkel J, et al. A trial of artemether or quinine in children with cerebral malaria. *N Engl J Med* 1996;**335**:69–75.
60. Huda SN, Shahab T, Ali SM, Afzal K, Khan HM. A comparative clinical trial of artemether and quinine in children with severe malaria. *Indian Pediatr* 2003;**40**(10):939–45.
61. Jacob A, Weinshenker BG, Violich I, McLinskey N, Krupp L, Fox RJ, et al. Treatment of neuromyelitis optica with rituximab: retrospective analysis of 25 patients. *Arch Neurol* 2008;**65**(11):1443–8.
62. Bedi GS, Brown AD, Delgado SR, Usmani N, Lam BL, Sheremata WA. Impact of rituximab on relapse rate and disability in neuromyelitis optica. *Mult Scler* 2011;**17**(10):1225–30.
63. Pellkofer HL, Krumbholz M, Berthele A, Hemmer B, Gerdes LA, Havla J, et al. Long-term follow-up of patients with neuromyelitis optica after repeated therapy with rituximab. *Neurology* 2011;**76**(15): 1310–5.
64. Soni S, Ruhela RK, Medhi B. Nanomedicine in central nervous system (CNS) disorders: a present and future prospective. *Adv Pharm Bull* 2016;**6**(3):319–35.
65. Ajetunmobi A, Prina-Mello A, Volkov Y, Corvin A, Tropea D. Nanotechnologies for the study of the central nervous system. *Prog Neurobiol* 2014;**123**:18–36.
66. Jain KK. Nanobiotechnology-based drug delivery to the central nervous system. *Neurodegener Dis* 2007;**4**(4):287–91.
67. Provenzale JM, Mohs AM. Nanotechnology in neurology current status and future possibilities. *US Neurol* 2010;**6**(1):12–7.
68. Costantino L. Drug delivery to the CNS and polymeric nanoparticulate carriers. *Future Med Chem* 2010;**2**(11):1681–701.
69. Nishikant D, Shiva S, Rishu T, Vikrant C. Nanotechnology as a targeted drug delivery system to brain. *Nanosci Nanotechnol* 2016;**10**(3):105.
70. Wong HL, Wu XY, Bendayan R. Nanotechnological advances for the delivery of CNS therapeutics. *Adv Drug Deliv Rev* 2012;**64**(7):686–700.
71. Desai MP, Labhasetwar V, Walter E, Levy RJ, Amidon GL. The mechanism of uptake of biodegradable microparticles in Caco-2 cells is size dependent. *Pharm Res* 1997;**14**(11):1568–73.

72. Kreuter J. Influence of the surface properties on nanoparticle-mediated transport of drugs to the brain. *J Nanosci Nanotechnol* 2004;**4**(5):484–8.
73. Mohanraj K, Sethuraman S, Krishnan UM. Development of poly(butylene succinate) microspheres for delivery of levodopa in the treatment of Parkinson's disease. *J Biomed Mater Res B Appl Biomater* 2013;**101**(5):840–7.
74. Vaghela V, Kesavadas C, Thomas B. Functional magnetic resonance imaging of the brain: a quick review. *Neurol India* 2010;**58**:879–85.
75. Nowostawska M, Corr S, Byrne S, Conroy J, Volkov Y, Gunko Y. Porphyrin-magnetite nanoconjugates for biological imaging. *J Nanobiotechnol* 2011;**9**:13.
76. Betke SC, Lammers T, Kiessling F. Applications of nanoparticles for diagnosis and therapy of cancer. *Br J Radiol* 2015;**88**(1054). https://doi.org/10.1259/bjr.20150207.
77. Varallyay CG, Nesbit E, Fu R, Gahramanov S, Moloney B, Earl E, et al. High-resolution steady-state cerebral blood volume maps in patients with central nervous system neoplasms using ferumoxytol, a super paramagnetic iron oxide nanoparticle. *J Cereb Blood Flow Metab* 2013;**33**:780–6.
78. Millward J, Schnorr J, Taupitz M, Wagner S, Wuerfel J, Infante-Duarte C. Iron oxide magnetic nanoparticles highlight early involvement of the choroid plexus in central nervous system inflammation. *ASN Neuro* 2013;**5**.
79. Farrell BT, Hamilton BE, Dósa E, Rimely E, Nasseri M, Gahramanov S, et al. Using iron oxide nanoparticles to diagnose CNS inflammatory diseases and PCNSL. *Neurology* 2013;**81**:256–63.
80. Skaat H, Corem-Slakmon E, Grinberg I, Last D, Goez D, Mardor Y, et al. Antibody-conjugated, dual-modal, near-infrared fluorescent iron oxide nanoparticles for antimyloidgenic activity and specific detection of amyloid-β fibrils. *Int J Nanomedicine* 2013;**8**:4063–76.
81. Akhtari M, Bragin A, Moats R, Frew A, Mandelkern M. Imaging brain neuronal activity using functionalized magneto particles and MRI. *Brain Topogr* 2012;**25**(4):374–88.
82. Schefer PW, Barak ER, Kamalian S, Gharai LR, Schwamm L, Gonzalez RJ, et al. Quantitative assessment of core/penumbra mismatch in acute stroke: CT and MR perfusion imaging are strongly correlated when sufficient brain volume is imaged. *Stroke* 2008;**39**(11):2986–92.
83. McAteer MA, Sibson NR, von Zur MC, Schneider JE, Lowe AS, Warrick N, et al. In vivo magnetic resonance imaging of acute brain inflammation using microparticles of iron oxide. *Nat Med* 2007;**13**(10):1253–8.
84. Tosi G, Bondioli L, Ruozi B, Badiali L, Severini G, Biffi S, et al. NIR-labeled nanoparticles engineered for brain targeting: in vivo optical imaging application and fluorescent microscopy evidences. *J Neural Transm* 2011;**118**:145–53.
85. Feng L, Long HY, Liu RK, Sun DN, Liu C, et al. A quantum dot probe conjugated with a β antibody for molecular imaging of Alzheimer's disease in a mouse model. *Cell Mol Neurobiol* 2013;**33**(6):759–65.
86. Minami S, Sun B, Popat K, Kauppinen T, Pleiss M, Zhou Y, et al. Selective targeting of microglia by quantum dots. *J Neuroinflammation* 2012;**9**:22.
87. Yang HW, Huang CY, Lin CW, Liu HL, Huang CW, Liao SS, et al. Gadolinium-functionalized nanographene oxide for combined drug and micro RNA delivery and magnetic resonance imaging. *Biomaterials* 2014;**35**(24):6534–42.
88. David-Pur M, Bareket-Keren L, Beit-Yaakov G, Raz-Prag D, Hanein Y. All carbonnano tube flexible multi-electrode array for neuronal recording and stimulation. *Biomed Microdevices* 2014;**16**:43–53.
89. Upadhyay RK. Drug delivery systems, CNS protection, and the blood brain barrier. *Biomed Res Int* 2014;**4**: https://doi.org/10.1155/2014/869269.
90. Yan F, Wang Y, He S, Ku S, Gu W, Ye L. Transferrin conjugated, fluorescein-loaded magnetic nanoparticles for targeted delivery across the blood-brain barrier. *J Mater Sci Mater Med* 2013;**24**(10):2371–9.
91. Cai Q, Wang L, Deng G, Liu J, Chen Q, Chen Z. Systemic delivery to central nervous system by engineered PLGA nanoparticles. *Am J Transl Res* 2016;**8**(2):749–64.
92. Zou LL, Ma JL, Wang T, Yang TB, Liu CB. Cell-penetrating peptide-mediated therapeutic molecule delivery into the central nervous system. *Curr Neuropharmacol* 2013;**11**:197–208.
93. Guo L, Ren J, Jiang X. Perspectives on brain-targeting drug delivery systems. *Curr Pharm Biotechnol* 2012;**13**:2310–8.

94. Zhang C, Chen J, Feng C, Shao X, Liu Q, Zhang Q, et al. Intranasal nanoparticles of basic fibroblast growth factor for brain delivery to treat Alzheimer's disease. *Int J Pharm* 2014;**461**(1):192–202.
95. Zhang C, Wan X, Zheng X, Zheng X, Shao X, Liu Q, et al. Dual-functional nanoparticles targeting amyloid plaques in the brains of Alzheimer's disease mice. *Biomaterials* 2014;**35**(1):456–65.
96. Wen Z, Yan Z, Hu K, Pang Z, Cheng X, Guo L, et al. Odorranalectin conjugated nanoparticles: preparation, brain delivery and pharmacodynamic study on Parkinson's disease following intranasal administration. *J Control Release* 2011;**151**(2):131–8.
97. Mathew A, Fukuda T, Nagaoka Y, Hasumura T, Morimoto H, Yasuhiko Y, et al. Curcumin loaded-PLGA nanoparticles conjugated with tet-1 peptide for potential use in Alzheimer's disease. *PLoS One* 2012;**7**(3):1–10.
98. Singh K, Ahmad Z, Shakya P, Ansari VA, Kumar A, Zishan M, et al. Nano formulation: a novel approach for nose to brain drug delivery. *J Chem Pharm Res* 2016;**8**(2):208–15.
99. Chaudhury A, Das S. Recent advancement of chitosan based nanoparticles for oral controlled delivery of insulin and other therapeutic agents. *AAPS PharmSciTech* 2011;**12**(1):10–20.
100. Bisson-Boutelliez C, Fontanay S, Finance C, Kedzierewicz F. Preparation and physicochemical characterization of amoxicillin-cyclodextrin complexes. *AAPS PharmSciTech* 2010;**11**(2):574–81.
101. Pilakka-Kanthikeel S, Atluri VSR, Sagar V, Nair M. Targeted brain derived neurotropic factors (BDNF) delivery across the blood-brain barrier for neuro-protection using magnetic nano carriers: an in vitro study. *PLoS One* 2013;**8**(4). https://doi.org/10.1371/journal.pone.0062241.
102. Chen J, Dai WT, He ZM, Gao L, Huang X, Gong JM, et al. Fabrication and evaluation of curcumin-loaded nanoparticles based on solid lipid as a new type of colloidal drug delivery system. *Indian J Pharm Sci* 2013;**75**(2):178–84.
103. Zhuang CY, Li N, Wang M, Zhang XN, Pan WS, Peng JJ, et al. Preparation and characterization of vinpocetine loaded nanostructured lipid carriers (NLC) for improved oral bioavailability. *Int J Pharm* 2010;**394**(1–2):179–85.
104. Bondi ML, Craparo EF, Giammona G, Drago F. Brain-targeted solid lipid nanoparticles containing riluzole: preparation, characterization and biodistribution. *Nanomedicine* 2009;**5**(1):25–32.
105. Wang BS, Wang H, Wei ZH, Song YY, Zhang L, Chen HZ. Efficacy and safety of natural acetylcholinesterase inhibitor huperzine a in the treatment of Alzheimer's disease: an updated meta-analysis. *J Neural Transm* 2009;**116**(4):457–65.
106. Kakkar V, Muppu SK, Chopra K, Kaur IP. Curcumin loaded solid lipid nanoparticles: an efficient formulation approach for cerebral ischemic reperfusion injury in rats. *Eur J Pharm Biopharm* 2013;**85**(3): 339–45.
107. Salvati E, Re F, Sesana S, Cambianica I, Sancini G, Masserini M, et al. Liposomes functionalized to overcome the blood-brain barrier and to target amyloid-β peptide: the chemical design affects the permeability across an *in vitro* model. *Int J Nanomedicine* 2013;**8**:1749–58.
108. Yang ZZ, Zhang YQ, Wang ZZ, Wu K, Lou JN, Qi XR. Enhanced brain distribution and pharmacodynamics of rivastigmine by liposomes following intranasal administration. *Int J Pharm* 2013;**452**(1): 344–54.
109. Yoo SI, Yang M, Brender JR, Kotov N. Inhibition of amyloid peptide fibril action by inorganic nanoparticles: functional similarities with protein. *Angew Chem Int Ed Eng* 2011;**50**:110–5.
110. Brooking J, Davis SS, Illum L. Transport of nanoparticles across the rat nasal mucosa. *J Drug Target* 2001;**9**:267–79.
111. Yang Z, Liu ZW, Allaker RP, Reip P, Oxford J, Ahmad Z, et al. A review of nanoparticle functionality and toxicity on the central nervous system. *J R Soc Interface* 2010;**7**:S411–22.
112. Cramer S, Tacke S, Bornhorst J, Klingauf J, Schwerdtle T, Galla HJ. The influence of silver nanoparticles on the blood-brain and the blood-cerebrospinal fluid barrier in vitro. *J Nanomed Nanotechnol* 2014;**5**: https://doi.org/10.4172/2157-7439.1000225.
113. Panyala NR, Pena-Mendez EM, Havel J. Silver or silver nanoparticles: a hazardous threat to the environment and human health? *J Appl Biomed* 2008;**6**:117–29.
114. Tang M, Xing T, Zeng J, Wang H, Li C, Yin S, et al. Unmodified CdSe quantum dots induce elevation of cytoplasmic calcium levels and impairment of functional properties of sodium channels in rat primary cultured hippocampal neurons. *Environ Health Perspect* 2008;**116**:915–22.

115. Au C, Mutkus L, Dobson A, Riffle J, Lalli J, Aschner M. Effects of nanoparticles on the adhesion and cell viability on astrocytes. *Biol Trace Elem Res* 2007;**120**:248–56.
116. Xu F, Piett C, Farkas S, Qazzaz M, Syed NI. Silver nanoparticles (AgNPs) caused regeneration of cytoskeleton and disrupt synaptic machinery of cultured cortical neurons. *Mol Brain* 2013;**6**:29. https://doi.org/10.1186/1756-6606-6-29.
117. Sharma HS, Sharma A. Nanoparticles aggravate heat stress induced cognitive deficits, blood-brain barrier disruption, edema formation and brain pathology. *Prog Brain Res* 2007;**162**:245–73.
118. Sun Z, Yathindranath V, Worden M, Thliveris JA, Chu S, Parkinson FE, et al. Characterization of cellular uptake and toxicity of aminosilane-coated iron oxide nanoparticles with different charges in central nervous system-relevant cell culture models. *Int J Nanomedicine* 2013;**8**:961–70.
119. Wang B, Wang Q, Chen H, Zhou X, Wang H, Wang H, et al. Size-dependent translocation pattern, chemical and biological transformation of nano- and submicron-sized ferric oxide particles in the central nervous system. *J Nanosci Nanotechnol* 2016;**16**(6):5553–61.
120. Niu Q, Zhang Q. Combined effect on immune and nervous system of aluminum nanoparticles. In: Otsuki T, Petratca C, Gioacchino MD, editors. *Allergy and immunotoxicology in occupational health.* Singapore: Springer; 2017. p. 115–28.
121. Shin SH, Ye MK, Kim HS, Kang HS. The effects of nano-silver on the proliferation and cytokine expression by peripheral blood mononuclear cells. *Int Immunopharmacol* 2007;**7**:1813–8.
122. Cha K, Hong HW, Choi YG, Lee MJ, Park JH, Chae HK, et al. Comparison of acute responses of mice livers to short-term exposure to nano-sized or micro-sized silver particles. *Biotechnol Lett* 2008;**30**:1893–9.
123. Kreuter J. Drug delivery to the central nervous system by polymeric nanoparticles: what do we know? *Adv Drug Deliv Rev* 2014;**71**:2–14.
124. Xu L, Dan M, Shao A, Cheng X, Zhang C, Yokel RA, et al. Silver nanoparticles induce tight junction disruption and astrocyte neurotoxicity in a rat blood-brain barrier primary triple coculture model. *Int J Nanomedicine* 2015;**10**:6105–61118.

CHAPTER 6

Natural Products as Immune System Modulators, and Against Infections of the Central Nervous System

Ali Parsaeimehr*, Hafiz M.N. Iqbal*, Roberto Parra-Saldívar*,†,‡,§
*Tecnologico de Monterrey, School of Engineering and Sciences, Campus Monterrey, Ave. Eugenio Garza Sada 2501, Monterrey, N. L., CP 64849, Mexico
†Microsystems Technologies Laboratories, MIT, Cambridge, MA 02139, USA
‡Division of Engineering in Medicine, Department of Medicine, Brigham and Women's Hospital, Harvard Medical School, Cambridge, MA02139, USA
§Harvard-MIT Division of Health Sciences and Technology, Massachusetts Institute of Technology, Cambridge, MA 02139, USA

1. INTRODUCTION

Nature has always generously provided necessary medicinal compounds for humankind. Plants, specifically, have mostly been the main traditional methodologies for curing disease, with the first historical evidence dating back to 2600 BCE with the use of nearly 1000 plant-derived medicines in Mesopotamia.[1,2] Plant-based materials are still performing essential roles in healthcare, and acetyldigoxin, aescin, khellin, agrimophol, anabasine, arecoline, berberine, betulinic acid, catechin, codeine, colchicine, curcumin, digoxin, irinotecan, α-lobeline, papain, picrotoxin, taxol, thymol, vinblastine, and vincristine are a few examples of valuable medicines derived from plants.[3,4]

Reports from the World Health Organization demonstrate the escalation in popularity and usage of traditional medicines around the world.[5] In China, medicinal plants are the sources of 30%–50% of the total medicinal intake. In Zambia, Ghana, Nigeria, and Mali, the initial action for 60% of children with high fever due to malaria is the consumption of plant-derived medicines at home. In North America, Europe, and other industrialized areas, more than 50% of people are consuming complementary or alternative medicine (i.e., herbalism, homeopathy, and acupuncture). In Germany, 90% of the population has consumed natural-based medicines once in their lives. From 1995 to 2000, the number of doctors in Germany who attempted to learn specific training in natural products nearly doubled. The current worldwide annual market for plant-based medicines is estimated to be more than $60 billion.[6,7]

Autoimmune disorders are an illness rising from an irregular immune response to a normal part of the body. Among the bioactive molecules derived from medicinal plants, a number of them have shown immune-stimulatory activities with the ability to affect the antigens-independent and unspecific defense system of an organism. The ability of herbal

The Microbiology of Central Nervous System Infections
https://doi.org/10.1016/B978-0-12-813806-9.00006-8

medicines to prevent cellular and humoral immune responses exhibits a promising approach in curing some immune system disorders. This immunomodulatory activity of medicinal plants offers an alternative approach to current chemotherapy.[8–10] The immune system of our body is separated into two classes: innate (nonspecific) and adaptive (acquired). The innate system responds fast and includes the initial defense line of our body, which includes the physical barriers (skin and endothelial cells). The second line of defense includes myeloid cells (macrophages, eosinophils, erythrocytes, dendritic cells, basophils, monocytes, neutrophils, and megakaryocytes). The adaptive system responds slowly, attacks the specific foreign substances, and works closely with the innate system.[11]

A weak immune system causes infectious diseases, cancer, metabolic disorders (obesity, type 2 diabetes mellitus, and malnutrition); conversely, an overactivated immune system (out-of-balance) causes cancer, cardiovascular disease, joint and tissue destruction, asthma, and skin disease.[12]

Infections of the central nervous system (CNS) can be life-threatening, particularly for people with a weak immune system. Common CNS infection sources are meningitis, encephalitis, and brain abscesses; they tend to cause more disability and mortality than other organ system infections. There are five main causes of CNS infections: (1) viruses (i.e., hepatitis C virus, Epstein-Barr virus, human cytomegalovirus virus, West Nile fever, herpes simplex virus, and the rabies virus), (2) bacteria (i.e., *Listeria monocytogenes*, *Mycobacterium tuberculosis*, *Borrelia burgdorferi*, *Bartonella henselae*, *Salmonella* spp., *Brucella* spp., and *Tropheryma whipplei*), (3) fungi (i.e., Cryptococcal meningitis, *Candida* spp., *Aspergillus* spp., and *Zygomycetes* spp), (4) protozoa (i.e., naegleriasis, malaria, and toxoplasmosis), and (5) prions (i.e., Creutzfeldt-Jakob disease, fatal familial insomnia, and Kuru).[13,14]

Evidence shows that natural products solely or in combination with chemical medicines provide a substantial opportunity to fight CNS infections. For example, medicinal plants such as *Geum japonicum*, *Rhus javanica*, *Syzygium aromaticum*, and *Terminalia chebula* Retzius in combination with acyclovir expressed stronger inhibitory effects on human herpes virus 1.[15] Similarly, medicinal plants, including *Chaerophyllum macropodum*, *Allium vineale*, and *Prangos ferulacea*, exhibited anti-*L. monocytogenes* activities.[16]

The aim of this paper is to present and discuss our current knowledge on using natural products as immune system modulators and against nervous systems infections.

2. NATURAL PRODUCTS AS IMMUNE MODULATORS

The immune system is known as a complex system linking to the various metabolic networks. The stimulation of the immune system using different drugs in order to reduce the threat of disease has been of interest for several decades. To date, two types of immune modulators have been introduced: (1) immunosuppressants such as cyclosporine, azathioprine, sirolimus, and prednisone, which suppress the immune response and can be applied for the control of immunologic and inflammatory diseases; and (2) immunostimulants such as glatiramer acetate, pegademase bovine, plerixafor, proleukin, and oprelvekin, which increase the body's resistance against infections and improve the basal levels of the immune response. Nevertheless, these chemotherapy drugs typically have side effects, including appetite changes, fatigue, hair loss, blood disorder, constipation, infection, anemia, nausea, and vomiting.[17] Alternatively, natural products and their derivatives could by a key to regulating and suppressing immune disorders. They could also have the potential to assist patients to have a more stabilized immune function without the side effects of chemotherapy drugs. For example, alprogen, a metabolite derived from *Aloe vera,* was able to reduce the leukotriene and histamine releases, inhibit the activation of mastocyte cells by blocking the calcium influx, and diminish the activity of the protein kinase C and phospholipase D in guinea pigs.[18]

A. vera polysaccharides were also identified as being able to raise the level of tumor necrosis factor alpha (TNF-α), interleukin-2 (IL-2), interferon gamma (IFN-γ), superoxide dismutase, (SOD), and immunoglobulin G (IgG) of immune-inhibited mice.[19] The secretion of proinflammatory cytokines such as TNF-α and IL-2 from cells of the innate immune system (i.e., mast cells, phagocytes, neutrophils, and natural killer cells) during the infection induces the release of acute phase proteins and influences the immune system.[20] In a report from Im et al.[21] the polysaccharides of *A. vera* were potent enough to decrease the proinflammatory cytokines; inhibit activity of the nuclear factor kappa-light-chain-enhancer of activated B cells (NF-κB) through the activation of proliferator-activated receptors (PPARγ); and upregulate the expression of CDX2, which was identified as a tumor suppressor in colon cancer.

Allium sativum is another medicinal plant with immunomodulatory activity. *A. sativum* has been shown to improve cellular immunity by upregulating the activity of natural killer (NK) cells and stimulate the activities of IFN-γ, IL-2, TNF-α, and interleukin 4 (IL-4).[21] Different bioactive molecules, such as alliin, γ-glutamyl cysteine, ajoene, and thioallyl compounds with immunomodulatory activities, are successfully isolated from *A. sativum.*[21] According to the literature, the aged *A. sativum* has more immunomodulatory properties than the fresh *A. sativum* because of the transformation of organosulfur compounds.[22] Fructans are the polymers of fructose molecules, and these molecules can be found in *A. sativum* extracts. Fructans have shown comparable immunostimulatory activity to the well-known polysaccharide immunomodulators such as mannan and zymosan. The literature also revealed that the *A. sativum* contains proteins such as lectins with mitogenic and immunostimulatory activity that exhibit mitogenic activity toward human peripheral blood lymphocytes, murine splenocytes, and thymocytes.[23]

Andrographis paniculata, belonging to the family of the Acanthaceae, is a medicinal plant that can be find in southern and southeastern parts of Asia. *A. paniculata* is expressed as a bitter plant with a sweet future, since it is a source of valuable diterpenes, such as andrographolide, isoandrographolide, 14-deoxyandrographolide, neoandrographolide, 14-deoxy-11,12-didehydroandrographolide, 19-*O*-acetylanhydroandrographolide, 14-deoxy-14,15-didehydroandrographolide, andrograpanin, and 14-acetylandrographolide with anticancer, antiinflammatory, immunomodulatory, antiinfection, antihepatotoxicity, and antiatherosclerosis activities.[24,25] Several immunomodulatory activities of andrographolide, such as reduction of nitric oxide synthases (iNOS), prostaglandin-endoperoxide synthase 2 (COX-2), prostaglandin E2 (PGE2), TNF-α, and interleukin 12, (IL-12) have been reported in in vitro studies.[26] A study by Kumar et al.[27] disclosed that the dichloromethane fraction of the methanolic extract of *A. paniculata* could considerably inhibit the proliferation of HT-29 (colon cancer) cells and increase the proliferation of human peripheral blood lymphocytes (HPBLs). Further studies disclosed the molecules as andrographolide, 14-deoxyandrographolide, 14-deoxy-11, and 12 didehydroandrographolide. In another report by Wang et al.[28] the andrographolide reduced the lipopolysaccharide-induced macrophage or the interleukin 4-induced macrophage activation and suppressed the alternatively activated macrophages cytokines expression (Fig. 1).

Asparagus racemosus belongs to the family of Liliaceae, and it is widely used in Nepal, India, the Himalayas, and Sri Lanka. The literature revealed that the roots of *A. racemosus* are diuretic, tonic, and galactagogue. Indeed, *A. racemosus* has exposed healing effects by

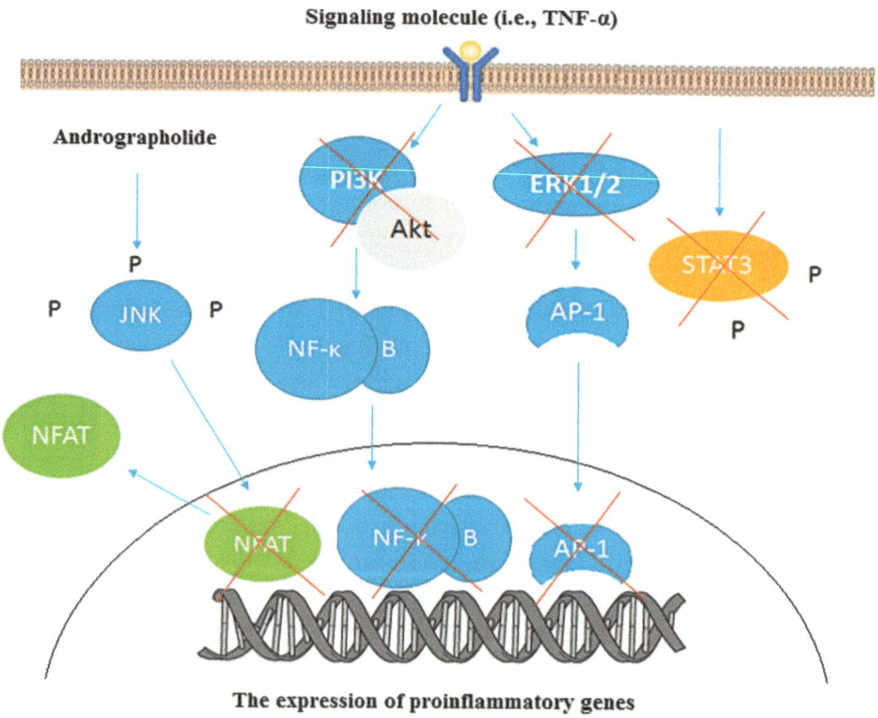

Fig. 1 Proposed role of andrographolide in inhibition of inflammation by Hidalgo et al.[29] *Note*: Andrographolide has inhibitory influence on the phosphatidylinositol 3-kinase (PI3K)/protein kinase B (Akt), extracellular signal-regulated kinases 1 and 2 mitogen-activated protein kinase (ERK1/2 MAPK), nuclear factor-κB (NF-κB), nuclear factor of activated T-cells (NFAT), activator protein 1 (AP-1), and signal transducer and activator of transcription 3 (STAT3), and increases the c-Jun N-terminal kinases (JNKs) phosphorylation.

improving the mucosal resistance, and it was used against inflammation, specific infections and nervous disorders while improving mental functions. *A. racemosus* contains a range of bioactive molecules, such as steroidal saponins, oligospirostanoside, polycyclic alkaloid-aspargamine A, Isoflavones-8-methoxy-5,6,4-trihydroxy isoflavone-7-*O*-beta-D-glucopyranoside, cyclic hydrocarbon-racemosol, dihydrophenantherene, sitosterol, 4,6-dihydryxy-2-*O* (-2-hydroxy isobutyl) benzaldehyde, and undecanyl cetanoate.[30,31]

A study by Gutam et al.[32] showed that the aqueous extract of *A. racemosus* at the dosage of 100mg/kg b.w. p.o. could stimulate the activation of T cells (T lymphocytes) by enhancing the cluster of differentiation 3 (CD3) and enhancing the cluster of differentiation 4 and cytotoxic T lymphocyte ratio. The analysis of the extracts from *A. racemosus* demonstrated the existence of 26.7% of 2→1 linked fructo-oligosaccharides with immunomodulatory activities. Sharma et al.[33] identified a new steroidal sapogenin molecule from the root of *A. racemosus* with an unusual 21 nor and C19 carboxylic acid functional group. This molecule was further tested on the immune system of normal and cyclosporine-A immune-suppressed-Swiss albino mice and it showed a dosage-dependent increase in CD3, B-lymphocyte antigen CD19, and the ratio of Th1/Th2 cells.

Azadirachta indica is another medicinal plant with immunostimulatory activity that grows in tropical and semitropical regions. *A. indica* is identified as a valuable source of bioactive compounds, such as nimbiol, nimbanene, nimbin, nimbandiol, 6-desacetylnimbinene, nimbolide, 17-hydroxyazadiradione, *n*-hexacosanol, 7-desacetyl-7-benzoylazadiradione, and 7-desacetyl-7-benzoylgedunin.[34] SaiRam et al.[35] isolated a volatile fraction named NIM-76 from *A. indica* oil, which was tested for immunomodulatory activity using rats as a model. A dosage-dependent immunomodulatory activity of NIM-78 was observed on polymorphonuclear leukocytes and enhanced macrophage activity and lymphocyte proliferation response, even though a drop in primary and secondary antibodies was recorded. The aqueous extract of *A. indica* flowers was also tested for the immunomodulatory activity using murine. The result indicated an increase in the synthesis of nitric oxide (NO) and super oxide in murine peritoneal macrophages, and a considerable rise in the phagocytic index representing the activation of the reticuloendothelial system by synthesis of mediators. The authors concluded that the *A. indica* extracts affect both specific and nonspecific immune responses.

In traditional medicine, the dried and fresh fruits of *Phyllanthus emblica* that belong to the family of Phyllanthaceae have also been used as immunostimulants.[36] Currently, different bioactive compounds, such as emblicanin A, phyllantine, phyllembein, triacontanol, kaempferol-3-*O*-α-L-(6″-methyl)-rhamnopyranoside, gallic acid, methyl gallate, ellagic acid, trigallayl glucose, β-amyrin ketone, betulonic acid, daucosterol, lupeol acetate, β-Amyrin-3-palmitate, gallic acid, betulinic acid, ursolic acid, oleanolic acid, quercetin, and rutin, are isolated from *P. emblica.*[37,38] The oral uptake of *P. emblica* in syngeneic BALB/c mice increased the NK activity and the antibody-dependent cellular cytotoxicity.[39] The phenolic compounds of *P. emblica,* including geraniin, quercetin 3-β-D-glucopyranoside, kaempferol 3-β-D-glucopyranoside, isocorilagin, quercetin, kaempferol, and rutin, were tested on the splenocyte proliferation and the cytotoxicity of the human breast cancer cell and human embryonic lung fibroblast cell. Among the mentioned phenolics, geraniin and isocorilagin showed the highest cytotoxicities against the breast cancer cell line. The authors revealed that the anticancer activity of these phenolics might be related to their immunomodulatory activities. Researchers also reported that the *P. emblica* could be applied as a nonsteroidal antiinflammatory drug. Moreover, the ethanolic extracts of *P. emblica* had a biphasic activity in ulcerated mice with a dosage-dependent manner. Also, the ethanolic extracts decreased the synthesis level of TNF-α and interleukin 1 beta (IL-1-β) and increased the synthesis of Interleukin 10 (human cytokine synthesis inhibitory factor).[39]

Among the limited number of medicinal plants with promising immunomodulatory properties, the *Echinacea* species from Asteraceae has attracted considerable attention. Recent activities in pharmaceutical applications of *Echinacea* have focused on its treatment and prevention activities toward respiratory tract infections, including tonsillitis, pharyngitis, laryngitis, sinusitis, the common cold, otitis media, and certain types of influenza.[40]

Researchers have isolated different compounds such as alkamides, caffeic acid esters, cichoric acid, polysaccharides, polyacetylenes, and cynarin from *Echinacea* species. A study by Guiotto et al.[41] disclosed that the dodeca-2E, 4E, 8Z, and the 10E/*Z*-tetraenoic acid isobutylamides (alkamides) from *Echinacea purpurea* have considerable effects on downregulation of the proinflammatory cytokines, such as IL-12, Interleukin 8 (IL-8), interleukin 6 (IL-6), Interleukin 10 (IL-10), and TNF. In another report, the immunomodulatory effect of three compounds, including cichoric acid, polysaccharides, and alkylamides, was tested on mice. Among them, the alkylamides at the concentration of 12 μg/kg body weight/day could considerably enhance the phagocytic activity and the phagocytic index of the alveolar macrophages. The authors also mentioned that the immunomodulatory activity of alkylamides seems to be more effective in lungs than the spleen. The Polysaccharides derived from *Echinacea* have also demonstrated immunostimulatory effects. The antiinflammatory activity of the extracts from the *E. purpurea* roots inhibited the synthesis of Pam3Csk4, a synthetic triacylated lipopeptide that regulates the production of TNFα by human monocytic THP-1 cells.[42]

Interestingly, the secondary metabolites, including alternariol, *O*-prenylporriolide, porritoxin, β-zearalenol, and (*S*)-zearalenone, that were derived from the fungal endophytes of *E. purpurea* had cytokine-suppressive activities with inhibitory influence on the synthesis of TNF-α.[43]

Panax ginseng is one of the most used and studied species of Ginseng. *P. ginseng* is a source of different bioactive compounds, such as dammarane, ginsenosi, protopanaxatriol, protopanaxadiol, panaxytriol, and panaxatriol.[44] Tran et al.[45] reported the isolation of three dammarane-type triterpene molecules, including 20(*S*)-ginsenoside Rg_2 (1), 27-demethyl-(*E*,*E*)-20(22),23-dien-3β,6α,12β-trihydroxydammar-25-one (2), and 3β,20(*S*)-dihydroxydammar-24-en-12β,23β-epoxy-20-*O*-β-D-glucopyranoside (3) from the *P. ginseng* with immunomodulatory activities. The authors claimed that all three isolated molecules upregulated the expression of IL-12 in lipopolysaccharide-induced activated mouse peritoneal macrophage at a dosage of 100 ng/mL. Nevertheless, the isolated molecule (2) had considerably upregulated the response of the Th1-type cytokines (produced the proinflammatory responses responsible for killing intracellular parasites) and downregulated the Th2 responses (humoral-mediated immunity).

NF-κB has vital roles in immune and inflammatory responses through controlling the genes responsible for the adhesion molecules (i.e., integrin, the immunoglobulin superfamily, selectins, and cadherins), proinflammatory cytokines (i.e., IL-1, IL-12, TNF-α, and IFN-γ), chemokines (i.e., chemokine (C—C motif) ligand 5), growth factors (i.e., adrenomedullin, angiopoietin, bone morphogenetic proteins, leukemia inhibitory factor, ephrins, and foetal bovine somatotrophin), and adaptive enzymes (i.e., COX-2).[46,47] Thus, the NF-κB pathway inhibitors could perform as antiinflammatory and immunostimulant drugs.

The bioactive molecules derived from the flowers of *P. ginseng* are identified as rich in ginsenosides and their derivative. A study by Cho et al.[48] revealed that the isolated ginsenosides Rk3, Rs4, SF, and Rg6 had strong-to-moderate inhibitory effects on NF-κB, and all these molecules downregulated the TNF-α-induced expression of interleukin 8, chemokine (C—X—C motif) ligand 1, inducible nitric oxide synthase, and the intercellular adhesion molecule 1 genes. Thus, the isolated ginsenosides and their derivative have the potential to be used as immunostimulants and antiinflammatory drugs.

Alkaloids are a class of nitrogen-based compounds with a wide range of pharmaceutical activities; microorganisms, plants and animals could produce alkaloids. Lunamarine (Punarnavine) is identified as a quinolone alkaloid existing in *Boerhavia diffusa*. Manu and Kuttan[49] studied the immunostimulatory effects of Lunamarine using Balb/c mice. They found out that the total white blood cells (WBC), the bone marrow cellularity, the number of α-esterase positive cells, and the proliferation of splenocytes and thymocytes were increased by injecting the lunamarine at the dosage of 40 mg/kg body weight. The higher dosages of lunamarine considerably decreased the proinflammatory activities of TNF-α, IL-6, and IL-1-ß induced by lipopolysaccharide (lipoglycans). The *Stephania cepharantha* (Menispermaceae) plant is a source of cepharanthine, which is a biscoclaurine alkaloid with antiinflammatory, antineoplastic, and immunomodulatory activities. Reports demonstrate that the cepharanthine has an inhibitory influence on the peroxidation of lipids in the plasma membrane, the production of TNF-α, and cytokine synthesis. Cepharanthine at the dosage of 1–5 μg/mL inhibits the dendritic cells (accessory cells), which connect the innate and adaptive immunity systems. The dendritic cell affected by cepharanthine is shown to have a low stimulatory role on allogeneic T cells and IFN-γ, which in turn demonstrates the potential of cepharanthine as an immunostimulant against autoimmune disease.[50]

Tylophora indica is a medicinal plant and a source of valuable phenanthroindolizidine alkaloids, such as tylophorine, tylophorinidine, septicine, Tylohirsutinidine, tyloindicine A, Tylohirsutinine, and tyloindicine G with antiinflammatory and antiallergic activities. The Tylophora alkaloids at the higher dosage (300 pg/mL) demonstrated an inhibitory influence on the proliferation of splenocytes, and suppressed the responses of the immune system by inhibiting the T cell proliferation and production of IL-2.[51]

Tetrandrine is a bisbenzylisoquinoline alkaloid (phenylalanine-derived metabolites) and a calcium channel blocker which is isolated from the *Stephania tetrandra* (Menispermaceae).[52] Evidence shows that tetrandrine has considerable immunomodulator activity, which can be used to treat rheumatic diseases, silicosis (miner's phthisis) and high-blood pressure.[53] Tetrandrine has a broad range of immune-modulating activities, including inhibitory effects on the production of TNF-α and formation of antitype II collagen antibodies, which are directly responsible for the cartilage and tissue destruction in autoimmune arthritis diseases such as rheumatoid arthritis.[54]

Noribogaine, ibogaine, tabernanthine, coronaridine, voacangine, ibogamine, and harmaline are iboga alkaloids, which are identified in *Tabernanthe iboga* (Apocynaceae). *T. iboga* can be found in west central parts of Africa (i.e., Congo, Gabon, and Cameroon).[55] A study by House et al.[56] showed that the alkaloids ibogaine and harmaline had a dosage-dependent suppression effect on the immune system at the concentration from 10 to 100 μmol/L.

Indigofera suffruticosa is a subtropical and tropical medicinal plant that is native to Mexico, northern Argentina, Central America, and the southern part of the United States. *Indigofera* genus are sources of valuable alkaloids, such as glucopyranose, coronarian, kirilowin G, endecaphyllin C1, and corynocarpin.[57] Lopes et al.[58] studied the effectiveness of *I. suffruticosa* alkaloids on the activation of macrophage and production of TNF-α, and they observed higher levels of NO and moderated levels of TNF-α release with a cytotoxic effect on adenocarcinoma cell lines (IC_{50} from 0.89 to 1.44 μg/mL).

Carbohydrates are biological molecules with either simple or complex structures. Even though the carbohydrates' immune-modulatory effects were discovered almost 50 years ago, the majority of the current literature has focused on the immunological activities of polysaccharides from the bacteria, which is related to the lipid A, an endotoxin responsible for the toxicity of Gram-negative bacteria. Carbohydrates can be classified into three major groups: monosaccharides, oligosaccharides, and polysaccharides. They are able to bind to other organic molecules such as lipids and proteins. Polysaccharides, such as mannans, arabinogalactans, galactans, fucoidans, glucans, fructans, hyaluronans, and xylans, are able to stimulate the immune system directly or indirectly through triggering several cellular and molecular pathways linking to the immune system. Evidence shows that the macrophages, neutrocytes, and monocytes are the primary targets of the polysaccharide immunomodulators.[59–61]

Glucans are D-glucose monomers, which are classified into two main alpha (i.e., dextran, glycogen, and pullulan) and beta (i.e., chrysolaminarin, curdlan, laminarin, lichenin,

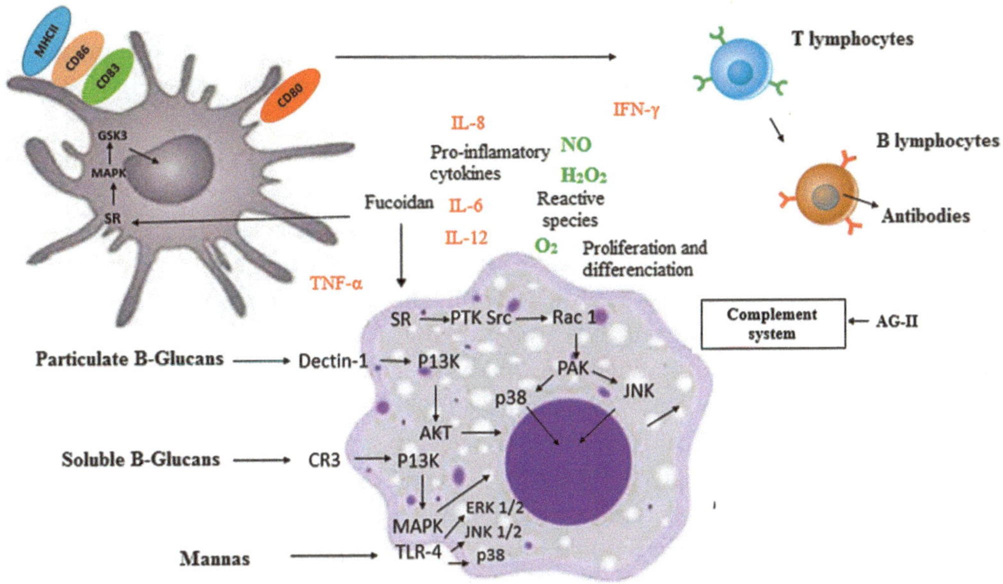

Fig. 2 A scheme reported by Ferreira et al.[62] on the activation process of the immune system by polysaccharides with immunostimulatory activities. *Note*: *AG-II*, type II arabinogalactans; *CD*, cluster of differentiation; *CR3*, complement receptor 3; *GSK3*, glycogen synthase kinase 3; *MHCII*, major histocompatibility complex class II; *PAK*, p21-activated kinase; *PI3K*, phosphatidylinositol-3 kinase; *PTK*, protein-tyrosine kinase; *Rac 1*, Ras-related C3 botulinum toxin substrate 1; *SR*, scavenger receptor; *Src*, proto-oncogene protein kinase; *TLR-4*, toll-like receptor 4.

and zymosan) groups. Glucans are identified in *Aconitum carmichaelii, Agaricus bisporus, Armillariella tabescens, Dendrobium huoshanense, Dictyophora indusiata, Cordyceps sinensis, E. purpurea, Ganoderma lucidum, Hordeum vulgare,* and *Hyriopsis cumingii*. The literature reveals that the immune-stimulatory activities of glucans are related to their effects on different immune receptors. This includes C-type lectin domain family 7 member A (Dectin-1), complement receptor, toll-like receptor 2–6 ligands, and upregulation of proinflammatory genes, increasing the production of the splenocyte antibody, NO, TNF-α, IL-1-ß, and IL-6[62–64] (Fig. 2).

The fungi *G. lucidum* is known as a source of different biological molecules.[65]

In a study by Guo et al.[66] the water-soluble polysaccharide from the spores of *G. lucidum* was tested for its biological activities using Murinae as a model. The authors found that the derived polysaccharide binds to dectin-1, and that it is an effective inducer of MAPKs, the spleen tyrosine kinase-dependent TNF-α, and IL-6.

Mannan is a linear polymer of mannose, and structurally, it has β (1–4) linked, branches in plants, and α (1–6), α (1–2), and α (1–3) linked branches in yeasts. *Aloe Vera, Coffea, Cordyceps militaris, Haematococcus lacustris, Hericium erinaceus, Peltigera canina, Tremella aurantialba,* and *Trigonella foenum-graecum* are some natural sources of mannan.[62,67]

Researchers have reported that mannan has mitogenic and comitogenic activities, and it is potent enough to increase the activation of the B lymphocyte and several prototype macrophages, enhance the secretion of IgM in human-human hybridoma HB4C5 cells, and upregulate the expression of COX-2 and iNOS.[62,67,68]

Galactans are oligosaccharide carbohydrates that can be found in plants (i.e., *Anogeissus latifolia*, *Atractylodes lancea,* and Acacia) and algae (i.e., *Porphyra vietnamensis*, *Solieria chordalis*, and *Chlorella pyrenoidosa*). The literature shows that galactans are potent enough to stimulate the immune system by increasing the activities of NK, the activities of alkaline phosphatase, the adhesion of neutrophil, and the proliferation of the lymphocyte.[69,70] Chemically, the sulfated and carboxymethylated derivatives of β (1,3) D-glucan are more soluble, have more influence on intramolecular and intermolecular hydrogen bonding, and are more efficient on the immune system.[71] Arabinogalactans are biopolymers containing arabinose and galactose molecules; they occur in a complex with proteins as proteoglycans or glycoproteins. The type I and II (AG-I, AG-II) of arabinogalactans have been commonly reported for their biological activities.[72] Arabinogalactans have been isolated from the *Anadenanthera colubrina*, *Artemisia tripartite*, *C. militaris*, *Entada africana*, *E. purpurea*, *Euterpe olerácea*, *Glinus oppositifolius*, *Juniperus scopolorum*, *Lycium barbarian,* and *Radix Astragali.*[62]

The immunostimulatory effects of arabinogalactans were reported to be due to the increase of macrophages and NK cells and the secretion of proinflammatory cytokines (i.e., IL-1 and TNF).[72,73]

A clinical trial on using the arabinogalactan isolated from the larch against the common cold virus showed a decrease in the occurrence of common cold incidences by 23%.[74]

A study by Taguchi et al.[75] on different types of arabinogalactans isolated from rhizomes of *A. lancea* showed that the arabino-3, 6-galactan compared to other types could considerably increase the production of cytokines from Peyer's patch cells (aggregated lymphoid nodules). β-D-(1 → 3)-Galactan chains are introduced as the main key factor for the immunostimulatory of in arabino-3, 6-galactan, since the digestion of β-D-(1 → 3)-galactan chains by *exo*-β-D-(1 → 3)-galactanase considerably decreased its immunostimulatory effects (Fig. 3).

Xylan is a hemicellulose with a backbone pentose sugar named xylose. Xylan can be found in plant cell walls.[77] A new water-soluble polysaccharide identified as 4-*O*-methyl-D-glucurono-D-xylan was isolated from the root of *Cudrania tricuspidata* by the Shi et al.[78] This bioactive compound in a concentration from 6.25–100 μg/mL was potent enough to stimulate the proliferation of splenocytes of a mouse, suggesting its mitogenic activity. The immunostimulatory of 4-*O*-methyl-D-glucurono-D-xylan was observed dependently or in combination with concanavalin A (a lectin) or lipopolysaccharide from *Salmonella abortus*.

Xylo-oligosaccharides (XOS) are the polymers of the xylose, which are isolated from the plant tissue. The structural analysis of isolated XOS from almond shells showed a mixture of neutral and acidic oligomers and the low-molecular weight polymers linked

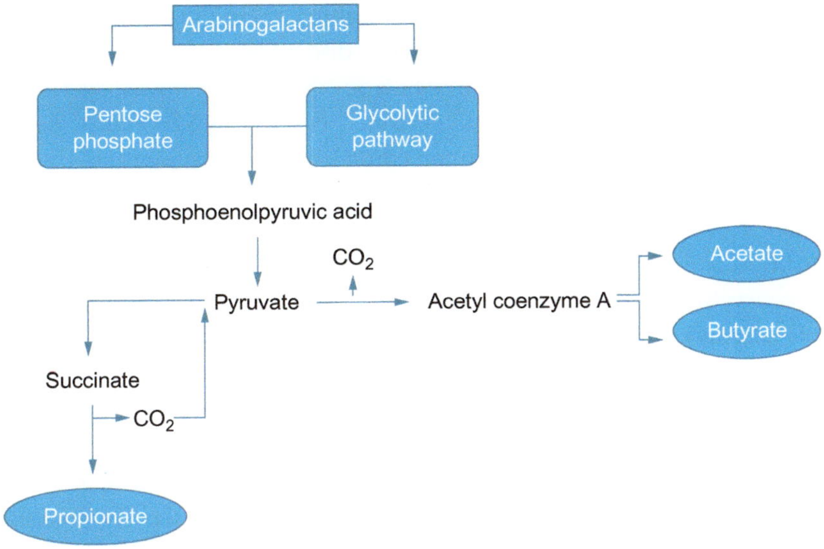

Fig. 3 The suggested mode of action of arabinogalactan on the immune system by Dion et al.[76]

to (4-O-methyl-D-glucurono)-D-xylan. XOS have shown immunostimulatory activity by increasing the T-mitogen-induced proliferation of rat thymocytes.[79]

A human clinical study established by Lecerf et al.[80] on the immunostimulatory effect of a new identified XOS and an inulin-XOS complex disclosed that the consumption of XOS (daily dosage of 5 g) enhanced the concentration of Bifidbacterium and the activities of α-glucosidase and β-glucuronidase while consumption of inulin-XOS complex (daily dosage of 3 g inulin + 1 g XOS) reduced the expression levels of interleukin 13 (the main mediator of the physiologic changes by inflammation).

Hyaluronic acid (Hyaluronans) is nonsulfated glycosaminoglycan with a molecular weight from 1050 to 1338 kDa. Hyaluronic acid is an essential part of articular cartilage, extracellular matrix, group A streptococcal, synovial fluid, and skin in mammalians.[81] Hyaluronic acid can also be found in *Streptococcus pyogenes*, a Gram-positive, aerotolerant bacterium.[82] Studies on Kunming mice showed that the hyaluronic acid is potent enough to increase the proliferation of splenocyte, the activity of acid phosphate, and the activity of lysozyme in serum.[83]

Flavonoids are bioactive compounds with a 15-backbone carbon structure and two pheyl rings. Flavonoids are categorized into flavonoids, isoflavonoids, and neoflavonoids classes.[84]

The primary effects of flavonoids on the immune system are due to their diverse modes of action, including active site interference and asymmetric activation, protein binding, and free radical scavenging and antioxidant effects.[85] A study by Akbay et al.[86] demonstrated that the three flavonoids—quercetin-3-O-rutinoside, kaempherol-3-O-rutinoside, and isorhamnetin-3-O-glucoside—isolated from the *Urtica dioica* had considerable intracellular killing activities, describing their potential as a drug against chronic granulomatous and neutrophil function disorders.

Silymarin is a priceless flavonoid complex from the *Silybum marianum* plant, which is composed of silibinin, silydianin, and silychrisin.[87] A study by Lee et al.[88] demonstrated that silibinin in a dose-dependent manner could considerably suppress the expression level of a cluster of differentiation 80 and 86 (CD80, CD86) and the major histocompatibility complex (MHC) molecules in dendritic cells. CD80 and CD86 play a vital role in the activation and survival of T cells while the MHC is a range of cell surface proteins that is necessary to recognize foreign molecules. Silibinin also suppressed the MAPKs and the p65 subunit of nuclear translocation of the NF-κB induced by lipopolysaccharides. In another report, the flavonoids kaempferol, quercetin, fisetin, and chrysin were demonstrated to be potent enough to suppress the production of IL-8 and the activation of TNF-α in human embryonic kidney cells 293. The generation of reactive oxygen species in response to TNF-α was also decreased by these flavonoids.[89] Cianidanol ((+)-catechin), *O*-methyl-derivative (+)-3- methoxy-5,7,3′,4′-tetrahydroxyflavan, and (+)-3- palmitoyl-ubiquinone 5,7,3′,4′-tetrahydroxyflavan are the other isolated flavonoids with immunomodulatory activities. These flavonoids were potent enough to enhance the spontaneous lymphocyte transformation.[90]

Terpenes are a diverse class of bioactive compounds that produce biosynthetically from isoprene and unites. Terpenes could be classified by the number of isoprene unites into (1) hemiterpenes (isoprene), (2) monoterpenes (i.e., limonene, linalool, and geraniol), (3) sesquiterpenes (i.e., humulene and farnesol), (4) diterpenes (i.e., cafestol, cembrene and taxadiene), sesterterpenes (i.e., geranylfarnesol), triterpenes (i.e., squalene), sesquarterpenes (i.e., tetraprenylcurcumene), tetraterpenes (i.e., carotenes and xanthophylls), polyterpenes (natural rubber), and norisoprenoids (i.e., megastigmane-3,9-diol, C13-norisoprenoids 3-oxo-α-ionol). Currently, material is scarce in the literature about the immunomodulatory influence of terpenes.[91]

Ganoderma is a polypore mushroom that can be found in tropical regions. Ganoderma has been traditionally used in Asian medicine. Among the Ganoderma species, *Ganoderma lucidum* and *Ganoderma applanatum* are recognized for their higher levels of terpenes and their derivatives (i.e., ganoderals, ganoderols, ganodermanontriol, ganodermanondiol, and lucidone) which are potent enough to stimulate the pathway of the nuclear factor (NF)-κB and the activation of MAPKs.[91] In a study by Raphael and G. Kuttan,[92] the potential for immunomodulation by some terpenes was assessed using Balb/c mice as a model. Uptaking monoterpenes, including carvone, limonene, and perillic acid, in a dosage from 50 to 100 μmol/kg body upregulated the production of the antibody, the WBC, bone marrow cellularity, and α-esterase positive cells. Geraniol trioxolane is a trioxolane derivative of a terpene with different biological activities; it is potent enough to be used against a range of diseases, including chronic fatigue syndrome, fungal infections of the skin, impetigo, herpes simplex viruses, psoriasis, and the human

Table 1 A list of natural sources for immunomodulatory proteins and their modes of action

Natural source	Immunomodulatory protein/mechanism of immunomodulatory action	Reference
Taiwanofungus camphoratus	FIP-Aca1/Induction of different cytokines (TNF-α, IL-1β, IL-12, IL-6) and signaling proteins (i.e., Chemokine (C—C motif) ligand 3)	97
Flammulina velutipes	FIP-Fve/Inducer of INF-γ, and IL-3 expression and reducer of the anaphylactic shock	98
Ganoderma sinensis	FIP-Gsi/Upregulator of IL-2, IL-4, IFN-γ, and inducer of cytokine production	99
Trametes versicolor	FIP-Tve/Enhancer of *peripheral blood lymphocyte activity, and the* generator of NO, TNF-α	100
Smilax glabra rhizomes (Sarsaparilla)	Smilaxin/Suppressor of [methyl-3H] thymidine uptake via mannose binding lectin 2 and 9-butyl-8-(3-methoxybenzyl)-9H-purin-6-amine in tumor cells. Higher production levels of NO	101
Tinospora cordifolia	Guduchi ImP/Enhancer of mitogenic activity and inducer of NO by macrophages	102

immunodeficiency virus. Clinical studies showed that using 200 mg daily of geraniol trioxolane was potent enough to reduce the symptoms of rheumatoid arthritis in a 2 weeks. Also, all the signs and the symptoms of the lupus erythematosus were cleared in 90 days by using Geraniol trioxolane.[93] Betulinic acid, oleanolic acid, α-amyrin acetate, ursolic acid, and lupeol acetate are a few other natural terpenes with antiinflammatory, immunostimulatory, and analgesic activities.[94]

The high-quality proteins are effective to enhance immune responses and regulate the activation of natural killer cells, T and B-lymphocytes, macrophages, gene expression, lymphocyte proliferation, production of cytokines, and antibodies.[95,96] A list of natural source for proteins as immunomodulators with their mode of action is presented in Table 1.

3. NATURAL PRODUCTS AGAINST CNS INFECTIONS

Infections of the CNS are always catastrophic and potentially lethal to patients. The treatment of the CNS infection needs knowledge, including prevention, epidemiology of the meningitis, vaccinology, the pharmacokinetics and pharmacodynamics of antimicrobial drugs, and the appearance of antimicrobial multidrug-resistant pathogens.[103] Considerably, natural products are potent enough to be used essentially to fight against CNS infection (independently or synergistically with antibiotics) beside the effectual chemical drugs like moxalactam and cefotaxime, which were used against CNS infection.[104] The laboratory diagnosis of CNS infection is vital and urgent treatment has to be initiated based on empirical therapy. Indeed, the selected drugs would preferably relate to the local

Table 2 A list of natural products against some of the main causes of central nervous system infections

Source of infection	Potent natural product	Bioactive compound/its mode of action	References
Tuberculosis	*Nocardia* sp.	Thiolactomycin/An inhibitor of β-ketoacyl ACP synthase I, and β-ketoacyl ACP synthase II responsible for the synthesis of very long chain meromycolic acids	105
	Cephalosporium caerulens	Cerulenin/An inhibitor of the enzyme fatty acid synthase, and blocker of the beta-ketoacyl-ACP synthase	106
	Pleurotus mutilus	Pleuromutilin/An inhibitor of the bacterial protein synthesis by influencing the prokaryotic ribosome	107
	Streptomyces coeruleorubidus	Caprazamycin/An inhibitor of the phospho-MurNAc-pentapeptide translocase (MraY) that catalyzes the synthesis of lipid I, which is a bacterial peptidoglycan precursor	108,109
	Streptomyces griseus 446-S3	Capuramycin/A robust inhibitor of translocase I (MurX/MraY)	110
Treponema pallidum	*Streptomyces parvulus*	Borrelidin/An inhibitor of threonyl-tRNA synthetase	111
	Penicillium	Penicillin/An inhibitor of peptidoglycan cross-links synthesis in the bacterial cell wall	112
Lyme disease	*Streptomyces roseosporus*	Daptomycin/A disrupter of ion exchange in the bacterial cell membrane, and inhibitor of protein, DNA, and RNA synthesis	113
	Streptomyces aureofaciens	Tetracycline/An inhibitor of the protein synthesis through blocking the attachment of charged aminoacyl-tRNA to the A site on the ribosome	114
	Saccharopolyspora erythraea	Erythromycin/An inhibitor of the protein synthesis through attaching to the 50S subunit of the bacterial rRNA complex	115,116
Bacterial meningitis	*Streptomyces orientalis*, *Acremonium chrysogenum*, *S. griseus*, *S. aureofaciens*	Vancomycin/An inhibitor of the bacterial cell wall synthesis by forming the hydrogen bond, which disrupts the terminal D-alanyl-D-alanine moieties of the NAM/NAG-peptides Cephalosporin/An inhibitor of peptidoglycan layer synthesis by binding to the D-Ala-D-Ala at the end of muropeptides Streptomycin/A protein synthesis blocker, which attaches to the 16S rRNA of the 30S subunit of the bacterial ribosome Tetracycline/A protein synthesis inhibitor by disrupting the binding of aminoacyl-tRNA to the A site on the ribosome	114,117–119

Cryptococcal meningitis	*Streptomyces nodosus*	Amphotericin B/A Disrupter of monovalent ions exchange, and a binder to ergosterol, a sterol in fungi and protozoa	120
Toxoplasma gondii	*S. ambofaciens*	Spiramycin/An inhibitor of protein synthesis in the bacterial cell	121
Malaria	*Cinchona* sp.	Quinine/Poses a fast schizonticidal action against intraerythrocytic malaria parasites	122
	Artemisia annua	Artemisinin/A producer of carbon-centered free radicals	123
	Dichroa febrifuga	Febrifugine/A disrupter of prolyl-tRNA synthetase, resulting in an intracellular accumulation of uncharged tRNA and mimicking reduced cellular proline accessibility	124

epidemiological data of occurrence of relevant organisms and its antibiotic sensitivity. Table 2 presents a list of natural products potent enough to be used against the pathogens responsible for CNS infections.

4. CONCLUSION AND PERSPECTIVES

Immunomodulators are becoming very popular worldwide in natural health pharmaceuticals because people have started to understand the essential role of a healthy immune system in health and disease prevention and recovery. For centuries, natural products have been the main source of drugs, and their dynamic position in future drug discovery is clear. As is discussed, chemical and synthetic drugs have diverse negative side effects on the patients (i.e., renal failure and toxicity) while natural products and their derivatives have fewer or even no side effects compared to chemotherapy drugs. Furthermore, natural products could be used synergistically with chemical drugs, though still more studies on kinetics, pharmacodynamics, their modes of action, and their concentration are still required. Indeed, even though some natural products have excellent bioactivity, they have poor bioavailability and absorption because of poor lipid solubility and improper molecular size. Considerably, phytosomes are a complex form of natural active ingredients and a phospholipid with a better bioavailability and absorption; they disclose an enhanced pharmacokinetic and pharmacodynamic profile versus conventional natural extracts. Phytosomes can be essentially used as an approach to manage and treat the CNS infection, where a higher bioavailability of the bioactive molecules is needed.

ACKNOWLEDGMENT

The authors appreciate the School of Engineering and Science, Tecnologico de Monterrey, Campus Monterrey, México, for supporting this study.

CONFLICT OF INTERESTS

The authors declare that they have no competing interests.

REFERENCES

1. Kellogg JJ, Todd DA, Egan JM, Raja HA, Oberlies NH, Kvalheim OM, et al. Biochemometrics for natural products research: comparison of data analysis approaches and application to identification of bioactive compounds. *J Nat Prod* 2016;**79**:376–86.
2. Parsaeimehr A, Sargsyan E, Jassbi AR. Perspectives and key factors affecting the use of herbal extracts against multidrug-resistant gram-negative bacteria. In: Rai M, Kon K, editors. *Fighting multidrug resistance with herbal extracts, essential oils and their components*. London: Academic Press; 2013. p. 181–90.
3. Fabricant D, Farnsworth N. The value of plants used in traditional medicine for drug discovery. *Environ Health Perspect* 2001;**109**:69–75.

4. Parsaeimehr A, Sargsyan E, Vardanyan A, Davtyan GS. Expression of secondary metabolites in plants and their useful perspective in animal health. *ABAH Bioflux* 2011;**3**:115–24.
5. Wachtel-Galor S, Benzie IFF. Herbal medicine: an introduction to its history, usage, regulation, current trends, and research needs. In: Raton B, editor. *Herbal medicine: biomolecular and clinical aspects*. Boca Raton, FL: CRC Press/Taylor & Francis; 2011.
6. Joos S, Glassen K, Musselmann B. Herbal medicine in primary healthcare in Germany: the patient's perspective. *Evid Based Complement Alternat Med* 2012;**2012**:1–10.
7. Kayne S. Introduction to traditional medicine. *Tradit Med A Glob Perspect* 2009;**1**:1–24.
8. Lall N, Kumar V, Meyer D, Gasa N, Hamilton C. In vitro and in vivo antimycobacterial, hepatoprotective and immunomodulatory activity of Euclea natalensis and its mode of action. *J Ethnopharmacol* 2016;**194**:740–8.
9. Deharo E, Baelmans R, Gimenez A, Quenevo C. In vitro immunomodulatory activity of plants used by the Tacana ethnic group in Bolivia. *Phytomedicine* 2004;**11**:516–22.
10. Ganju L, Karan D, Chanda S, Srivastava K. Immunomodulatory effects of agents of plant origin. *Biomed Pharmacother* 2003;**57**:296–300.
11. Delves P, Roitt I. The immune system. *N Engl J Med* 2000;**343**:37–49.
12. Abbas A, Lichtman A, Pillai S. *Basic immunology: functions and disorders of the immune system*. 4th ed. Philadelphia: Elsevier; 2014.
13. Luby J. Infections of the central nervous system. *Am J Med Sci* 1992;**304**:379–91.
14. Weddle G. Infections of the central nervous system. In: Cartwright CC, Wallace DC, editors. *Nursing care of the pediatric neurosurgery patient*. Springer; 2017. p. 515–28.
15. Kurokawa M, Nagasaka K, Hirabayashi T, Uyama S. Efficacy of traditional herbal medicines in combination with acyclovir against herpes simplex virus type 1 infection in vitro and in vivo. *Antivir Res* 1995;**27**:19–37.
16. Sagun E, Durmaz H, Tarakci Z. Antibacterial activities of the extracts of some herbs used in Turkish herby cheese against *Listeria monocytogenes* serovars. *Int J Food Prop* 2006;**9**:255–60.
17. Bascones-Martinez A, Mattila R. Immunomodulatory drugs: oral and systemic adverse effects. *Med Oral Patol Oral Cir Bucal* 2014;**19**:24.
18. Ro J, Lee B, Kim J, Chung Y, Chung M. Inhibitory mechanism of aloe single component (alprogen) on mediator release in guinea pig lung mast cells activated with specific antigen-antibody reactions. *J Pharmacol Exp Ther* 2000;**292**:114–21.
19. Yu Z, Jin C, Xin M, JianMin H. Effect of *Aloe vera* polysaccharides on immunity and antioxidant activities in oral ulcer animal models. *Carbohydr Polym* 2009;**75**:307–11.
20. Arango Duque G, Descoteaux A. Macrophage cytokines: involvement in immunity and infectious diseases. *Front Immunol* 2014;**5**:491.
21. Im SA, Kim JW, Kim HS, Park CS, Shin E, Do SG, et al. Prevention of azoxymethane/dextran sodium sulfate-induced mouse colon carcinogenesis by processed Aloe vera gel. *Int Immunopharmacol* 2016;**40**:428–35.
22. Morioka N, Sze L, Morton D, Irie R. A protein fraction from aged garlic extract enhances cytotoxicity and proliferation of human lymphocytes mediated by interleukin-2 and concanavalin A. *Cancer Immunol Immunother* 1993;(5):316–22.
23. Chandrashekar PM, Prashanth KVH, Venkatesh YP. Isolation, structural elucidation and immunomodulatory activity of fructans from aged garlic extract. *Phytochemistry* 2011;**72**:255–64.
24. Rajagopal S, Kumar R, Deevi D. Andrographolide, a potential cancer therapeutic agent isolated from *Andrographis paniculata*. *J Exp Ther Oncol* 2003;**3**:147–58.
25. Zhang X, Tan B. Antihyperglycaemic and antioxidant properties of *Andrographis paniculata* in normal and diabetic rats. *Clin Exp Pharmacol* 2000;**27**:358–63.
26. Chao W, Kuo Y, Lin B. Anti-inflammatory activity of new compounds from *Andrographis paniculata* by NF-κB transactivation inhibition. *J Agric Food Chem* 2010;**58**:2505–12.
27. Kumar R, Sridevi K, Kumar N, Nanduri S. Anticancer and immunostimulatory compounds from *Andrographis paniculata*. *J Ethnopharmacol* 2004;**92**:291–5.
28. Wang W, Wang J, Dong S, Liu C, Italiani P. Immunomodulatory activity of andrographolide on macrophage activation and specific antibody response. *Acta Pharmacol Sin* 2010;**31**:191–201.

29. Hidalgo M, Hancke J, Bertoglio J. Andrographolide a new potential drug for the long term treatment of rheumatoid arthritis disease. In: Matsuno H, editor. *Innovative rheumatology*. Hampshire: InTech; 2013. p. 247–70.
30. Alok S, Jain S, Verma A, Kumar M, Mahor A. Plant profile, phytochemistry and pharmacology of *Asparagus racemosus* (Shatavari): a review. *Asian Pac J Trop Dis* 2013;**3**:242–51.
31. Bhatnagar M, Sisodia S. Antisecretory and antiulcer activity of *Asparagus racemosus* Willd. Against indomethacin plus pyloric ligation-induced gastric ulcer in rats. *J Herb Pharmacother* 2006;**6**:13–20.
32. Gautam M, Saha S, Bani S, Kaul A, Mishra S, Patil D, et al. Immunomodulatory activity of *Asparagus racemosus* on systemic Th1/Th2 immunity: implications for immunoadjuvant potential. *J Ethnopharmacol* 2009;**121**:241–7.
33. Sharma P, Chauhan PS, Dutt P, Amina M, Suri KA, Gupta BD, et al. A unique immuno-stimulant steroidal sapogenin acid from the roots of *Asparagus racemosus*. *Steroids* 2011;**76**:358–64.
34. Alzohairy M. Therapeutics role of *Azadirachta indica* (Neem) and their active constituents in diseases prevention and treatment. *Evid Based Complement Alternat Med* 2016;**2016**:1.
35. SaiRam M, Sharma S, Ilavazhagan G, Kumar D. Immunomodulatory effects of NIM-76, a volatile fraction from neem oil. *J Ethnopharmacol* 1997;**55**:133–9.
36. Yahayo W, Supabphol A. Suppression of human fibrosarcoma cell metastasis by *Phyllanthus emblica* extract in vitro. *Asian Pac J Cancer Prev* 2013;**14**:6863–7.
37. Luo W, Zhao M, Yang B, Shen G, Rao G. Identification of bioactive compounds in *Phyllenthus emblica* L. fruit and their free radical scavenging activities. *Food Chem* 2009;**114**:499–504.
38. Li B, Huang G, Lu R, Wei J, Zhong Z. Study on chemical composition of *Phyllanthus emblica*. *J Chin Med Mater* 2015;**38**:290–3.
39. Suresh K, Vasudevan DM. Augmentation of murine natural killer cell and antibody dependent cellular cytotoxicity activities by *Phyllanthus emblica*, a new immunomodulator. *J Ethnopharmacol* 1994;**44**:55–60.
40. Barnes J, Anderson L, Gibbons S. Echinacea species (*Echinacea angustifolia* (DC.) Hell., *Echinacea pallida* (Nutt.) Nutt., *Echinacea purpurea* (L.) Moench): a review of their chemistry, pharmacology. *J Pharm Pharmacol* 2005;**8**:929–54.
41. Guiotto P, Woelkart K, Grabnar I, Voinovich D, Perissutti B, Invernizzi S, et al. Pharmacokinetics and immunomodulatory effects of phytotherapeutic lozenges (bonbons) with *Echinacea purpurea* extract. *Phytomedicine* 2008;**15**:547–54.
42. Manayi A, Vazirian M, Saeidnia S. *Echinacea purpurea*: pharmacology, phytochemistry and analysis methods. *Pharmacogn Rev* 2015;**9**(17):63.
43. Kaur A, Oberhofer M, Juzumaite M, Raja HA, Gulledge TV, Kao D, et al. Secondary metabolites from fungal endophytes of *Echinacea purpurea* suppress cytokine secretion by macrophage-type cells. *Nat Prod Commun* 2016;**11**:1143–6.
44. Shin B-K, Kwon SW, Park JH. Chemical diversity of ginseng saponins from *Panax ginseng*. *J Ginseng Res* 2015;**39**:287–98.
45. Tran T-L, Kim Y-R, Yang J-L, Oh D-R, Dao T-T, Oh W-K. Dammarane triterpenes from the leaves of *Panax ginseng* enhance cellular immunity. *Bioorg Med Chem* 2014;**22**:499–504.
46. Zhang G, Ghosh S. Toll-like receptor-mediated NF-κB activation: a phylogenetically conserved paradigm in innate immunity. *J Clin Invest* 2001;**107**:13–9.
47. Ghosh S, May M, Kopp E. NF-κB and rel proteins: evolutionarily conserved mediators of immune responses. *Annu Rev Immunol* 1998;**16**:225–60.
48. Cho K, Song SB, Tung NH, Kim KE, Kim YH. Inhibition of TNF-α-mediated NF-κB transcriptional activity by dammarane-type Ginsenosides from steamed flower buds of *Panax ginseng* in HepG2 and SK-Hep1 cells. *Biomol Ther* 2014;**22**:55–61.
49. Manu K, Kuttan G. Immunomodulatory activities of Punarnavine, an alkaloid from *Boerhaavia diffusa*. *Immunopharmacol Immunotoxicol* 2009;**31**:377–87.
50. Uto T, Nishi Y, Toyama M, Yoshinaga K. Inhibitory effect of cepharanthine on dendritic cell activation and function. *Int Immunopharmacol* 2011;**11**:1932–8.
51. Ganguly T. Immunomodulatory effect of on Con A induced lymphoproliferation. *Phytomedicine* 2001;**8**:431–7.
52. Wang G, Lemos J, Iadecola C. Herbal alkaloid tetrandrine: from an ion channel blocker to inhibitor of tumor proliferation. *Trends Pharmacol Sci* 2004;**25**(3):120–3.

53. Xiao-Hui C, Shuai W, Bao-An C. Research advances on the pharmacological effects of tetrandrine. *Chin J Nat Med* 2011;**6**:473–80.
54. Araujo GR, Fonseca JE, Fujimura PT, Cunha-Junior JP, Silva CHM, Mourão AF, et al. Anti-type II collagen antibodies detection and avidity in patients with oligoarticular and polyarticular forms of juvenile idiopathic arthritis. *Immunol Lett* 2015;**165**:20–5.
55. Zhao G, Xie X, Sun H, Yuan Z, Zhong Z, Tang S. Bioinspired collective syntheses of iboga-type indole alkaloids. *Org Lett* 2016;**18**:2447–50.
56. House R, Thomas P, Bhargava H. Comparison of the hallucinogenic indole alkaloids ibogaine and harmaline for potential immunomodulatory activity. *Pharmacology* 1995;**51**:56–65.
57. Southon I, Buckingham J. *Dictionary of alkaloids, with CD-ROM*. Chapman Hall, London: CRC Press; 1989.
58. Lopes F, Calvo T, Colombo L. Immunostimulatory and cytotoxic activities of *Indigofera suffruticosa* (Fabaceae). *Nat Prod Res* 2011;**25**:1796–809.
59. Yang L, Zhang L. Chemical structural and chain conformational characterization of some bioactive polysaccharides isolated from natural sources. *Carbohydr Polym* 2009;**76**:349–61.
60. Schepetkin I, Quinn M. Botanical polysaccharides: macrophage immunomodulation and therapeutic potential. *Int Immunopharmacol* 2006;**31**:317–33.
61. Sharma S, Erickson K, Troutman J. Complete tetrasaccharide repeat unit biosynthesis of the immunomodulatory *Bacteroides fragilis* capsular polysaccharide A. *ACS Chem Biol* 2016;**12**:92–101.
62. Ferreira S, Passos C, Madureira P, Vilanova M. Structure–function relationships of immunostimulatory polysaccharides: a review. *Carbohydr Polym* 2015;**132**:378–96.
63. Geijtenbeek T, Gringhuis S. C-type lectin receptors in the control of T helper cell differentiation. *Nat Rev Immunol* 2016;**16**:433–48.
64. Vannucci L, Krizan J, Sima P, Stakheev D, Caja F, Rajsiglova L, et al. Immunostimulatory properties and antitumor activities of glucans. *Int J Oncol* 2013;**43**:357–64 [Review].
65. Ćilerdžić J, Vukojević J, Stajić M, Stanojković T. Biological activity of *Ganoderma lucidum* basidiocarps cultivated on alternative and commercial substrate. *J Ethnopharmacol* 2014;**155**:312–9.
66. Guo L, Xie J, Ruan Y, Zhou L, Zhu H, Yun X. Characterization and immunostimulatory activity of a polysaccharide from the spores of *Ganoderma lucidum*. *Int Immunopharmacol* 2009;**9**:1175–82.
67. Moreira L. An overview of mannan structure and mannan-degrading enzyme systems. *Appl Microbiol Biotechnol* 2008;**79**:165–78.
68. Mikami T, Nagase T, Matsumoto T, Suzuki M, Suzuki S, Kumano N. Mitogenic effect of the mannans from *Saccharomyces cerevisiae* on mouse spleen lymphocytes. *Microbiol Immunol* 1982;**26**:913–22.
69. Paulsen B. Plant polysaccharides with immunostimulatory activities. *Curr Org Chem* 2001;**5**:939–50.
70. Wongprasert K, Rudtanatip T, Praiboon J. Immunostimulatory activity of sulfated galactans isolated from the red seaweed *Gracilaria fisheri* and development of resistance against white spot syndrome virus (WSSV) in shrimp. *Fish Shellfish Immunol* 2014;**36**:52–60.
71. Lehne G, Haneberg B, Gaustad P, Johansen PW, Preus H, Abrahamsen TG. Oral administration of a new soluble branched beta-1,3-D-glucan is well tolerated and can lead to increased salivary concentrations of immunoglobulin a in healthy volunteers. *Clin Exp Immunol* 2006;**143**:65–9.
72. Ellis M, Egelund J, Schultz C, Bacic A. Arabinogalactan-proteins: key regulators at the cell surface? *Plant Physiol* 2010;**153**:403–19.
73. Nair P, Rodriguez S, Ramachandran R. Immune stimulating properties of a novel polysaccharide from the medicinal plant *Tinospora cordifolia*. *Int Immunopharmacol* 2004;**4**:1645–59.
74. Riede L, Grube B, Gruenwald J. Larch arabinogalactan effects on reducing incidence of upper respiratory infections. *Curr Med Res Opin* 2013;**29**:251–8.
75. Taguchi I, Kiyohara H, Matsumoto T, Yamada H. Structure of oligosaccharide side chains of an intestinal immune system modulating arabinogalactan isolated from rhizomes of *Atractylodes lancea* DC. *Carbohydr Res* 2004;**339**:763–70.
76. Dion C, Chappuis E, Ripoll C. Does larch arabinogalactan enhance immune function? A review of mechanistic and clinical trials. *Nutr Metab (Lond)* 2016;**13**:28.
77. Rennie E, Scheller H. Xylan biosynthesis. *Curr Opin Biotechnol* 2014;**26**:100–7.
78. Shi L, Dong Q, Ding K. Structure elucidation and immunomodulatory activity in vitro of a xylan from roots of *Cudrania tricuspidata*. *Food Chem* 2014;**152**:291–6.

79. Nabarlatz D, Montané D, Kardošová A, Bekešová S. Almond shell xylo-oligosaccharides exhibiting immunostimulatory activity. *Carbohydr Res* 2007;**342**:1122–8.
80. Lecerf J, Dépeint F, Clerc E. Xylo-oligosaccharide (XOS) in combination with inulin modulates both the intestinal environment and immune status in healthy subjects, while XOS alone only shows. *Br J Nutr* 2012;**108**:1847–58.
81. Kogan G, Šoltés L, Stern R, Gemeiner P. Hyaluronic acid: a natural biopolymer with a broad range of biomedical and industrial applications. *Biotechnol Lett* 2007;**29**:17–25.
82. Bernish B, van de Rijn I. Characterization of a two-component system in *Streptococcus pyogenes* which is involved in regulation of hyaluronic acid production. *J Biol Chem* 1999;**274**:4786–93.
83. Ke C, Wang D, Sun Y, Qiao D, Ye H, Zeng X. Immunostimulatory and antiangiogenic activities of low molecular weight hyaluronic acid. *Food Chem Toxicol* 2013;**58**:401–7.
84. Havsteen B. The biochemistry and medical significance of the flavonoids. *Pharmacol Ther* 2002;**96**:67–202.
85. Middleton E, Kandaswami C. Effects of flavonoids on immune and inflammatory cell functions. *Biochem Pharmacol* 1992;**43**:1167–79.
86. Akbay P, Basaran A, Undeger U. In vitro immunomodulatory activity of flavonoid glycosides from *Urtica dioica* L. *Phytother Res* 2003;**17**:34–7.
87. Chlopčíková Š, Psotová J, Miketová P. Chemoprotective effect of plant phenolics against anthracycline-induced toxicity on rat cardiomyocytes. Part I. Silymarin and its flavonolignans. *Phytother Res* 2004;**18**:107–10.
88. Lee J, Kim S, Kim H, Lee T. Silibinin polarizes Th1/Th2 immune responses through the inhibition of immunostimulatory function of dendritic cells. *J Cell Physiol* 2007;**210**:385–97.
89. Lee S, Kim Y, Kwon S, Lee Y, Choi S. Inhibitory effects of flavonoids on TNF-α-induced IL-8 gene expression in HEK 293 cells. *BMB Rep* 2009;**42**:265–70.
90. Brattig NW, Diao G-J, Berg PA. Immunoenhancing effect of flavonoid compounds on lymphocyte proliferation and immunoglobulin synthesis. *Int J Immunopharmacol* 1984;**6**:205–15.
91. Gershenzon J, Kreis W. *Biochemistry of plant secondary metabolism*. Sheffield: Sheffieled Academic Press; 1999.
92. El Enshasy H, Hatti-Kaul R. Mushroom immunomodulators: unique molecules with unlimited applications. *Trends Biotechnol* 2013;**31**:668–77.
93. Herman S, Inventor. Method of treating a systemic disorder using trioxolane and diperoxide compounds. United States Patent US 5,270,344; 1993.
94. Dzubak P, Hajduch M, Vydra D, Hustova A. Pharmacological activities of natural triterpenoids and their therapeutic implications. *Nat Prod Rep* 2006;**23**:394–411.
95. Wu G. *Amino acids: biochemistry and nutrition*. Boca Raton, FL: CRC Press; 2013.
96. Gokhale AS, Satyanarayanajois S. Peptides and peptidomimetics as immunomodulators. *Immunotherapy* 2014;**6**:755–74.
97. Sheu F, Chien P, Hsieh K, Chin K. Purification, cloning, and functional characterization of a novel immunomodulatory protein from *Antrodia camphorata* (bitter mushroom) that exhibits TLR2-dependent. *J Agric Food Chem* 2009;**57**:4130–41.
98. Wang P, Hsu C, Tang S, Huang Y. Fungal immunomodulatory protein from *Flammulina velutipes* induces interferon-γ production through p38 mitogen-activated protein kinase signaling pathway. *J Agric Food Chem* 2004;**52**:2721–5.
99. Li Q, Wang X, Chen Y, Lin J, Zhou X. Cytokines expression induced by *Ganoderma sinensis* fungal immunomodulatory proteins (FIP-gsi) in mouse spleen cells. *Appl Biochem Biotechnol* 2010;**162**: 1403–13.
100. Li F, Wen H, Zhang Y, Aa M, Liu X. Purification and characterization of a novel immunomodulatory protein from the medicinal mushroom *Trametes versicolor*. *Sci China Life Sci* 2011;**54**:379–85.
101. Chu KT, Ng TB. Smilaxin, a novel protein with immunostimulatory, antiproliferative, and HIV-1-reverse transcriptase inhibitory activities from fresh Smilax glabra rhizomes. *Biochem Biophys Res Commun* 2006;**340**:118–24.
102. Aranha I, Clement F, Venkatesh YP. Immunostimulatory properties of the major protein from the stem of the Ayurvedic medicinal herb, guduchi (*Tinospora cordifolia*). *J Ethnopharmacol* 2012;**139**: 366–72.
103. Parikh V, Tucci V, Galwankar S. Infections of the nervous system. *Int J Crit Illn Inj Sci* 2012;**2**:82.

104. Garvey G. Current concepts of bacterial infections of the central nervous system: bacterial meningitis and bacterial brain abscess. *J Neurosurg* 1983;**59**:735–44.
105. Bommineni GR, Kapilashrami K, Cummings JE, Lu Y, Knudson SE, Gu C, et al. Thiolactomycin-based inhibitors of bacterial β-ketoacyl-ACP synthases with in vivo activity. *J Med Chem* 2016;**59**:5377–90.
106. Nomura S, Horiuchi T, Omura S. The action mechanism of cerulenin. *J Biochem* 1972;**71**:783–96.
107. Wilson D. The A–Z of bacterial translation inhibitors. *Crit Rev Biochem Mol Biol* 2009;**44**:393–433.
108. Igarashi M, Nakagawa N, Doi N, Hattori S. Caprazamycin B, a novel anti-tuberculosis antibiotic, from *Streptomyces* sp. *J Antibiot* 2003;**56**:580–3.
109. de Souza M. Promising candidates in clinical trials against multidrug-resistant tuberculosis (MDR-TB) based on natural products. *Fitoterapia* 2009;**80**:453–60.
110. Yamaguchi H, Sato S, Yoshida S, Takada K. Capuramycin, a new nucleoside antibiotic. Taxonomy, fermentation, isolation and characterization. *J Antibiot* 1986;**39**:1047–53.
111. Olano C, Wilkinson B, Sánchez C, Moss S. Biosynthesis of the angiogenesis inhibitor borrelidin by *Streptomyces parvulus* Tü4055: cluster analysis and assignment of functions. *Chem Biol* 2004;**11**:87–97.
112. Izaki K, Matsuhashi M, Strominger J. Biosynthesis of the peptidoglycan of bacterial cell walls XIII. Peptidoglycan transperptid and d-alanine carboxypeptidase: Penicillin. *J Biol Chem* 1968;**243**:3180–92.
113. Ng I, Ye C, Zhang Z, Lu Y, Jing K. Daptomycin antibiotic production processes in fed-batch fermentation by Streptomyces roseosporus NRRL11379 with precursor effect and medium optimization. *Bioprocess Biosyst Eng* 2014;**37**:415–23.
114. Chopra I, Roberts M. Tetracycline antibiotics: mode of action, applications, molecular biology, and epidemiology of bacterial resistance. *Microbiol Mol Biol Rev* 2001;**65**:232–60.
115. Wu J, Zhang Q, Deng W, Qian J, Zhang S, Liu W. Toward improvement of erythromycin A production in an industrial *Saccharopolyspora erythraea* strain via facilitation of genetic manipulation with an artificial attB site for specific recombination. *Appl Environ Microbiol* 2011;**77**:7508–16.
116. Cortes J, Haydock S, Roberts G, Bevitt D. An unusually large multifunctional polypeptide in the erythromycin-producing polyketide synthase of *Saccharopolyspora erythraea*. *Nature* 1990;**348**:176–7.
117. Schnappinger D, Hillen W. Tetracyclines: antibiotic action, uptake, and resistance mechanisms. *Arch Microbiol* 1996;**165**:359–69.
118. Weil J, Miramonti J, Ladisch M. Cephalosporin C: mode of action and biosynthetic pathway. *Enzym Microb Technol* 1995;**17**:85–7.
119. Hammond S, Williamson M, Williams D. On the biosynthesis of the antibiotic vancomycin. *J Chem Soc Chem Commun* 1982;(6):344–6.
120. Lemke A, Kiderlen A, Kayser O. Amphotericin B. *Appl Microbiol Biotechnol* 2005;**68**:151–62.
121. Vasquez D. Binding to ribosomes and inhibitory effect on protein synthesis of the spiramycin antibiotics. *Life Sci* 1967;**15**:845–53.
122. Iwu M, Duncan A, Okunji C. *New antimicrobials of plant origin*. Alexandria, VA: ASHS Press; 1999.
123. Olliaro P, Haynes R, Meunier B, Yuthavong Y. Possible modes of action of the artemisinin-type compounds. *Trends Parasitol* 2001;**17**:122–6.
124. Pines M, Spector I. Halofuginone—the multifaceted molecule. *Molecules* 2015;**20**:573–94.

FURTHER READING

1. Choi E, Kim A, Kim Y. Immunomodulating activity of arabinogalactan and fucoidan in vitro. *J Med Food* 2005;**8**:446–53.

SECTION II

Bacterial and Viral Infections of the CNS

CHAPTER 7

Pneumococcal Meningitis

Benard W. Kulohoma[*,†]
[*]University of Nairobi, Nairobi, Kenya
[†]University of Oxford, Oxford, United Kingdom

1. PNEUMOCOCCAL DISEASE

Streptococcus pneumoniae (the pneumococcus) asymptomatically colonizes the human nasopharynx alongside multiple other commensal bacteria.[1] *S. pneumoniae* may invade the host's sterile sites and cause severe life-threatening invasive pneumococcal disease (IPD). Invasion of the lungs, blood, middle of the ear, and cerebral spinal fluid (CSF) leads to pneumonia, bacteremia (bloodstream infection), acute otitis media, and meningitis, respectively (Fig. 1).[2–4] The populations most at risk of IPD are children under the age of 5; the elderly (>65 years); diabetics; individuals with malignancy, chronic liver disease, malnutrition and sickle cell disease; and those with compromised immunity such as splenic dysfunction and immunosuppressive illness, particularly those infected with human immunodeficiency virus (HIV).[5,6] Asymptomatic colonization is most prevalent among young children.[7,8] Each year, 1 in 8 children under the age of 5 dies from IPD in sub-Saharan Africa, a childhood mortality rate nearly 20 times higher than that in developed nations (1 in 143).[9,10]

1.1 The Bacterium

S. pneumoniae is a Gram-positive, lancet-shaped coccus. It is a bile soluble, α-haemolytic when cultured on blood agar, and usually occurs in pairs or as chains. They grow optimally when cultured in 5% CO_2 at 37°C and have a doubling time of 20–30 min. They typically have a polysaccharide capsule but may be uncapsulated.[11] Pilus, a cell surface organelle that extends beyond the polysaccharide capsule, has been reported to be present in up to 30% of pneumococci.[12–15] Pneumococci are classified into >92 distinct serotypes based on the immunohistochemistry of their polysaccharide capsules, with several serotypes just recently identified.[16–18] The prevalence of carriage serotypes is much higher compared to those associated with invasive disease, and consequently pneumococcal carriage is more common.[19] Although all pneumococci are capable of causing invasive disease, only a small proportion of pneumococcal serotypes are most frequently associated with the majority of severe invasive disease; these serotypes are rarely carried at the nasopharynx.[19–21]

The Microbiology of Central Nervous System Infections
https://doi.org/10.1016/B978-0-12-813806-9.00007-X

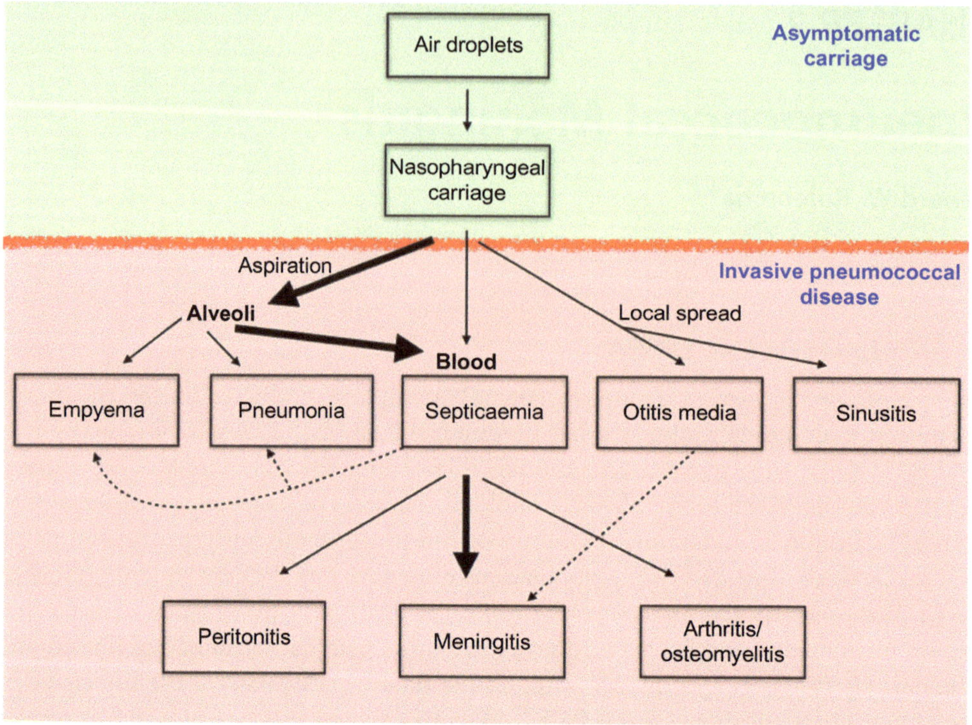

Fig. 1 Pathogenic routes of pneumococcal infection. The different host tissue sites are infected through the airborne and haematogenous routes. This may result in different disease outcomes (highlighted in *boxes*). The main route for pneumococcal meningitis infections is also highlighted using *bold arrows*.

1.2 Diagnosis and Laboratory Typing Schemes

Pneumococcal disease may be presumed on clinical (e.g. features of a lobar pneumonia or typical CSF findings, particularly in the context of otitis media) or radiological grounds (for which there are standardized WHO criteria[22]) but requires laboratory investigation for confirmation. Otitis media may be confirmed by middle ear aspiration and culture; pneumonia may be confirmed by lung or pleural fluid aspirate and/or blood culture; bacteremia may be confirmed by blood culture or pneumococcal DNA detection using polymerase chain reaction (PCR);[23] and meningitis may be confirmed by CSF Gram stain, culture or pneumococcal PCR, or blood culture in the context of typical CSF findings. Sputum culture is poorly predicative of pneumococcal pneumonia but pneumococcal DNA quantification of nasopharyngeal samples by PCR is currently attracting considerable attention, particularly as a research tool.[23–26] Pneumococci retrieved from blood, CSF and other sterile fluids are grown on selective media: blood agar with gentamycin (an antibiotic that allows only pneumococci to grow), at 37°C and 5% CO_2

overnight. Their identity is further confirmed by optochin sensitivity. Pneumococci are soluble in bile, a feature that distinguishes them from genetically closely related competitive nasopharyngeal commensal *Streptococcus viridans*. A Quellung reaction that is considered the "gold standard" is used to characterize pneumococci into serotypes.[27] It is based on the binding of the polysaccharide capsule to a homologous monoclonal antibody. More recently, multiplex PCR reactions have been used for pneumococcal serotyping.[28] Multilocus sequence typing (MLST) is used to classify pneumococci into evolutionarily related clones referred to as sequence types (STs).[29,30] MLST has contributed greatly in molecular epidemiology and provided a better understanding of bacterial diversity and evolution.[31–36]

2. HIGHLIGHTS OF THE EPIDEMIOLOGY OF PNEUMOCOCCAL DISEASE

S. pneumoniae is the most common cause of bacterial meningitis and substantially contributes to bacteremia among adults and children.[37,38] These bacterial infections exhibit seasonality with high prevalence in the cold and dry seasons followed by a reduction in the rainy season.[39] Treatment and prevention of pneumococcal infection in the developing world, which has the majority of cases, is paramount, particularly with the high burden of immunocompromising conditions such as malnutrition, malaria, and HIV.[40,41] Limited healthcare resources and access to suitable treatment, late presentation for treatment, and increasing antimicrobial resistance have further complicated the situation, resulting in high case fatality regardless of HIV status.[42]

Each year, an estimated 14.5 million episodes of severe IPD occur globally.[43,44] Approximately 1 million deaths occur in children under 5, a 10th of whom are HIV positive, and >61% of the majority of deaths in HIV-negative children occur in 10 countries in Africa and Asia.[43] Individuals with HIV infection are 30–100 times more at risk for IPD compared to age-matched HIV-negative counterparts, with rates increasing with decrease in CD4 T cell counts.[4,42,45–47] Highly active antiretroviral therapy (HAART) has led to a dramatic reduction in the incidence of IPD and prolonged the time to the first IPD incident in HIV-infected individuals compared to the period before HAART or in the absence of HAART.[45,47] Close exposure to children is a risk factor for IPD in HIV-infected adults, who commonly have serogroups that are most common in children compared to HIV uninfected adults.[48] Approximately 50% of children under the age of 9 have been reported to be asymptomatic carriers of pneumococcus in East Africa.[7] In West Africa, about half the cases of pneumococcal disease in children have been reported to occur in the first year of life, with 20% occurring before the age of 6 months.[49,50] Moreover, clinical surveillance in developing countries greatly underestimates the cases of pneumococcal disease. A large number of deaths are due to poor diagnosis and are often not fully investigated because most health centers lack the capacity to perform microbiology assays, and there is a paucity of data on invasive disease. Incidence is also underestimated due to the poor sensitivity of diagnostic tests such as blood cultures, improper handling and transportation of samples, and use of unsuitable media.[51] Antibiotic use also greatly compromises the sensitivity of blood cultures.[6,51,52] There is limited access to care, given that most health institutions are remotely located and not all patients present to hospitals.[6,43,53] This is especially true in rural areas where most infant deaths occur outside the hospital.[6]

The introduction of the 13 valent pneumococcal conjugate vaccine (PCV13) into the childhood immunization schedule of most developing countries is anticipated to provide coverage of up to 90% in children, based on serogrouping data from invasive isolates.[54] This distribution of pneumococcal serotypes associated with IPD varies both by age of individuals and geography.[20,21] Understanding the geographic distribution of pneumococcal disease is critical for the development of optimal vaccine formulations with increased coverage. These vaccine formulations are capable of protecting against additional invasive serotypes that are most prevalent in the host populations.[20,54,55]

3. PNEUMOCOCCAL DISEASE INTERVENTIONS

Interventions that inhibit specific invasive bacterial strain growth and colonization may disturb the niche cocolonization balance, thereby allowing increased niche occupation by previously outcompeted but highly virulent strains.[56] Interventions that target a particular bacterial species may also elicit nonspecific or cross-reactive immune responses, resulting in inhibition of other nontargeted bacteria.[1] Other interventions, such as antibiotics, may result in alteration of bacterial defense mechanisms resulting in more-tolerant bacterial strains.[57,58] Careful consideration must be made when administering interventions to ensure they are highly effective but do not result in drastic detrimental alterations in microbial population structures.

3.1 Antibiotics

Antibiotics have long been used for treatment against pneumococcal infection. This has led to a marked worldwide increase in the prevalence of pneumococci with decreased susceptibility to multiple drugs (penicillin, erythromycin, tetracycline, chloramphenicol, ceftriaxone, clindamycin, and cotrimoxazole).[59–63] Acquisition of drug resistance provides a selective advantage among pneumococci, making antibiotic prophylaxis difficult and highlighting the need for new antibiotic agents. The global spread of multidrug-resistant strains necessitates the urgent development of novel intervention strategies against highly invasive, multidrug-resistant pneumococci. Functional characterization of pneumococcal genes is necessary to understand the process of resistance acquisition, which greatly challenges the ability to treat infection. Consequently, more efforts are being directed toward disease prevention interventions.

3.2 Vaccines

Vaccines are an effective intervention strategy used for prevention of the first episode of disease and disease recurrence, and, if efficacious, to significantly reduce disease incidence and severity. Currently, two classes of pneumococcal vaccines are licensed: the 23-valent polysaccharide pneumococcal vaccine (PPV) and the pneumococcal conjugate vaccine (PCV). The 23-valent PPV is based on the polysaccharide capsule of 23 serotypes most commonly associated with disease in the United States.[64] However, PPV fails to induce an immune response and to reduce pneumococcal carriage and transmission among children under the age of 2 and the elderly (>65 years), who constitute a sizeable proportion of the populations most at risk of IPD.[4] More recently, PCVs have been developed. The recommended vaccine is a 13-valent conjugate pneumococcal vaccine (PCV13) that is based on 13 serotypes (1, 3, 4, 5, 6A, 6B, 7F, 9V, 14, 18C, 19A, 19F, and 23F) that are commonly associated with disease, conjugated to a diphtheria carrier protein.[65,66] The efficacy of these vaccines has been established and those that provide better coverage have been recommended, especially in regions of high disease burden.[65–70] Development of new protein vaccines capable of eliciting a protective immune response regardless of the pneumococcal capsular type is paramount. A better understanding of protein antigens that are universal to all pneumococci and play a significant role during the disease process is required for vaccine development.

4. IMPORTANT VIRULENCE FACTORS FOR PNEUMOCOCCAL CSF INVASION

Invasive pneumococcal disease is preceded by nasopharyngeal colonization and is a complex, multistep process that involves the interaction of specific bacterial virulence factors with the host. Pneumococcal polysaccharide capsules are major virulence factors that are capable of eliciting specific protective antibody-dependent immune responses in humans; different serotype capsules have been used in various vaccine formulations.[71] The antigenicity of each capsule is serotype-specific, but cross-protective immunity may be elicited between capsules with similar polysaccharides.[71] Pneumococcal proteins are also major contributors to the pathogen's disease process.[72–77] These proteins are able to act indirectly as mediators of inflammation or directly by invading host tissues.[72,73,76,78] Proteins elicit a T cell-mediated immune response and have been used as carriers in conjugate vaccines.[3,79] There are major efforts to identify next-generation pneumococcal protein vaccine candidates that induce immunologic memory and are present in all pneumococci, thus eliciting a serotype-independent immune response.[8,71,77,80–82] A variety of proteins that enable pneumococci to effectively invade and colonize hosts have been characterized (Fig. 2). Some of the most well-studied proteins can be broadly categorized as cytoplasmic proteins, choline-binding proteins, metal-binding lipoproteins, and sortase-dependent surface proteins.

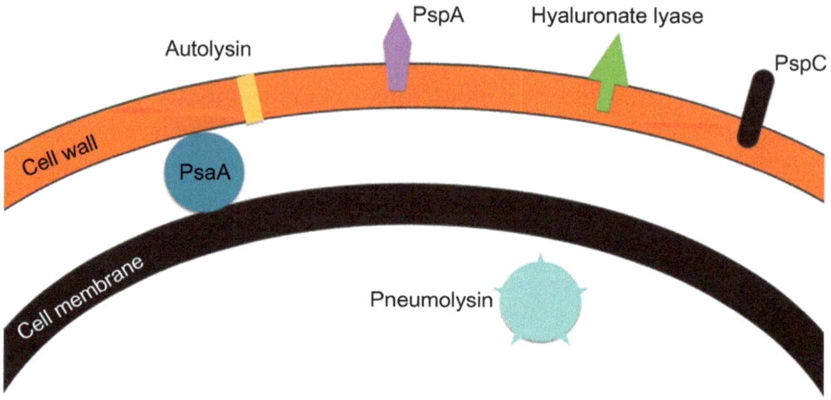

Fig. 2 Some key virulent pneumococcal antigens that have been considered as potential protein vaccine candidates. These include choline-binding proteins (choline-binding protein A (CbpA or PspC), autolysin, and pneumococcal surface protein A (PspA)); cytoplasmic proteins (pneumolysin); metal-binding lipoproteins (pneumococcal surface antigen A (PsaA)); and sortase-dependent surface proteins (Neuraminidase and hyaluronate lyase).

4.1 Choline-Binding Proteins

Choline-binding proteins are surface proteins that are attached to the pneumococcal cell wall.[73,83] Examples include pneumococcal surface protein A (PspA), pneumococcal surface protein C (PspC or CbpA or SpsA), choline-binding protein D (CbpD), choline-binding protein E (CbpE), and choline-binding protein G (CbpG), autolysin (also known as *N*-acetylmuramoyl-L-alanine amidase) (LytA), β-*N*-acetylglucosamidase (LtyB), β-*N*-acetylmuramidase (LytC).[73,76,84–88]

Choline-binding protein A (CbpA or PspC) binds to and blocks the polymeric immunoglobulin receptor (pIgR) that transports secretory IgA across the respiratory epithelium.[89] CbpA facilitates pneumococcal adhesion to human epithelial cells during initial colonization, and has also been shown to be essential in penetration across the blood-brain barrier (BBB) during CSF invasion.[84,85,90,91] CbpA facilitates bacteraemia by binding to the pIgR receptor on the host's epithelium, thereby allowing pneumococcal transcytosis across the mucosal epithelium into blood.[76] Once pneumococci are in blood circulation on reaching the BBB, PspC binds to endothelial laminin receptors, facilitating transcytosis across the BBB into the CSF.[76] CbpA plays a major role in pneumococcal pathogenesis and is under consideration as a potential vaccine candidate.

Pneumococcal surface protein A (PspA) is found on the cell surface of virtually all pneumococcal strains[73,92] and elicits an immune response in mice that protects against invasive infection >100-fold the 50% lethal dose.[73,93–96] PspA subverts immune responses by inhibiting complement binding.[73] However, PspA exhibits high polymorphism at both the DNA and protein level,[97] which undermines its use as a universal immunogenic antigen.

Autolysin (LytA) is normally present in an inactive form and is activated when cell wall biosynthesis stops as a result of nutrient starvation or antibiotic treatment.[72] LytA is involved in separation of daughter cells and cell lysis during the stationary phase and in the presence of antibiotics. It also plays a role in virulence by activating the release of cytoplasmic protein pneumolysin, which causes cell lysis releasing pneumococcal antigens into the host and further aggravates the disease.[72,73] LytA also is involved in the induction of cytokine release and cellular immunity, and LytA mutants have been associated with reduced severity of host tissue inflammation.[98] LytB is also involved in daughter cell separation while LytC has lysozyme activity at 30°C; mutants of LytB and LytC poorly colonize the nasopharynx in murine models.[74]

CbpG facilitates adherence to mucosal surfaces and virulence at both mucosal surfaces and blood, and immunization with recombinant CbpG is protective in murine models of colonization and bacteremia.[99]

4.2 Cytoplasmic Proteins

Pneumolysin (Pln) is the best-characterized virulent cytoplasmic protein and considered one of the most promising potential protein vaccine candidates.[100] It is released by autolysin activity.[73,87] Pln forms transmembrane pores in the target membranes, resulting in cell lysis, which impedes the host's ability to opsonize, phagocytose, and eliminate invading pneumococci. Pln also disrupts the humoral immune responses to infection by binding to the Fc region of IgG.[72,73,84,101–103]

4.3 Metal-Binding Lipoproteins

Metal-binding lipoproteins pneumococcal surface antigen A (PsaA), pneumococcal iron acquisition A (PiaA), pneumococcal iron uptake A (PiuA), and zinc metalloproteinases play important roles in bacterial transport and fitness.[73,104,105] PsaA is bound to an ABC transport protein complex that transports manganese ions and plays a role in host cell adhesion while PiaA and PiuA mediate pneumococcal iron uptake.[104,105] Vaccination with PiaA and PiuA elicits a protective immune response, and Pia-Piu double mutant strains have significantly reduced growth on an iron-deficient medium as well as reduced virulence in animal models of pneumonia and bacteremia.[89]

4.4 Sortase-Dependent Surface Proteins

Sortase-dependent surface proteins are surface proteins and include Neuramidases (Nan), immunoglobulin A1 protease (Iga1), hyaluronidase (Hyl), and putative proteinase maturation protein A (PpmA).[73,76,106] Hyl lyses hyaluronic acid is mostly associated with host connective tissue and the extracellular matrix, thereby allowing host tissue invasion.[72,73,76] NanA and NanB deglycosylate mucus glycoconjugates, consequently preventing pneumococcal mucus entrapment and prolonging nasopharyngeal colonization.[76]

Invasive pneumococcal disease relies on a combination of multiple key virulent proteins as opposed to individual virulent protein contributions. Identification of virulent pneumococcal proteins that are essential for disease and always present in all pneumococci is important for the development of novel interventions.

5. PATHOGENESIS OF PNEUMOCOCCAL MENINGITIS

Nasopharyngeal colonization is the prerequisite for the complex multistep process that leads to pneumococcal bloodstream invasion and, subsequently, meningitis.[8] The BBB normally protects the brain from most bacteria that reach the bloodstream through very tight cellular junctions which allow very low rates of pinocytosis. This restricts the etiology of bacterial meningitis to a few, predominantly encapsulated extracellular pathogens, including the pneumococcus.[76]

S. pneumoniae is highly adapted to colonization of humans and onward transmission to susceptible hosts.[8,107] This is facilitated both by mutations and by frequent transfers of genetic material (horizontal gene transfer) between strains and across bacterial species.[44,57,108] This highly active process results in considerable redundancy in the range of tools available to *S. pneumoniae* that targets host receptors, enabling the pneumococcus to overcome the mucosal barrier and survive in the nasopharynx (Fig. 3).[73,74,106,109] Pneumococci are able to evade mucus entrapment by breaking down mucus using neuraminidase A (NanA), beta-galactosidase A (BgaA), beta-*N*-acetylglucosaminidase (StrH), and neuraminidase B (NanB) while pneumolysin (Ply) hinders mucosal epithelium ciliary

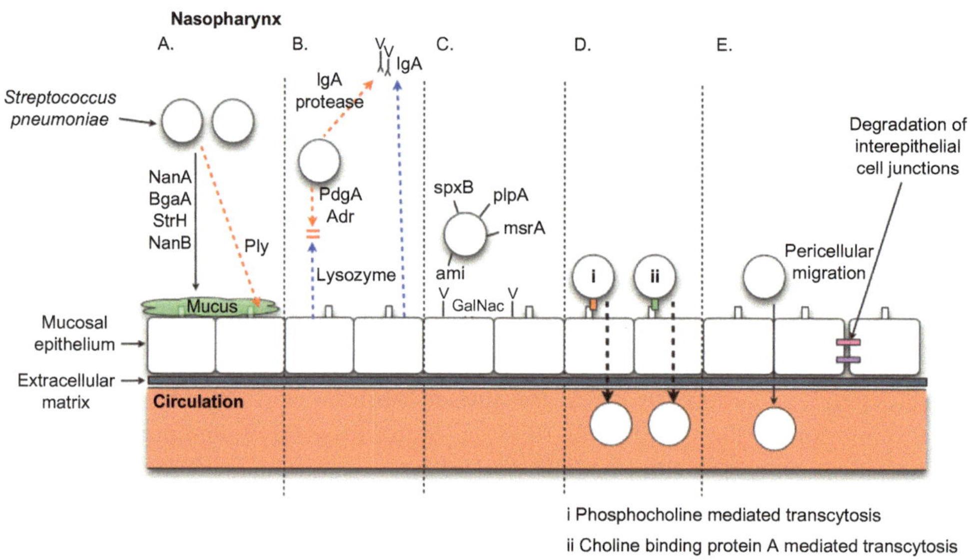

Fig. 3 *S. pneumoniae* evasion of host barriers to colonization at the nasopharynx.

beating, thereby facilitating nasopharyngeal colonization (Fig. 3A). IgA protease subverts IgA antigen specificity, enhancing persistence on the mucosal surface, while *N*-acetyl-glucosamine-deacetylase A (PdgA) and *O*-acetyltransferase (Adr) deacetylate pneumococcal surface peptidoglycan molecules, rendering them resistant to lysozyme activity (Fig. 3B). Pneumococcal receptors pyruvate oxidase (spxB), Oligopeptide ABC transporter, oligopeptide-binding protein (ami), macrolide efflux system encoded by mefE/mel (msrA), and permease-like protein (plpA) facilitate adherence to host mucosal epithelium via GalNac receptors enhancing colonization (Fig. 3C). Pneumococci are transcytosed across the mucosal epithelium into the blood through interactions of their surface antigens (phosphocholine (ChoP) and choline-binding protein A (PspC or CbpA)) with host receptors (platelet-activating factor (PAFr) and polymeric immunoglobulin receptor (pIgR), respectively) (Fig. 3D). Pneumococci translocate across the mucosal epithelium through intercellular and pericellular routes antigens-plasminogen complexes that facilitate degradation of interepithelial cell junctions (Fig. 3E).

Pneumococcal meningitis is an intense host inflammatory process triggered by pneumococci and bacterial products that is focused on the subarachnoid space (Fig. 4).[74,84,85,91] Although meningitis is largely a dead end for the pneumococcus in terms of colonization and onward transmission, meningeal invasion presumably results from the utilization of one or more of the mechanisms evolved for these purposes. The ability of the pneumococcus to cross the BBB either through host endothelial receptor engagement, cytoskeletal rearrangement and transcytosis, or by infecting migrating phagocytes is crucial for CSF invasion.[76,85,91,110] The avoidance of immune surveillance and the capacity of *S. pneumoniae* to grow and survive in the relatively hostile meningeal space facilitate this process.[76,106]

6. GENETICS OF PNEUMOCOCCAL MENINGITIS

The confined environment of the nasopharynx allows close interactions and exchange of genetic material between pneumococci and with other species. The result is a huge spectrum of mosaic pneumococcal genomes. Recent advances in genome sequencing technologies have allowed exploration of the genetic diversity of pneumococcal genomes, and multiple tools exist to analyze these huge amounts of data.[111] There is increased appreciation of the population genetic structure of pneumococci.[112] However, the role of genetic changes and the selective forces that drive them is not fully understood. Comparison of the pneumococcal genomes recovered from bloodstream infection and meningitis has revealed high similarity in their genetic composition.[107,113] This analysis also identified a subset of genes ($n = 32$) present in the bloodstream core genome that is absent in the meningitis core genome, and a subset of genes ($n = 29$) that is present in the meningitis core genome but absent in the bloodstream core genome. Elucidating the underlying selective forces that determine the likelihood and direction of gene transfer in

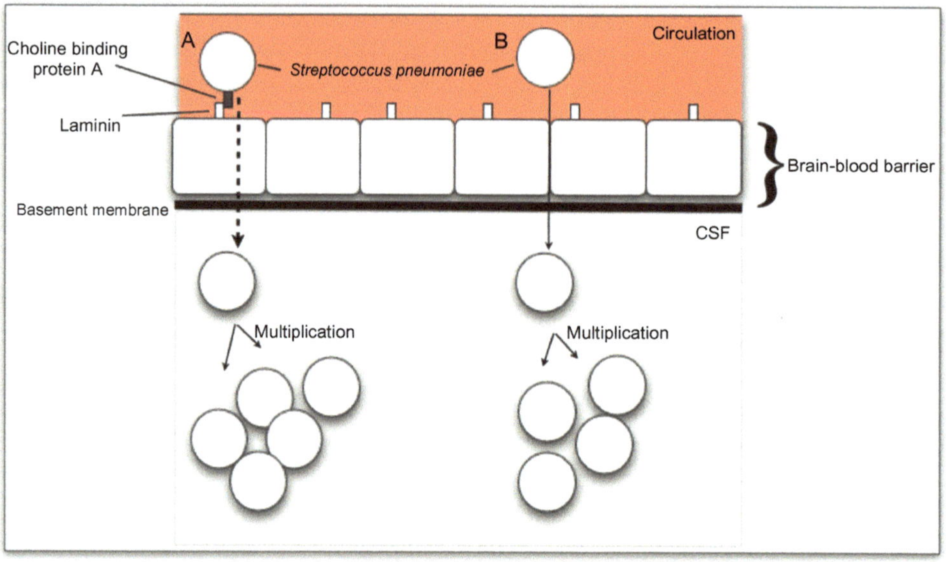

(A)

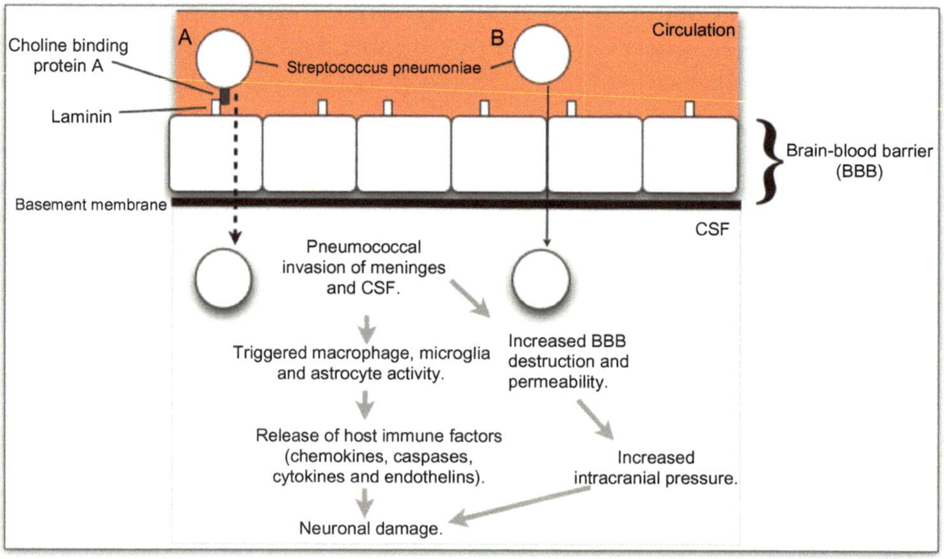

(B)

Fig. 4 Pneumococci crossing the blood-brain barrier (BBB). Shows hematogenous pneumococcal meningitis. (A) Bacteremic pneumococci cross the BBB by host endothelial receptor engagement and transcytosis, (B) thereby triggering host immune responses.

bloodstream and meningitis pneumococcal isolates is important for understanding the pathogenesis process, and may also highlight potential therapeutic and diagnostic targets. Phyletic (gene presence or absence) patterns robustly resolve the population structure of multilineage hyperrecombinant bacteria such as pneumococci and strongly suggest the presence of a phylogenetic signal.[113] Overall, bloodstream pneumococci exhibit more pronounced gene transfer events, and gene loss events exceed gene gain events in both the bloodstream and CSF isolates. These findings demonstrate the likely extent and direction of gene fluctuations at different niches during pneumococcal pathogenesis, highlighting some of the evolutionary dynamics required for tissue-specific host invasion and survival.

Future investigations of more detailed, well-sampled longitudinal datasets of carriage, bloodstream, and CSF isolates under different immunological pressures (particularly immunocomprising conditions such as HIV) will improve understanding of the pathogenesis of pneumococcal meningitis. This will help provide a clear understanding of the pathogenesis process by elaborating its transition from a largely commensal relationship at the nasopharynx to pneumococcal meningitis.

REFERENCES

1. Dagan R, Lipsitch M. *Changing the ecology of pneumococci with antibiotics and vaccines.* Washington, DC: ASM Press; 2004. p. 283–313.
2. Brouwer MC, Read RC, van de Beek D. Host genetics and outcome in meningococcal disease: a systematic review and meta-analysis. *Lancet Infect Dis* 2009;**10**(4):262–74. https://doi.org/10.1016/S1473-3099(10)70045-1 [Published Online First: 2010/03/26].
3. Eskola J, Kilpi T, Palmu A, et al. Efficacy of a pneumococcal conjugate vaccine against acute otitis media. *N Engl J Med* 2001;**344**(6):403–9 [Published Online First: 2001/02/15].
4. Janoff EN, Rubins JB. Invasive pneumococcal disease in the immunocompromised host. *Microb Drug Resist* 1997;**3**(3):215–32 [Published Online First: 1997/10/01].
5. Williams TN, Uyoga S, Macharia A, et al. Bacteraemia in Kenyan children with sickle-cell anaemia: a retrospective cohort and case control study. *Lancet* 2009. https://doi.org/10.1016/S0140-6736(09)61374-X [Published Online First: 2009/09/15].
6. Berkley JA, Lowe BS, Mwangi I, et al. Bacteremia among children admitted to a rural hospital in Kenya. *N Engl J Med* 2005;**352**(1):39–47. https://doi.org/10.1056/NEJMoa040275 [Published Online First: 2005/01/07].
7. Abdullahi O, Nyiro J, Lewa P, et al. The descriptive epidemiology of *Streptococcus pneumoniae* and *Haemophilus influenzae* nasopharyngeal carriage in children and adults in Kilifi district, Kenya. *Pediatr Infect Dis J* 2008;**27**(1):59–64. https://doi.org/10.1097/INF.0b013e31814da70c [Published Online First: 2007/12/29].
8. Bogaert D, De Groot R, Hermans PW. *Streptococcus pneumoniae* colonisation: the key to pneumococcal disease. *Lancet Infect Dis* 2004;**4**(3):144–54. https://doi.org/10.1016/S1473-3099(04)00938-7 [Published Online First: 2004/03/05].
9. UNICEF. *Levels and trends in child mortality*. Available from: http://www.unicef.org/media/files/UNICEF_Child_mortality_for_web_0831.pdf; 2010.
10. UNICEF. *Levels and trends in child mortality*. Available from: http://www.childinfo.org/files/Child_Mortality_Report_2011.pdf; 2011.

11. Hoskins J, Alborn Jr WE, Arnold J, et al. Genome of the bacterium *Streptococcus pneumoniae* strain R6. *J Bacteriol* 2001;**183**(19):5709–17. https://doi.org/10.1128/JB.183.19.5709-5717.2001 [Published Online First: 2001/09/07].
12. Barocchi MA, Ries J, Zogaj X, et al. A pneumococcal pilus influences virulence and host inflammatory responses. *Proc Natl Acad Sci U S A* 2006;**103**(8):2857–62.
13. Moschioni M, De Angelis G, Melchiorre S, et al. Prevalence of pilus-encoding islets among acute otitis media *Streptococcus pneumoniae* isolates from Israel. *Clin Microbiol Infect* 2010;**16**(9):1501–4. https://doi.org/10.1111/j.1469-0691.2009.03105.x [Published Online First: 2009/11/06].
14. Moschioni M, Donati C, Muzzi A, et al. *Streptococcus pneumoniae* contains 3 rlrA pilus variants that are clonally related. *J Infect Dis* 2008;**197**(6):888–96. https://doi.org/10.1086/528375 [Published Online First: 2008/02/14].
15. Kulohoma BW, Gray K, Kamng'ona A, et al. Piliation of invasive *Streptococcus pneumoniae* isolates in the era before pneumococcal conjugate vaccine introduction in Malawi. *Clin Vaccine Immunol* 2013;**20**(11):1729–35. https://doi.org/10.1128/CVI.00403-13 [Published Online First: 2013/09/13].
16. Bratcher PE, Kim KH, Kang JH, et al. Identification of natural pneumococcal isolates expressing serotype 6D by genetic, biochemical and serological characterization. *Microbiology* 2010;**156**(Pt 2):555–60. https://doi.org/10.1099/mic.0.034116-0 [Published Online First: 2009/11/28].
17. Park IH, Pritchard DG, Cartee R, et al. Discovery of a new capsular serotype (6C) within serogroup 6 of *Streptococcus pneumoniae*. *J Clin Microbiol* 2007;**45**(4):1225–33. https://doi.org/10.1128/JCM.02199-06 [Published Online First: 2007/02/03].
18. Nahm MH, Lin J, Finkelstein JA, et al. Increase in the prevalence of the newly discovered pneumococcal serotype 6C in the nasopharynx after introduction of pneumococcal conjugate vaccine. *J Infect Dis* 2009;**199**(3):320–5. https://doi.org/10.1086/596064 [Published Online First: 2008/12/23].
19. Brueggemann AB, Peto TE, Crook DW, et al. Temporal and geographic stability of the serogroup-specific invasive disease potential of *Streptococcus pneumoniae* in children. *J Infect Dis* 2004;**190**(7):1203–11. https://doi.org/10.1086/423820 [Published Online First: 2004/09/04].
20. Hausdorff WP, Bryant J, Paradiso PR, et al. Which pneumococcal serogroups cause the most invasive disease: implications for conjugate vaccine formulation and use, part I. *Clin Infect Dis* 2000;**30**(1):100–21. https://doi.org/10.1086/313608 [Published Online First: 2000/01/05].
21. Scott JA, Hall AJ, Dagan R, et al. Serogroup-specific epidemiology of *Streptococcus pneumoniae*: associations with age, sex, and geography in 7,000 episodes of invasive disease. *Clin Infect Dis* 1996;**22**(6):973–81 [Published Online First: 1996/06/01].
22. Scott JA. The preventable burden of pneumococcal disease in the developing world. *Vaccine* 2007;**25**(13):2398–405. https://doi.org/10.1016/j.vaccine.2006.09.008 [Published Online First: 2006/10/10].
23. Carrol ED, Guiver M, Nkhoma S, et al. High pneumococcal DNA loads are associated with mortality in Malawian children with invasive pneumococcal disease. *Pediatr Infect Dis J* 2007;**26**(5):416–22. https://doi.org/10.1097/01.inf.0000260253.22994.61 [Published Online First: 2007/05/01].
24. Yu J, Carvalho Mda G, Beall B, et al. A rapid pneumococcal serotyping system based on monoclonal antibodies and PCR. *J Med Microbiol* 2008;**57**(Pt 2):171–8. https://doi.org/10.1099/jmm.0.47549-0 [Published Online First: 2008/01/19].
25. Jin P, Xiao M, Kong F, et al. Simple, accurate, serotype-specific PCR assay to differentiate *Streptococcus pneumoniae* serotypes 6A, 6B, and 6C. *J Clin Microbiol* 2009;**47**(8):2470–4. https://doi.org/10.1128/JCM.00484-09 [Published Online First: 2009/06/19].
26. Werno AM, Murdoch DR. Medical microbiology: laboratory diagnosis of invasive pneumococcal disease. *Clin Infect Dis* 2008;**46**(6):926–32. https://doi.org/10.1086/528798 [Published Online First: 2008/02/12].
27. Arai S, Konda T, Wad A, et al. Use of antiserum-coated latex particles for serotyping *Streptococcus pneumoniae*. *Microbiol Immunol* 2001;**45**(2):159–62 [Published Online First: 2001/04/11].
28. Azzari C, Moriondo M, Indolfi G, et al. Realtime PCR is more sensitive than multiplex PCR for diagnosis and serotyping in children with culture negative pneumococcal invasive disease. *PLoS One* 2010;**5**(2). https://doi.org/10.1371/journal.pone.0009282 [Published Online First: 2010/02/23].
29. Enright MC, Spratt BG. A multilocus sequence typing scheme for *Streptococcus pneumoniae*: identification of clones associated with serious invasive disease. *Microbiology* 1998;**144**(Pt 11):3049–60 [Published Online First: 1998/12/10].

30. Maiden MC, Bygraves JA, Feil E, et al. Multilocus sequence typing: a portable approach to the identification of clones within populations of pathogenic microorganisms. *Proc Natl Acad Sci U S A* 1998;**95**(6):3140–5 [Published Online First: 1998/04/18].
31. Feil EJ, Enright MC. Analyses of clonality and the evolution of bacterial pathogens. *Curr Opin Microbiol* 2004;**7**(3):308–13. https://doi.org/10.1016/j.mib.2004.04.002 [Published Online First: 2004/06/16].
32. Feil EJ, Enright MC, Spratt BG. Estimating the relative contributions of mutation and recombination to clonal diversification: a comparison between *Neisseria meningitidis* and *Streptococcus pneumoniae*. *Res Microbiol* 2000;**151**(6):465–9. https://doi.org/10.1016/S0923-2508(00)00168-6 [Published Online First: 2000/08/29].
33. Feil EJ, Li BC, Aanensen DM, et al. eBURST: inferring patterns of evolutionary descent among clusters of related bacterial genotypes from multilocus sequence typing data. *J Bacteriol* 2004;**186**(5):1518–30 [Published Online First: 2004/02/20].
34. Feil EJ, Maiden MC, Achtman M, et al. The relative contributions of recombination and mutation to the divergence of clones of Neisseria meningitidis. *Mol Biol Evol* 1999;**16**(11):1496–502 [Published Online First: 1999/11/11].
35. Feil EJ, Smith JM, Enright MC, et al. Estimating recombinational parameters in *Streptococcus pneumoniae* from multilocus sequence typing data. *Genetics* 2000;**154**(4):1439–50 [Published Online First: 2000/04/04].
36. Hanage WP, Kaijalainen T, Herva E, et al. Using multilocus sequence data to define the pneumococcus. *J Bacteriol* 2005;**187**(17):6223–30. https://doi.org/10.1128/JB.187.17.6223-6230.2005 [Published Online First: 2005/08/20].
37. Gordon SB, Chaponda M, Walsh AL, et al. Pneumococcal disease in HIV-infected Malawian adults: acute mortality and long-term survival. *AIDS* 2002;**16**(10):1409–17 [Published Online First: 2002/07/20].
38. Gordon SB, Kanyanda S, Walsh AL, et al. Poor potential coverage for 7-valent pneumococcal conjugate vaccine, Malawi. *Emerg Infect Dis* 2003;**9**(6):747–9 [Published Online First: 2003/06/05].
39. Gordon SB, Walsh AL, Chaponda M, et al. Bacterial meningitis in Malawian adults: pneumococcal disease is common, severe, and seasonal. *Clin Infect Dis* 2000;**31**(1):53–7. https://doi.org/10.1086/313910 [Published Online First: 2000/07/29].
40. UNAIDS. *AIDS epidemic update: UNAIDS*. Available from: http://www.unaids.org/en/KnowledgeCentre/HIVData/EpiUpdate/EpiUpdArchive/2009/default.asp; 2009.
41. Glennie SJ, Nyirenda M, Williams NA, et al. Do multiple concurrent infections in African children cause irreversible immunological damage? *Immunology* 2012;**135**(2):125–32. https://doi.org/10.1111/j.1365-2567.2011.03523.x [Published Online First: 2011/11/03].
42. Hart CA, Beeching NJ, Duerden BI, et al. Infections in AIDS. *J Med Microbiol* 2000;**49**(11):947–67 [Published Online First: 2000/11/10].
43. O'Brien KL, Wolfson LJ, Watt JP, et al. Burden of disease caused by *Streptococcus pneumoniae* in children younger than 5 years: global estimates. *Lancet* 2009;**374**(9693):893–902. https://doi.org/10.1016/S0140-6736(09)61204-6 [Published Online First: 2009/09/15].
44. Golubchik T, Brueggemann AB, Street T, et al. Pneumococcal genome sequencing tracks a vaccine escape variant formed through a multi-fragment recombination event. *Nat Genet* 2012;**44**(3):352–5. https://doi.org/10.1038/ng.1072 [Published Online First: 2012/01/31].
45. Feikin DR, Feldman C, Schuchat A, et al. Global strategies to prevent bacterial pneumonia in adults with HIV disease. *Lancet Infect Dis* 2004;**4**(7):445–55. https://doi.org/10.1016/S1473-3099(04)01060-6 [Published Online First: 2004/06/29].
46. French N, Gordon SB, Mwalukomo T, et al. A trial of a 7-valent pneumococcal conjugate vaccine in HIV-infected adults. *N Engl J Med* 2010;**362**(9):812–22 [Published Online First: 2010/03/05].
47. Kapogiannis BG, Soe MM, Nesheim SR, et al. Trends in bacteremia in the pre- and post-highly active antiretroviral therapy era among HIV-infected children in the US Perinatal AIDS Collaborative Transmission Study (1986–2004). *Pediatrics* 2008;**121**(5):e1229–39. https://doi.org/10.1542/peds.2007-0871 [Published Online First: 2008/05/03].
48. Klugman KP, Madhi SA, Feldman C. HIV and pneumococcal disease. *Curr Opin Infect Dis* 2007;**20**(1):11–5. https://doi.org/10.1097/QCO.0b013e328012c5f1 [Published Online First: 2007/01/02].

49. O'Dempsey TJ, McArdle TF, Lloyd-Evans N, et al. Pneumococcal disease among children in a rural area of west Africa. *Pediatr Infect Dis J* 1996;**15**(5):431–7 [Published Online First: 1996/05/01].
50. Usen S, Adegbola R, Mulholland K, et al. Epidemiology of invasive pneumococcal disease in the Western Region, The Gambia. *Pediatr Infect Dis J* 1998;**17**(1):23–8 [Published Online First: 1998/02/20].
51. WHO. 23-Valent pneumococcal polysaccharide vaccine. WHO position paper. *Wkly Epidemiol Rec* 2008;**83**(42):373–84 [Published Online First: 2008/10/22].
52. Shann F. Bacterial pneumonia: commoner than perceived. *Lancet* 2001;**357**(9274):2070–2. https://doi.org/10.1016/S0140-6736(00)05226-0 [Published Online First: 2001/07/11].
53. Mudhune S, Wamae M. Report on invasive disease and meningitis due to *Haemophilus influenzae* and *Streptococcus pneumoniae* from the network for surveillance of pneumococcal disease in the East African region. *Clin Infect Dis* 2009;**48**(Suppl 2):S147–52. https://doi.org/10.1086/596494 [Published Online First: 2009/02/24].
54. Cornick JE, Everett DB, Broughton C, et al. Invasive *Streptococcus pneumoniae* in children, Malawi, 2004–2006. *Emerg Infect Dis* 2011;**17**(6):1107–9. https://doi.org/10.3201/eid/1706.101404 [Published Online First: 2011/07/14].
55. Ritchie ND, Mitchell TJ, Evans TJ. What is different about serotype 1 pneumococci? *Future Microbiol* 2012;**7**(1):33–46. https://doi.org/10.2217/fmb.11.146 [Published Online First: 2011/12/24].
56. Lipsitch M, Samore MH. Antimicrobial use and antimicrobial resistance: a population perspective. *Emerg Infect Dis* 2002;**8**(4):347–54 [Published Online First: 2002/04/25].
57. Hanage WP, Fraser C, Tang J, et al. Hyper-recombination, diversity, and antibiotic resistance in pneumococcus. *Science* 2009;**324**(5933):1454–7. https://doi.org/10.1126/science.1171908 [Published Online First: 2009/06/13].
58. Lipsitch M. The rise and fall of antimicrobial resistance. *Trends Microbiol* 2001;**9**(9):438–44. https://doi.org/10.1016/S0966-842X(01)02130-8 [Published Online First: 2001/09/13].
59. Casal J. Antimicrobial susceptibility of *Streptococcus pneumoniae*: serotype distribution of penicillin-resistant strains in Spain. *Antimicrob Agents Chemother* 1982;**22**(2):222–5 [Published Online First: 1982/08/01].
60. Munoz R, Coffey TJ, Daniels M, et al. Intercontinental spread of a multiresistant clone of serotype 23F *Streptococcus pneumoniae*. *J Infect Dis* 1991;**164**(2):302–6 [Published Online First: 1991/08/01].
61. Koornhof HJ, Wasas A, Klugman K. Antimicrobial resistance in *Streptococcus pneumoniae*: a South African perspective. *Clin Infect Dis* 1992;**15**(1):84–94 [Published Online First: 1992/07/01].
62. McGee L, McDougal L, Zhou J, et al. Nomenclature of major antimicrobial-resistant clones of *Streptococcus pneumoniae* defined by the pneumococcal molecular epidemiology network. *J Clin Microbiol* 2001;**39**(7):2565–71. https://doi.org/10.1128/JCM.39.7.2565-2571.2001 [Published Online First: 2001/06/28].
63. Everett DB, Mukaka M, Denis B, et al. Ten years of surveillance for invasive *Streptococcus pneumoniae* during the era of antiretroviral scale-up and cotrimoxazole prophylaxis in Malawi. *PLoS One* 2011;**6**(3). https://doi.org/10.1371/journal.pone.0017765 [Published Online First: 2011/03/23].
64. Makela H, Butler JC. *History of pneumococcal immunization*. Washington, DC: ASM Press; 2008. p. 19–29.
65. ClinicalTrials.gov. *Safety & immunogenicity study of 10-valent pneumococcal conjugate vaccine when administered as a 2-dose schedule*. National Institutes of Health; 2006. [Updated 23 October 2008]. Available from: http://clinicaltrials.gov/ct2/show/NCT00307034.
66. ClinicalTrials.gov. *Study evaluating 13-valent pneumococal conjugate vaccine in healthy infants*. National Institute of Health; 2007. [Updated 6 May 2009]. Available from: http://clinicaltrials.gov/ct2/show/NCT00475033.
67. Cutts FT, Zaman SM, Enwere G, et al. Efficacy of nine-valent pneumococcal conjugate vaccine against pneumonia and invasive pneumococcal disease in The Gambia: randomised, double-blind, placebo-controlled trial. *Lancet* 2005;**365**(9465):1139–46. https://doi.org/10.1016/S0140-6736(05)71876-6 [Published Online First: 2005/03/30].
68. Saaka M, Okoko BJ, Kohberger RC, et al. Immunogenicity and serotype-specific efficacy of a 9-valent pneumococcal conjugate vaccine (PCV-9) determined during an efficacy trial in The Gambia. *Vaccine* 2008;**26**(29–30):3719–26. https://doi.org/10.1016/j.vaccine.2008.04.066 [Published Online First: 2008/06/03].

69. Klugman KP. Efficacy of pneumococcal conjugate vaccines and their effect on carriage and antimicrobial resistance. *Lancet Infect Dis* 2001;**1**(2):85–91. https://doi.org/10.1016/S1473-3099(01)00063-9 [Published Online First: 2002/03/02].
70. Klugman KP, Madhi SA, Huebner RE, et al. A trial of a 9-valent pneumococcal conjugate vaccine in children with and those without HIV infection. *N Engl J Med* 2003;**349**(14)):1341–8. https://doi.org/10.1056/NEJMoa035060 [Published Online First: 2003/10/03].
71. Bogaert D, Hermans PW, Adrian PV, et al. Pneumococcal vaccines: an update on current strategies. *Vaccine* 2004;**22**(17–18):2209–20. https://doi.org/10.1016/j.vaccine.2003.11.038 [Published Online First: 2004/05/20].
72. Paton JC, Andrew PW, Boulnois GJ, et al. Molecular analysis of the pathogenicity of *Streptococcus pneumoniae*: the role of pneumococcal proteins. *Annu Rev Microbiol* 1993;**47**:89–115. https://doi.org/10.1146/annurev.mi.47.100193.000513 [Published Online First: 1993/01/01].
73. Jedrzejas MJ. Pneumococcal virulence factors: structure and function. *Microbiol Mol Biol Rev* 2001;**65**(2):187–207. https://doi.org/10.1128/MMBR.65.2.187-207.2001 [Published Online First: 2001/05/31].
74. Bergmann S, Hammerschmidt S. Versatility of pneumococcal surface proteins. *Microbiology* 2006;**152**(Pt 2):295–303. https://doi.org/10.1099/mic.0.28610-0 [Published Online First: 2006/01/27].
75. Hammerschmidt S. Adherence molecules of pathogenic pneumococci. *Curr Opin Microbiol* 2006;**9**(1):12–20. https://doi.org/10.1016/j.mib.2005.11.001 [Published Online First: 2005/12/13].
76. Mook-Kanamori BB, Geldhoff M, van der Poll T, et al. Pathogenesis and pathophysiology of pneumococcal meningitis. *Clin Microbiol Rev* 2011;**24**(3):557–91. https://doi.org/10.1128/CMR.00008-11 [Published Online First: 2011/07/08].
77. Ogunniyi AD, Grabowicz M, Briles DE, et al. Development of a vaccine against invasive pneumococcal disease based on combinations of virulence proteins of *Streptococcus pneumoniae*. *Infect Immun* 2007;**75**(1):350–7. https://doi.org/10.1128/IAI.01103-06 [Published Online First: 2006/11/08].
78. Ogunniyi AD, LeMessurier KS, Graham RM, et al. Contributions of pneumolysin, pneumococcal surface protein A (PspA), and PspC to pathogenicity of *Streptococcus pneumoniae* D39 in a mouse model. *Infect Immun* 2007;**75**(4):1843–51. https://doi.org/10.1128/IAI.01384-06 [Published Online First: 2007/01/31].
79. Kirkham LA, Kerr AR, Douce GR, et al. Construction and immunological characterization of a novel nontoxic protective pneumolysin mutant for use in future pneumococcal vaccines. *Infect Immun* 2006;**74**(1):586–93. https://doi.org/10.1128/IAI.74.1.586-593.2006 [Published Online First: 2005/12/22].
80. Cockeran R, Anderson R, Feldman C. The role of pneumolysin in the pathogenesis of *Streptococcus pneumoniae* infection. *Curr Opin Infect Dis* 2002;**15**(3):235–9 [Published Online First: 2002/05/17].
81. Briles DE. Protection of the elderly from pneumococcal pneumonia with a protein-based vaccine? *Mech Ageing Dev* 2004;**125**(2):129–31. https://doi.org/10.1016/j.mad.2003.11.008 [Published Online First: 2004/03/24].
82. Briles DE, Hollingshead SK, Swiatlo E, et al. PspA and PspC: their potential for use as pneumococcal vaccines. *Microb Drug Resist* 1997;**3**(4):401–8 [Published Online First: 1997/01/01].
83. Rigden DJ, Galperin MY, Jedrzejas MJ. Analysis of structure and function of putative surface-exposed proteins encoded in the *Streptococcus pneumoniae* genome: a bioinformatics-based approach to vaccine and drug design. *Crit Rev Biochem Mol Biol* 2003;**38**(2):143–68. https://doi.org/10.1080/713609215 [Published Online First: 2003/05/17].
84. Orihuela CJ, Gao G, Francis KP, et al. Tissue-specific contributions of pneumococcal virulence factors to pathogenesis. *J Infect Dis* 2004;**190**(9):1661–9. https://doi.org/10.1086/424596 [Published Online First: 2004/10/13].
85. Ring A, Weiser JN, Tuomanen EI. Pneumococcal trafficking across the blood-brain barrier. Molecular analysis of a novel bidirectional pathway. *J Clin Investig* 1998;**102**(2):347–60. https://doi.org/10.1172/JCI2406 [Published Online First: 1998/07/17].
86. Weiser JN. The pneumococcus: why a commensal misbehaves. *J Mol Med* 2009. https://doi.org/10.1007/s00109-009-0557-x [Published Online First: 2009/11/10].
87. Lock RA, Paton JC, Hansman D. Comparative efficacy of pneumococcal neuraminidase and pneumolysin as immunogens protective against *Streptococcus pneumoniae*. *Microb Pathog* 1988;**5**(6):461–7 [Published Online First: 1988/12/01].

88. Lock RA, Paton JC, Hansman D. Purification and immunological characterization of neuraminidase produced by *Streptococcus pneumoniae*. *Microb Pathog* 1988;**4**(1):33–43. https://doi.org/10.1016/0882–4010(88)90046-0 [Published Online First: 1988/01/01].
89. Kadioglu A, Weiser JN, Paton JC, et al. The role of *Streptococcus pneumoniae* virulence factors in host respiratory colonization and disease. *Nat Rev Microbiol* 2008;**6**(4):288–301. https://doi.org/10.1038/nrmicro1871 [Published Online First: 2008/03/15].
90. Gillespie SH, Balakrishnan I. Pathogenesis of pneumococcal infection. *J Med Microbiol* 2000;**49**(12):1057–67 [Published Online First: 2000/12/29].
91. Orihuela CJ, Mahdavi J, Thornton J, et al. Laminin receptor initiates bacterial contact with the blood brain barrier in experimental meningitis models. *J Clin Investig* 2009;**119**(6):1638–46. https://doi.org/10.1172/JCI36759 [Published Online First: 2009/05/14].
92. Morsczeck C, Prokhorova T, Sigh J, et al. *Streptococcus pneumoniae*: proteomics of surface proteins for vaccine development. *Clin Microbiol Infect* 2008;**14**(1):74–81. https://doi.org/10.1111/j.1469-0691.2007.01878.x [Published Online First: 2007/11/24].
93. Briles DE, Ades E, Paton JC, et al. Intranasal immunization of mice with a mixture of the pneumococcal proteins PsaA and PspA is highly protective against nasopharyngeal carriage of *Streptococcus pneumoniae*. *Infect Immun* 2000;**68**(2):796–800 [Published Online First: 2000/01/20].
94. Briles DE, Tart RC, Wu HY, et al. Systemic and mucosal protective immunity to pneumococcal surface protein A. *Ann N Y Acad Sci* 1996;**797**:118–26 [Published Online First: 1996/10/25].
95. King QO, Lei B, Harmsen AG. Pneumococcal surface protein A contributes to secondary *Streptococcus pneumoniae* infection after influenza virus infection. *J Infect Dis* 2009;**200**(4):537–45. https://doi.org/10.1086/600871 [Published Online First: 2009/07/10].
96. McDaniel LS, Sheffield JS, Delucchi P, et al. PspA, a surface protein of *Streptococcus pneumoniae*, is capable of eliciting protection against pneumococci of more than one capsular type. *Infect Immun* 1991;**59**(1):222–8 [Published Online First: 1991/01/01].
97. Gor DO, Ding X, Briles DE, et al. Relationship between surface accessibility for PpmA, PsaA, and PspA and antibody-mediated immunity to systemic infection by *Streptococcus pneumoniae*. *Infect Immun* 2005;**73**(3):1304–12. https://doi.org/10.1128/IAI.73.3.1304-1312.2005 [Published Online First: 2005/02/26].
98. Canvin JR, Marvin AP, Sivakumaran M, et al. The role of pneumolysin and autolysin in the pathology of pneumonia and septicemia in mice infected with a type 2 pneumococcus. *J Infect Dis* 1995;**172**(1):119–23 [Published Online First: 1995/07/01].
99. Mann B, Orihuela C, Antikainen J, et al. Multifunctional role of choline binding protein G in pneumococcal pathogenesis. *Infect Immun* 2006;**74**(2):821–9. https://doi.org/10.1128/IAI.74.2.821-829.2006 [Published Online First: 2006/01/24].
100. Abry MF, Kimenyi KM, Osowo FO, et al. Genetic diversity of the Pneumococcal CbpA: implications for next-generation vaccine development. *Hum Vaccin Immunother* 2015;**0**. https://doi.org/10.1080/21645515.2015.1021521 [Published Online First: 2015/04/23].
101. Rubins JB, Paddock AH, Charboneau D, et al. Pneumolysin in pneumococcal adherence and colonization. *Microb Pathog* 1998;**25**(6):337–42. https://doi.org/10.1006/mpat.1998.0239 [Published Online First: 1999/01/23].
102. Quin LR, Moore 3rd QC, McDaniel LS. Pneumolysin, PspA, and PspC contribute to pneumococcal evasion of early innate immune responses during bacteremia in mice. *Infect Immun* 2007;**75**(4):2067–70. https://doi.org/10.1128/IAI.01727-06 [Published Online First: 2007/01/16].
103. Etuwewe OM, Swann N, Hollingshead S, et al. Effect of recurrent invasive pneumococcal disease on serum anti-pneumolysin IgG titres in HIV infected adults. *Vaccine* 2009;**27**(29):3881–4. https://doi.org/10.1016/j.vaccine.2009.04.026 [Published Online First: 2009/04/21].
104. Rajam G, Anderton JM, Carlone GM, et al. Pneumococcal surface adhesin A (PsaA): a review. *Crit Rev Microbiol* 2008;**34**(3–4):131–42. https://doi.org/10.1080/10408410802275352 [Published Online First: 2008/08/30].
105. Jomaa M, Terry S, Hale C, et al. Immunization with the iron uptake ABC transporter proteins PiaA and PiuA prevents respiratory infection with *Streptococcus pneumoniae*. *Vaccine* 2006;**24**(24):5133–9. https://doi.org/10.1016/j.vaccine.2006.04.012 [Published Online First: 2006/05/19].

106. Weiser JN, Bae D, Fasching C, et al. Antibody-enhanced pneumococcal adherence requires IgA1 protease. *Proc Natl Acad Sci U S A* 2003;**100**(7):4215–20. https://doi.org/10.1073/pnas.0637469100 [Published Online First: 2003/03/19].
107. Kulohoma BW, Cornick JE, Chaguza C, et al. Comparative genomic analysis of meningitis- and bacteremia-causing pneumococci identifies a common core genome. *Infect Immun* 2015;**83**(10):4165–73. https://doi.org/10.1128/IAI.00814-15 [Published Online First: 2015/08/12].
108. Croucher NJ, Harris SR, Fraser C, et al. Rapid pneumococcal evolution in response to clinical interventions. *Science* 2011;**331**(6016):430–4. https://doi.org/10.1126/science.1198545 [Published Online First: 2011/01/29].
109. Selva L, Viana D, Regev-Yochay G, et al. Killing niche competitors by remote-control bacteriophage induction. *Proc Natl Acad Sci U S A* 2009;**106**(4):1234–8. https://doi.org/10.1073/pnas.0809600106 [Published Online First: 2009/01/15].
110. Tuomanen E. Entry of pathogens into the central nervous system. *FEMS Microbiol Rev* 1996;**18**(4):289–99. https://doi.org/10.1111/j.1574-6976.1996.tb00245.x [Published Online First: 1996/07/01].
111. Kulohoma BW. BMX: a tool for computing bacterial phyletic composition from orthologous maps. *BMC Res Notes* 2015;**8**:51. https://doi.org/10.1186/s13104-015-1017-z [Published Online First: 2015/03/11].
112. Brueggemann AB, Muroki BM, Kulohoma BW, et al. Population genetic structure of *Streptococcus pneumoniae* in Kilifi, Kenya, prior to the introduction of pneumococcal conjugate vaccine. *PLoS One* 2013;**8**(11). https://doi.org/10.1371/journal.pone.0081539 [Published Online First: 2013/11/28].
113. Kulohoma BW, Gray K, Bentley SD, et al. Genetic antigen diversity and gene flux among meningitic and bacteraemia-associated pneumococci from Malawi [PhD thesis]; 2012.

CHAPTER 8

Infections of the Central Nervous System Caused by Nontuberculous Mycobacteria

Carlos Franco-Paredes*,†
*University of Colorado Denver, Denver, CO, United States
†Hospital Infantil de México, Federico Gómez, Mexico City, Mexico

1. INTRODUCTION

Mycobacteria are ubiquitous microorganisms frequently found in engineered water systems, soils, and natural waters.[1] There are a few reasons for their success in thriving in these environments, including: (a) microbiologically, a lipid-rich outer membrane provides them with a selective advantage to live in harsh environments; (b) they have an overall decreased growth rate compared to other bacterial groups, which gives them a decreased susceptibility to antibiotics; (c) they are considered oligotrophs by their ability to grow at low carbon levels, thus making them effective competitors in low nutrient environments; (d) they have the ability to degrade and metabolize a variety of complex hydrocarbons and therefore are considered as agents of nutrient cycles; and (e) their hydrophobic membrane is a major determinant of their surface adherence, biofilm formation, their ability to become aerosolized, and their antibiotic/disinfectant resistance.[1,2] Based on these environmental aspects, the most common form of acquisition of NTM includes inhalation of aerosols, cutaneous inoculation, postsurgical, iatrogenic through the use of indwelling medical devices or prosthetic implants, and probably other modes.

A few mycobacterial species seem to have evolved from an environmental pool into major human pathogens (Table 1). Of these, *Mycobacterium* species comprising the nontuberculous mycobacteria (NTM) are considered important human pathogens that predominantly affect severely immunocompromised human hosts, including those with primary immunodeficiencies such as those with GATA2 deficiency or among those with secondary causes of immunosuppression including HIV/AIDS, transplant-associated immunosuppression, individuals receiving anti-TNF inhibitors, and those with hematologic malignancies.[1,3] The NTM group constitutes mycobacterial species other than those belonging to the MTB complex and that do not cause leprosy. It includes two major groups defined by their ability to grow in solid culture media: (a) rapidly growing mycobacteria (RGM) and (b) slow-growing mycobacteria (SGM)[2] (Fig. 1).

The Microbiology of Central Nervous System Infections
https://doi.org/10.1016/B978-0-12-813806-9.00008-1

Table 1 Major mycobacterial opportunistic pathogens

Mycobacterial species	Disease—clinical manifestation	Neurologic manifestations
Mycobacterium tuberculosis	Intrathoracic (intrapulmonary) Extrathoracic (extrapulmonary) Disseminated disease with involvement of the reticuloendothelial system	Meningoencephalitis Arachnoiditis Tuberculoma Hydrocephalus
Mycobacterium leprae	Leprosy	Neuritis Peripheral nerve dysfunction due to demyelination
Mycobacterium ulcerans	Buruli ulcer	Nonreported
Mycobacterium marinum	Cutaneous disease	Nonreported
Mycobacterium avium complex	Pulmonary Cutaneous Lymphatic Disseminated disease with involvement of the reticuloendothelial system	Meningitis, meningoencephalitis[a] Parenchymal forms (focalized abscess)
Mycobacterium haemophilum	Pulmonary Cutaneous Lymphatic Disseminated	Meningoencephalitis[a] Parenchymal forms (focalized abscess)
Mycobacterium abscessus complex	Pulmonary Cutaneous Disseminated	Meningoencephalitis[a] Parenchymal forms (focalized abscess)
Mycobacterium fortuitum	Cutaneous (direct inoculation such as dermal piercing, acupuncture, etc.) Pulmonary Disseminated Intravascular devices	Meningitis
Mycobacterium chelonei	Cutaneous Pulmonary Disseminated	Nonreported
Mycobacterium kansasii	Pulmonary Cutaneous	Meningoencephalitis[a]

[a]Most cases are in the setting of disseminated disease and in patients with advanced immunosuppression states.

In the United States and other countries, recent reports have suggested that the prevalence of NTM disease, particularly RGM and especially among older adults.[3,4] RGM may cause skin and soft tissue infections after trauma or surgery, pulmonary disease in patients with chronic pulmonary diseases or those with some degree of esophageal dysfunction or gastroesophageal reflux disease, and disseminated disease in immunocompromised hosts.[3,5,6] Mycobacterial species in the RGM group grow on solid media within 7 days and possess an inherent resistance to first-line antituberculosis drugs. RGM are

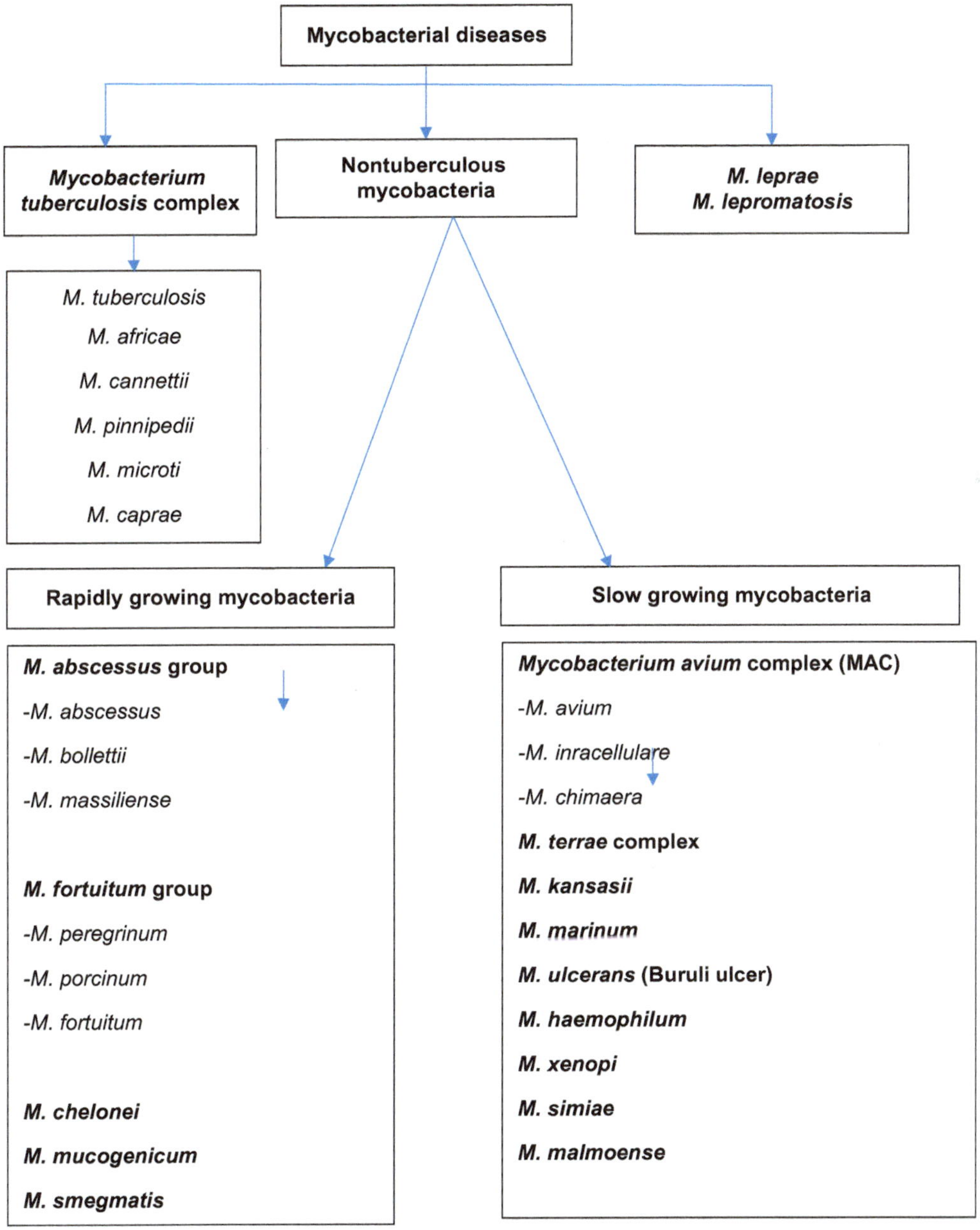

Fig. 1 Mycobacterial species associated with human disease.

notorious for causing skin and soft tissue infections with the increase in use of vascular catheters, prosthetic devices,[7,8] and disseminated disease among immunocompromised hosts. Ninety-percent of clinical disease is caused by *Mycobacterium fortuitum*, *M. chelonae*, and *M. abscessus*. RGM are the third most common cause of pulmonary disease caused by

NTM; 80% of the disease is caused by *M. abscessus* and 15% by *M. fortuitum*. The most important risk factors for pulmonary disease are aspiration of gastric contents and cystic fibrosis.[9] Other clinical presentations of RGM include endocarditis and ocular forms (keratitis). Most RGM are capable of producing a biofilm, and several species can grow at high temperatures (45°C). *M. abscessus*, *M. chelonae*, and *M. mucogenicum* have the ability to resist the activity of common disinfectants such as chlorine. RGM are generally not transmitted person to person but there are recent reports of some species of the *M. abscessus* group potentially been transmitted from person to person among patients with cystic fibrosis.[9] Among the SGM, the two most frequently isolated pathogens include species of the *Mycobacterium avium complex* (MAC) and *Mycobacterium kansasii*. Members of the MAC include *M. avium*, *M. intracellulare*, *M. chimaera*, *M. vulneris*, *M. marseilense*, and many other species.[5,6,9] Clinical syndromes associated with MAC include pulmonary disease among patients with underlying chronic pulmonary diseases are among postmenopausal women of Caucasian ancestry who normally suppress the cough reflex (Lady Windermere syndrome). Disseminated forms are often seen in patients with primary immunodeficiencies (i.e., GATA 2 deficiency) or among those with advanced HIV-associated immunosuppression. In addition, there are some isolated reports of cutaneous disease.[10–18] *M. kansasii* may cause diffuse pulmonary disease in patients with underlying chronic pulmonary diseases, patients with a history of smoking, lung carcinomas, and among those with HIV/AIDS. Interestingly, *M. kansasii* may cause cutaneous disease, sometimes in patients with concomitant pulmonary disease.

2. CENTRAL NERVOUS SYSTEM INFECTIONS DUE TO NTM

Of the >150 species of mycobacterial species within the *Mycobacterium* genus, the two most important species that may affect the nervous system are *Mycobacterium tuberculosis* complex (MTB) and *Mycobacterium leprae*. However, there are increasing reports of NTM identified as pathogens capable of causing meningoencephalitis and, sometimes, parenchymal disease in the form of a localized abscesses. *M. tuberculosis* is a major human pathogen that may infect every human organ or system including the central nervous system (CNS) with potentially devastating consequences. The most frequent clinical neurologic syndromes associated with MTD include meningoencephalitis, arachnoiditis and their associated complications including hydrocephalus, cerebrovascular accidents, and others. In turn, the Schwann cell of the peripheral nerve is the target cell of the *M. leprae* infection leading to important demyelination with severe peripheral nerve damage and dysfunction.

While CNS infections caused by NTM are rare (Table 1), there are an increasing number of reports in patients with AIDS.[5–23] In addition, other important risk factors for the acquisition of NTM include history of neurosurgery, trauma, otomastoiditis, disseminated mycobacterial infection, and intracranial implant or CSF shunts.[6] The use of an anti-TNR receptor blockade appears to be an increasing risk factor for the acquisition of NTM including *Mycobacterium avium-intracellulare* and *M. haemophilum*. NTM involvement of the CNS is often associated with high mortality.[5] From a pathogenetic perspective, it is important to consider that it is likely that *M. tuberculosis* enters the CNS by hematogenous dissemination, seeds the brain parenchyma and forms rich nodules, which may eventually open into the subarachnoid space to cause meningitis, meningoencephalitis, or arachnoiditis. This successful pathogen has the molecular ability to enter the brain, likely by crossing the blood-brain-barrier. It is tempting to suggest that the entry of NTM into the CNS is either through the hematogenous route or through direct inoculation (trauma, neurosurgical, and others). Gaining access into the CNS may be due to high-grade bacteremia associated with disseminated disease forms of NTM infections.

The most frequently reported clinical manifestation of CNS involvement with NTM is meningitis consisting of headache, fever, and neck stiffness.[5–23] A few patients also presented with altered mental status or focal neurologic signs. Some patients presented predominantly with parenchymal disease in the form of single or multiple brain abscesses, otomastoiditis with subdural empyema, and some with regional spread from infected devices such as epidural catheters. Of the cases reported in the medical literature, cerebrospinal fluid findings demonstrated hypoglycorrachia (mean $<$ 30 mg/dL), high protein (mean 180 mg/dL), and an elevated cell count (mean $\geq$500 cells/μL) with polymorphonuclear predominance in some reports and mononuclear in others. Examination of a cerebrospinal fluid smear with acid-fast staining is rarely positive, which often results in a delay in diagnosis. Occasionally the Gram stain may show Gram-positive rods, which may provide a clue for the possibility of *Listeria*, *Nocardia*, or mycobacterial infection. One of the important principles in managing CNS disease with NTM is identifying the source of infection. Magnetic Resonance Imaging of the brain or spinal cord is warranted; adequate drainage of abscesses and source removal including the removal of foreign material and catheters are considered crucial interventions for sterilization.

Similar to other neuroinfections, an important principle in the treatment of NTM CNS disease is the use of drugs with the ability to cross the blood-brain barrier and the blood-cerebrospinal fluid barrier to achieve high bactericidal concentrations.[5,6] In some cases, intraventricular or intrathecal aminoglycosides have been an addition to oral therapy with macrolides (i.e., clarithromycin or zithromycin) and other intravenous antibiotic choices depending on the organism and the susceptibilities (i.e., meropenem, imipenem linezolid, moxifloxacin, or others). Repeat lumbar punctures to prove eradication of infection may help guide therapy. Patients should probably be treated for no less than 1 year and therapy should be continued for several months after a negative CSF culture is obtained. The duration of therapy among cases reported in the literature has been variable. However, it is important to consider that other serious infections and clinical syndromes caused by NTM, such as lung disease or disseminated disease, are usually treated for a minimum of 1 year; therefore CNS forms may require more prolonged courses of antimicrobials.[5,6] In patients who are immunocompromised or who have a persistent focus of infection, the duration of therapy needs to be longer. The role of corticosteroids in patients with CNS infections with NTM has not been elucidated but there are increasing reports suggesting that the use of corticosteroids may be beneficial for those with severe meningoencephalitis.[24]

3. CONCLUSION

NTM can cause a diverse range of infections including infection of the CNS, most commonly in the form of meningitis or meningoencephalitis. Some patients develop a chronic meningitis syndrome requiring prolonged courses of antimicrobials. Many of these patients have disseminated mycobacterial disease with likely high-grade bacteremia as the crucial determinant in allowing the entrance of mycobacteria into the CNS. In addition to the increasing number of reports of NTM disease in patients with advanced AIDS, with the increasing use of invasive neurosurgical procedures, invasive catheters and devices, and increasing use of highly immunosuppressive drugs such anti-TNF-alpha receptor blockers, the rate of these infections is bound to rise. In patients with NTM-associated meningitis with substantial meningeal involvement, the use of steroids should be considered just as it is recommended in the case of tuberculous meningitis.

REFERENCES

1. Falkinham JO. Surrounded by mycobacteria: nontuberculous mycobacteria in the human environment. *J Appl Microbiol* 2009;**107**:356–67.
2. Brown-Elliott BA, Wallace Jr. RJ. Clinical and taxonomic status of pathogenic nonpigmented or late-pigmenting rapidly growing mycobacteria. *Clin Microbiol Rev* 2002;**15**(4):716–46.
3. Wu U, Holland SM. Host susceptibility to non-tuberculous mycobacterial infections. *Lancet Infect Dis* 2015;**15**:968–80.
4. Talati NJ, Rouphael N, Kuppalli K, Franco-Paredes C. Spectrum of CNS disease caused by rapidly growing mycobacteria. *Lancet Infect Dis* 2008;**8**:390–8.
5. Wang SH, Pancholi P. Mycobacterial skin and soft tissue infection. *Curr Infect Dis Rep* 2014;**16**:438.
6. Ryu YJ, Koh WJ, Daley CL. Diagnosis and treatment of nontuberculous mycobacterial lung disease: clinicians' perspectives. *Tuberc Respir Dis* 2016;**79**:74–84.
7. Midani S, Rathore MH. *Mycobacterium fortuitum* infection of ventriculoperitoneal shunt. *South Med J* 1999;**92**(7):705–7.
8. Chan KH, Mann KS, Seto WH. Infection of a shunt by *Mycobacterium fortuitum:* case report. *Neurosurgery* 1991;**29**(3):472–4.
9. Lee MR, Cheng A, Lee YC, Yang CY, Lai CC, Huang YT, et al. CNS infections caused by *Mycobacterium abscessus* complex: clinical features and antimicrobial susceptibilities of isolates. *J Antimicrob Chemother* 2012;**67**:222–5.
10. O'Dwyer JP, O'Connor JG, McDermott H, Sheehan S, Francis Fanning N, Corcoran GD. Meningoencephalitis associated with non-tuberculous mycobacteria. *BMJ Case Rep* 2009;**2009**. https://doi.org/10.1136/bcr.03.2009.1696.
11. Flor A, Capdevila JA, Martin N, Gavalda J, Pahissa A. Nontuberculous mycobacterial meningitis: report of two cases and review. *Clin Infect Dis* 1996;**23**(6):1266–73.

12. Carson LA, Bland LA, Cusick LB, et al. Prevalence of nontuberculous mycobacteria in water supplies of hemodialysis centers. *Appl Environ Microbiol* 1988;**54**(12):3122–5.
13. O'Brien RJ, Geiter LJ, Snider Jr. DE. The epidemiology of nontuberculous mycobacterial diseases in the United States. Results from a national survey. *Am Rev Respir Dis* 1987;**135**(5):1007–14.
14. Rentian C, Qi T, Lu H. Central nervous system infection with non-tuberculous mycobacteria: a report of that infection in two patients with AIDS. *Drug Discov Ther* 2014;**8**(6):276–9.
15. Sardar P, Bandyopadhyay D, Roy D, Guha P, Guha G, Banerjee AK. Nontuberculous mycobacteria and toxoplasma co-infection of the central nervous system in a patient with AIDS. *Braz J Infect Dis* 2009;**13**(6):449–51.
16. Jacob CN, Henein SS, Heurich AE, Kamholz S. Nontuberculous mycobacterial infection of the central nervous system in patients with AIDS. *South Med J* 1993;**86**(6):638–40.
17. Smith MB, Boyars MC, Woods GL. Fatal *Mycobacterium fortuitum* meningitis in a patient with AIDS. *Clin Infect Dis* 1996;**23**(6):1327–8.
18. Buppajarmtham A, Apisarnthanarak A, Rutjanawech S, Khawcharoenporn T. Central nervous system infection due to *Mycobacterium haemophilum* in a patient with acquired immunodeficiency syndrome. *Int J STD AIDS* 2015;**26**(4):288–90.
19. Marie I, Heron F, Lecomte F, et al. Multiple cerebral abscesses as a complication of *Mycobacterium fortuitum* infection. *Eur J Intern Med* 2003;**14**(6):386–9.
20. Liebeskind DS, Ostrzega N, Wasterlain CG, Buttner EA. Neurologic manifestations of disseminated infection with *Mycobacterium abscessus*. *Neurology* 2001;**56**(6):810–3.
21. Madaras-Kelly KJ, DeMasters TA, Stevens DL. *Mycobacterium fortuitum* meningitis associated with an epidural catheter: case report and a review of the literature. *Pharmacotherapy* 1999;**19**(5):661–6.
22. Santamaria-Jauregui J, Sanz-Hospital J, Berenguer J, Munoz D, Gomez-Mampaso E, Bouza E. Meningitis caused by *Mycobacterium fortuitum*. *Am Rev Respir Dis* 1984;**130**(1):136–7.
23. Maniu CV, Hellinger WC, Chu SY, Palmer R, Alvarez-Elcoro S. Failure of treatment for chronic *Mycobacterium abscessus* meningitis despite adequate clarithromycin levels in cerebrospinal fluid. *Clin Infect Dis* 2001;**33**(5):745–8.
24. Thwaites GE, Nguyen DB, Nguyen HD, et al. Dexamethasone for the treatment of tuberculous meningitis in adolescents and adults. *N Engl J Med* 2004;**351**(17):1741–51.

FURTHER READING

1. Adjemian J, Olivier KN, Seitz AE, Holland SM, Prevots DR. Prevalence of nontuberculous mycobacterial lung disease in U.S. Medicare beneficiaries. *Am J Respir Crit Care Med* 2012;**185**(8):881–6.

CHAPTER 9

Central Nervous System Tuberculosis

Saeed Soleiman-Meigooni
AJA University of Medical Sciences, Tehran, Iran

1. INTRODUCTION

The tuberculosis (TB) pandemic status is larger than it was previously thought. In 2015, the World Health Organization (WHO) reported >10.4 million new cases of tuberculosis worldwide, which consisted of 580,000 multidrug-resistant (MDR) patients, 400,000 patients with HIV coinfections, and 1.4 million deaths related to tuberculosis. During 2000–15, 49 million people died from the disease. Although there was a 22% reduction in the mortality rate during previous years, it nevertheless remains one of the 10 most fatal diseases worldwide. Now, the WHO is targeting an end strategy with a 35% decrease in the mortality rate and a 20% decrease in new patients by 2020.[1]

2. EPIDEMIOLOGY

Few epidemiologic studies on central nervous system tuberculosis (CNS-TB) have been performed in recent years. In a 30-year cohort study from 1971 to 2000 in Canada, 824 out of about 83,000 patients with tuberculosis were diagnosed with CNS-TB. The risk factors of CNS-TB in this study consisted of the female gender, lower ages, and birth outside Canada. CNS-TB resulted in mortality rates that were more than four times those compared to non-CNS-TB.[2] In another cohort study from 1993 to 2001 in California, the risk factors of CNS involvement consisted of lower ages (under 5 years), birth in the United States, and being Hispanic. A CSF positive culture was associated with increased risk of death. The drug resistance and mortality rate were higher in CNS-TB patients, but closed contact, HIV coinfection, and positive skin tests (PPD) were more prevalent in non-CNS-TB.[3] Unlike the California study, one cohort study in China from 1995 to 2004 showed that about one-third of patients with CNS-TB also had HIV coinfection. The prevalence of CNS involvement in this study was 2.5% and the mortality rate was 17%. Older age, hydrocephalus, and positive CSF cultures were associated with higher mortality rates.[4] Finally, CNS-TB risk factors are a lower age, HIV coinfection, malnutrition, measles, alcoholism, malignancy, immunosuppressive medication, and living in high endemic regions.[2,3,5]

The Microbiology of Central Nervous System Infections
https://doi.org/10.1016/B978-0-12-813806-9.00009-3

3. MICROBIOLOGY

Mycobacterium SPP is an aerobic, rod-shaped, spore-free bacterium with natural immunity to alcohol and acid (acid-fast). The doubling times for these SPP are longer than many bacteria. More than 200 species of these microorganisms have been identified, and many of them are saprophyte. Humans and guinea pigs are naturally sensitive to Mycobacterium SPP. The Ziehl-Neelsen technique is widely utilized for identification of acid-fast organisms. Three formulations are used in the culture of mycobacteria:

(1) Semisynthetic agar media (Example: middle brook 7H10, 7H11), which is utilized with a larger quantity of the sample that grows after several weeks.

(2) Inspissated egg media (Example: Lowenstein Jensen), which is utilized by a small inoculation that grows after 3–6 weeks.

(3) Broth media (Example: middle brook 7H9, 7H12), which acts by small inoculation and faster growth compared with the two media mentioned above and is widely utilized for diagnosis.

The main entry portal of the organism is inhalation (for *M. tuberculosis*) or digestion (for *M. bovis*). The Mycobacterium cell wall consists of many factors that motivate immune responses. These parts are labeled.

(1) Lipids: the Mycobacterium cell wall is reached by lipids consisting of mycolic acid, VAX, and phosphates. A complex between mycolic acid and muramyl dipeptide produces granuloma.

(2) Proteins: there are many declared proteins in Mycobacterium, which result in tuberculin reaction and antibody secretion.

(3) Polysaccharides: they can produce immediate hypersensitivity, although their role in pathogenesis is not well recognized.[6]

4. PATHOGENESIS AND PATHOLOGY

Mycobacterium SPP consists of very small particles <25 μm that propagate by coughing, sneezing, and talking; they are deposited in microalveoli. An immune system reaction consists of cytokine releases, mononuclear inflammatory cell aggregation, and engorgement. This process may be dropped by killing the microorganism by macrophages, or continued, which results in two clinical forms of the pathologic lesions in the alveoli, 1–2 months after inhalation:

(1) The exudative type, consisting of edema and mycobacteria surrounded by acute inflammatory cells. Clinical presentation of this lesion resembles bacterial pneumonitis. This lesion may be resolved or produce a necrotic or productive lesion.

(2) Productive (proliferative) lesion: chronic inflammation occurs in this lesion, consisting of three zones: a central zone with multinuclear giant cells engorging mycobacteria, a middle zone consisting of pale epithelioid cells, and a peripheral zone consisting of fibroblasts, lymphocytes, and monocytes.

After primary infection, the organism may pervade via the lymphatic route, blood circulation, the gastrointestinal (GI) tract, and the bronchus. The most common process occurring after primary infection is a direct extension to the adjacent lymph node and remote metastasis by blood circulation.[7] Some genotypes of *Mycobacterium tuberculosis* such as RV0311, RV0805, RV0931c, RV0986, and MT3280 have more CNS affinity compared to others. Whatever primary infection and bacteremia are severe, CNS metastasis is more dominant. The blood-brain barrier (BBB) provides an intense protective wall for CNS against microorganisms. *M. tuberculosis* may transmit across the BBB by infected monocytes or neutrophils and deposit in any place of the CNS, called Rich foci. Months to years after primary infection, these focuses may enlarge and rupture, which produces a variety of clinical presentations.[8]

In most patients with CNS-TB, leptomeningeal involvement occurs after primary bacteremia. CNS-TB may present as meningitis, hard or soft granuloma, cerebritis, arachnoiditis, and spinal involvement. Cerebral parenchyma involvement occurs as tuberculoma with or without meningitis. Tuberculomas are round, oval-shaped or lobular lesions distributed in potentially any site of the CNS; tubercular granuloma are small round lesions with a size of 0.5–2 mm consisting of epithelioid cells containing *M. tuberculosis* bacilli and a peripheral collar with mononuclear inflammatory cells. Characteristic Langhans' giant cells may be found in granuloma. In a study of 32 children with TBM with a mean age of 4.6 years, 88% of the patients had a miliary pattern. All the latter patients had leptomeningeal nodules; 18% had deep parenchymal nodules and 39% had concomitant tuberculoma. These findings suggested that miliary tuberculosis was a part of the TBM pathogenesis.[8] In tuberculous meningitis, thick gelatinous exudates develop around the Sylvian fissure, basilar cisterns, the brainstem, and the cerebellum. Hydrocephalus may occur after an obstruction in CSF flow and is related to the prognosis. Edema, perivascular infiltration, and microglial reaction developed in the cerebral parenchyma next to the tuberculous exudates. Basilar exudate appears more severe around Willis's circle and may result in vascular changes, vasculitis-like syndromes, and thrombotic changes. The most commonly involved vessels are the internal carotid artery (ICA), the middle cerebral artery (MCA), and perforating vessels. A cerebral infarct occurs predominantly around Sylvian fissures. In an autopsy study on 31 patients with tuberculous meningitis, the most common pathologic changes were basilar exudates, hydrocephalus, caseating granuloma, ventriculitis, infarct, edema, and abscess formation, respectively.[9,10]

5. CLINICAL PRESENTATION

Some known presentations of CNS-TB have been defined in the literature, which consists of meningitis, brain tuberculoma, tuberculous brain abscess, and spinal involvement. The most common clinical presentation in almost all these papers is meningitis.

5.1 Tuberculous Meningitis

It occurs through leptomeningeal involvement by *M. tuberculosis*. Mycobacterium produces a thick exudate that deposits, mostly on the basilar meaning. Inflammatory cytokines have a crucial role in the pathogenesis of TBM. After the entrance of Mycobacterium into the subarachnoid space, local T cells induce an inflammatory reaction that leads to caseous granuloma formation. TNF-α has a crucial role in granuloma formation and tissue necrosis. In animal models, a high concentration of TNF-α in CSF correlated with a poor prognosis. Other cytokines that increased in TBM are IL-8, IFN-γ, and IFN-α. There are also increasing levels of IL-10, matrix metalloproteinases, and the tissue inhibitor of matrix metalloproteinases, which lead to the destruction of the BBB, high proteins and lactates, and low glucose concentration of the CSF. Nitric oxide (NO) is another cytokine that increases in the CSF of patients with TBM and induces vascular and perivascular inflammation; it may have a role in tuberculous encephalopathy.[8]

Most patients with TBM experienced a recent history of nonspecific prodromal symptoms such as loss of appetite, fever, malaise, fatigue, and headache lasting 2–8 weeks. After this period, specific CNS symptoms present. In adults, meningismus symptoms such as neck stiffness, photophobia, headache, vomiting, and lethargy usually develop at the time of presentation. In children, seizure, fever, abdominal pain, and neck stiffness are common presenting symptoms, but a headache is not common. Very young children may present with fever, poor feeding, lethargy, and bulging of the fontanelle. Less prevalent symptoms of TBM are visual disturbances, choroid tubercle (pathognomonic for miliary tuberculosis), focal neurologic deficits, seizures, apathy, lethargy, and coma. Many of the later symptoms develop in the advanced course of the disease. About 10% of the patients have a history of tuberculosis and one-third to half of the patients have concomitant pulmonary tuberculosis. TBM is commonly attributed to miliary tuberculosis, and it's suggested that miliary tuberculosis has a critical role in the pathogenesis of TBM.[11]

The worldwide accepted three stages for TBM are:

(1) Stage I: alert with no focal neurologic deficit.

(2) Stage II: lethargic or Glasgow's coma scale (GCS) between 11 and 14 and focal neurological deficit.

(3) Stage III: GCS ≤ 10 with or without a focal neurological deficit.

Cranial nerve paresis occurs in 20%–30% of patients, and the sixth nerve paresis is the most affected nerve involved.[5,12,13] In a report of two patients with TBM, acute polyradiculo neuritis mimicking Guillian Barre syndrome, which is unresponsive to IVIG and corticosteroids, occurs in the initial course of the disease. These two patients were cured with long-term antituberculosis treatment (ATT).[14] (Figs. 1 and 2).

Some studies report the frequency of the symptoms in TBM, which is listed in Table 1.

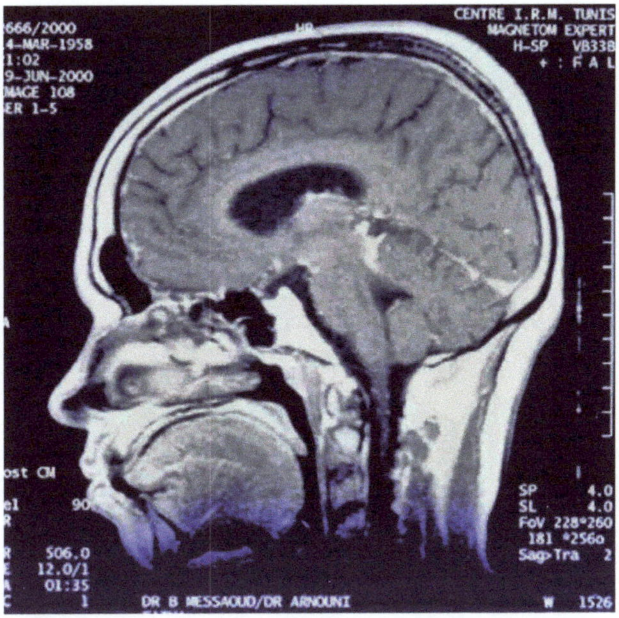

Fig. 1 Peripheral and basilar meningeal enhancement with hydrocephaly.[32]

5.2 Tuberculous Encephalopathy

Tuberculous encephalopathy (TE) is an uncommon presentation of CNS-TB that occurs at all ages, especially young children. Generalized brain edema and demyelination are characteristic findings of TE. Microscopic findings consist of microvascular necrosis, perivascular reaction by macrophages, demyelination, focal glial nodule, and loss of neurons in the gray matter. Hemorrhagic lesions rarely occur. The patients may present by seizure, meningismus, loss of consciousness, and disseminated intravascular coagulation.[5,13]

5.3 Tuberculoma

Tuberculoma is a well-defined nonvessel round granulomatous firm lesion larger than 2 cm. Most affected patients are children and young adults in developing countries. They usually present concomitant with meningitis or that may be an accidental finding upon imaging. In a large cross-sectional study consisting of 404 patients with TBM, about half the patients had concomitant brain tuberculoma, which consisted of supratentorial tuberculoma in 82.7%, infratentorial in 7.1%, and 10.2% in both sites.[15]

In another study in China on 23 patients with a mean age of 30 with CNS tuberculoma, common presentations were sensory-motor involvement (such as paresthesia), weakness, and paraparesis in 82.6%, urinary incontinence in 65.2%, and constitutional symptoms such as fever and headache in 34.8%. The most common sites of tuberculomas

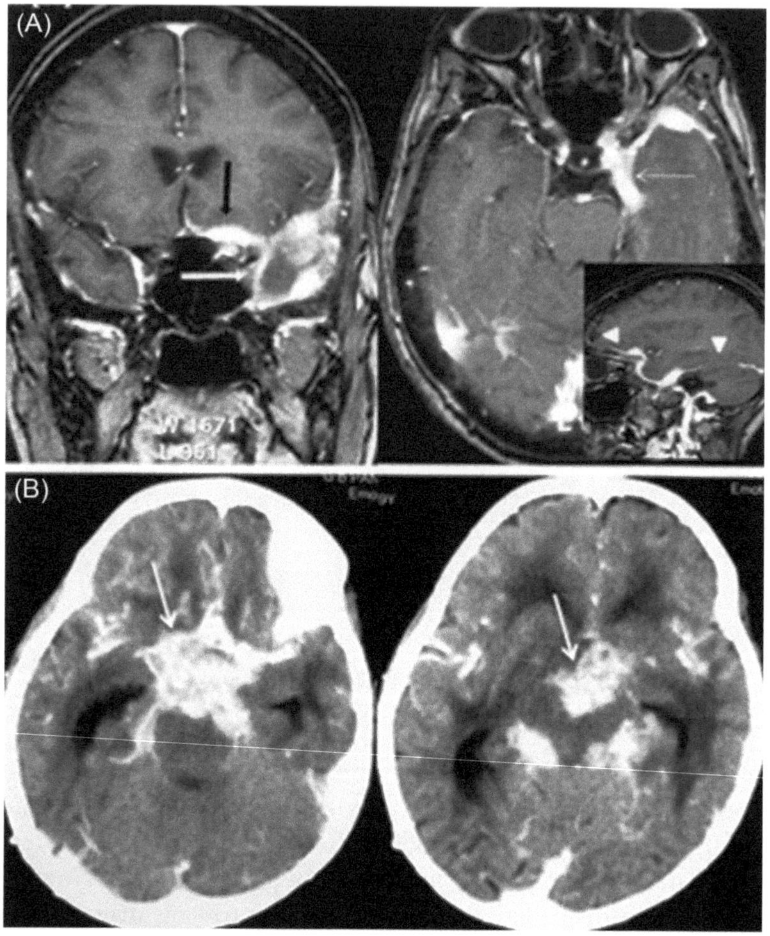

Fig. 2 Patchy meningitis and granulomatous basal meningitis. (A) Pachymeningitis. Contrast-enhanced T1-weighted coronal and sagittal images of brain show intensely enhancing, thickened pachymeninges (*dura mater*) encircling the left temporal lobe (*thick white arrow*), floor of ipsilateral anterior cranial fossa (*thick black arrow*), and the left cavernoussinus (*thin white arrow*). (B) Granulomatousbasal meningitis. Axial contrast-enhanced computed tomography images at pons and midbrain level show irregular lumpy masses (*arrows*) superimposed on dense basal enhancement. The patient was on treatment fortuberculous meningitis, and later presented with progressive loss of vision in both eyes.[11]

were spinal thoracic (43.5%) and intracranial and cervical spinal, each with 26.1%, cervico-thoracic at 13%, and thoraco-lumbar at 4.3%.[18]

There are some different clinical presentations with CNS tuberculoma related to its anatomical position. A significant proportion of the patients have no clinical symptoms until mass effect and elevated intracranial pressure results in some CNS symptoms such as headache, vomiting, seizure, focal neurological deficit, and papilledema. These symptoms are referred to as supratentorial tuberculoma. Infratentorial tuberculoma is more

Table 1 Comparison of TBM pediatrics patients in four studies

Study	Philippines[12]	Pakistan[15]	Turkey[16]	United Kingdom[17]
Study design				
Study period	1989–98	2002–11	1988–2000	1977–97
Number of patients	405	404	61	38
Mean age (years)	3.8	43	34.5	3.7
M/F ratio	1.3/1	1.1/1	1.8/1	NA
Clinical symptoms (%)				
Fever	88	78.5	NA	66
Headache	28	57	NA	37
Meningismus	11	35	NA	47
Alteration in consciousness	42	60	56	79
Seizure	62	22	16	53
Focal neurological deficit	14	29	56	66
Nausea and vomiting	57	53	NA	55
Mood changes	16	NA	39	16
Concomitant pulmonary tuberculosis	55	32	54	39.5
Stage I	4	33	0	NA
Stage II	39	40	44	NA
Stage III	57	27	56	NA
Complication (%)				
Mortality	16	18	28	13
Severe neurological deficit	35	21	31	47
Surgical intervention	36	47	NA	76
CSF positive smear or culture	0	13	20	51.5
Hydrocephaly	90	29	23	94
Basilar enhancement	100	32	NA	93
Infarct	10	25	21	33
Ventriculitis	19	0.2	5	NA

NA, not available.

common in children and presents with brainstem symptoms, cerebellum dysfunction, and cranial nerve paresis.

A CT scan with IV contrast is a valuable diagnostic procedure that shows round or lobular ring-enhanced lesions with an irregular wall and surrounding edema. The target sign is a tuberculoma pathognomonic finding in CT that consists of central calcification and a surrounding ring-enhanced area after intravenous contrast injection. The other differential diagnosis of brain tuberculoma consists of cysticercosis, pyogenic abscess, and neoplastic lesions such as lymphoma. Unlike these differential diagnoses, a tuberculoma lesion is >2 cm in size and has an irregular wall and surrounding edema. Brain MRI findings of tuberculoma are hypo intense or iso intense in tuberculomas with caseous necrosis.[13] (Figs. 3–5).

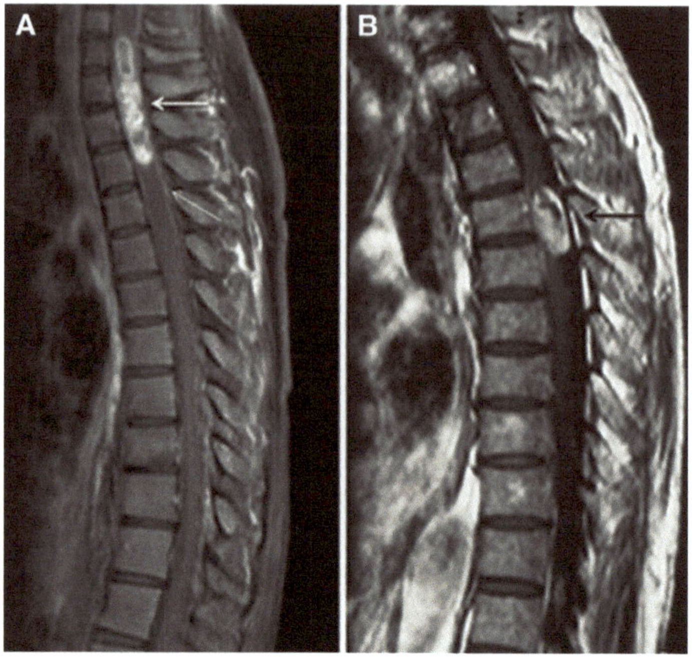

Fig. 3 Intramedulary and extramedulary tuberculoma.[11]

5.4 Tuberculous Brain Abscess

A tuberculous brain abscess is an uncommon presentation of CNS-TB, even in the endemic region. The frequency of this presentation is about 4% in non-HIV patients, and may reach up to 20% in HIV-infected patients. It is suggested that necrosis may develop as a result of an inflammatory response and may appear as a single or multiple uniloculate or multiloculated lesion with a thin surrounding wall on imaging.[19] A tuberculous brain abscess is a solitary multiloculated mass larger than tuberculoma with acute progressive CNS symptoms such as partial seizures, focal neurological deficits, hemiparesis, loss of consciousness, and signs of elevated intracranial pressure.[13] Other symptoms such as chronic headaches, fatigue, chills, and weight loss have also been reported.[20] A 5-year study in Mexico showed that the frequency of the tuberculous brain abscess was six out of 149 patients with CNS-TB (4%). The mean age of the patients was 42 years, and common complaints were hemiparesis, meningismus, and intracranial hypertension. One study compared 15 cases of tuberculous brain abscess. The mean age of the patients was 25 years. Eight of 15 patients had HIV coinfection and 10 of 15 patients had previously had tuberculosis. Common symptoms were seizures, headaches, intracranial hypertension, hemiparesis, and loss of consciousness. All patients received antituberculosis drugs and seven of 15 cases underwent surgical drainage. Surgery and drainage have diagnostic and therapeutic value in the management of these patients.[19] (Fig. 6).

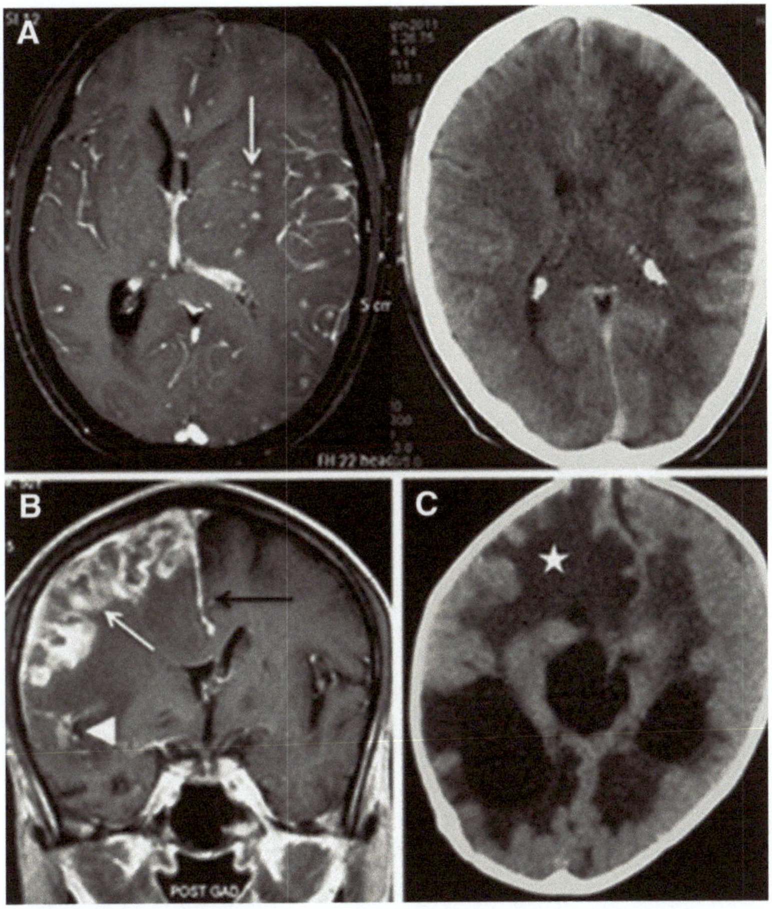

Fig. 4 Miliary tuberculosis, tuberculoma, and tuberculous encephalopathy.[11]

5.5 Spinal Tuberculosis

Less than 1% of the patients with tuberculosis developed spinal tuberculosis, which is one of the paraplegia in the developing world. There has been some known presentation of spinal tuberculosis:

(1) *Tuberculous spondylitis (Pott's disease)*: the vertebral body adjacent to the intervertebral disk has been affected, and neural arcs will be intact. It may cause vertebral body destruction and anterior wedge. The common affected vertebra are the lower thoracic and upper lumbar spine. Usually, some adjacent vertebra are affected together. It may extend to paraspinal spaces or result in paravertebral or epidural tuberculous abscess and cord compression syndrome.

(2) *Spinal tuberculoma*: common sites of spinal tuberculoma are extradural and intramedullary. Cord edema, syrinx, or arachnoiditis may be present.

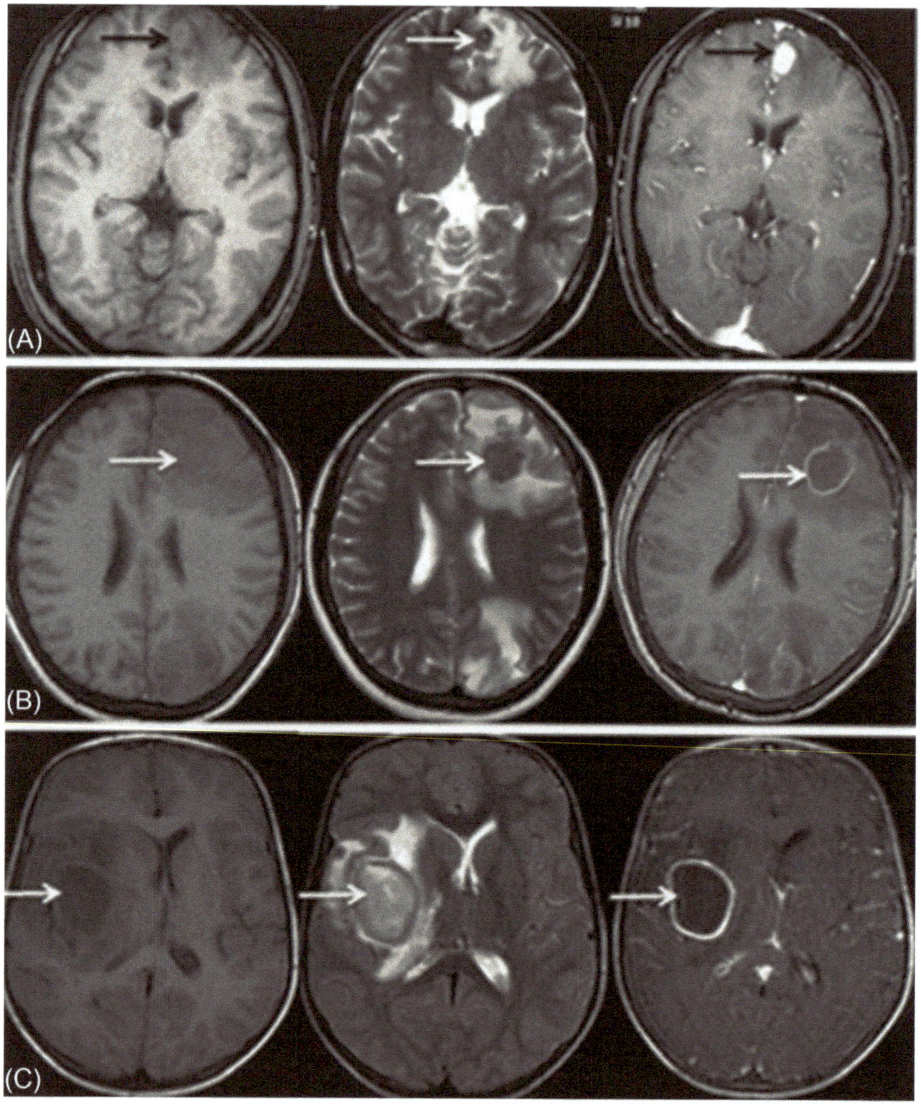

Fig. 5 Staging of tuberculoma, noncaseating, caseating, and liquificated (abscess formation).[11]

(3) *Tuberculous myelitis*: it involves a common cervicomedullary segment of cord. Paraparesis is a common symptom, but diffuse cord swelling and enhancement in MRI may last several months. It's on a differential diagnosis of cord infarction, acute transverse myelitis, and demyelination disease.[11,13,21] (Figs. 7–9).

6. COMPLICATIONS

6.1 Hydrocephaly

Hydrocephaly is the most prominent complication of CNS-TB, occurring in up to 80% of children and 40% of adults with TBM. The most common presentation symptoms are

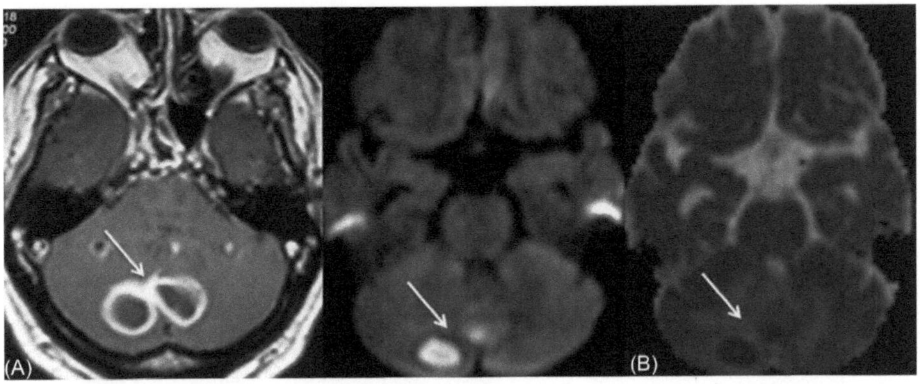

Fig. 6 Tuberculous abscess and perilesional edema.[11]

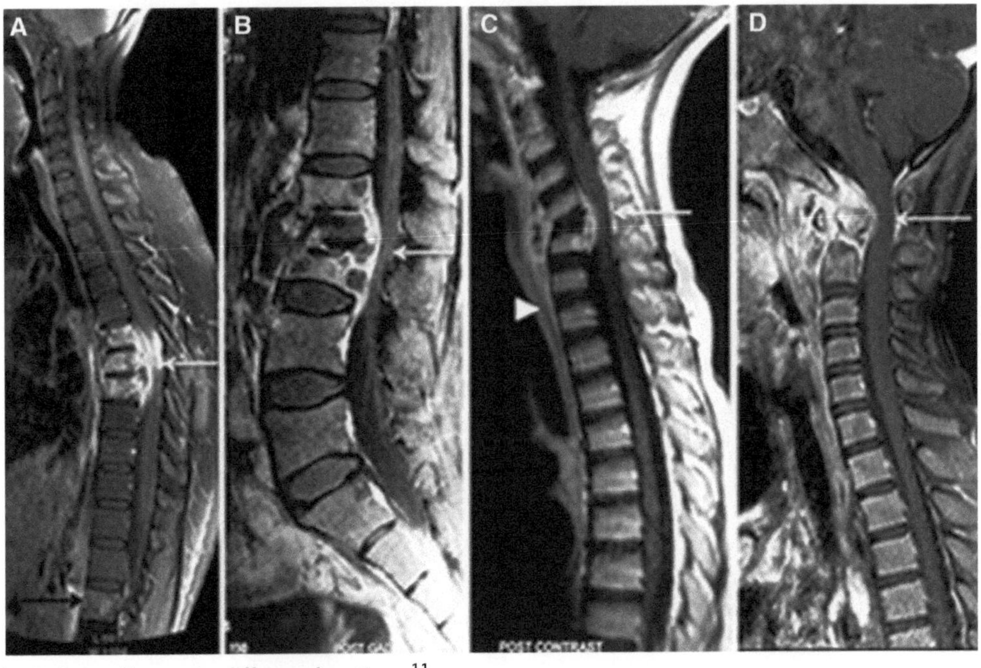

Fig. 7 Potts disease in different locations.[11]

loss of consciousness and rising intracranial pressure. A large proportion of the patients are managed with antituberculosis treatment (ATT) and adjunctive corticosteroids. However, there may be a need for surgical intervention and insertion of ventricular-peritoneal shunting. In a report from South Africa, it was found that the triad of basilar enhancement, hydrocephalous, and infarct has 41% sensitivity and 100% specificity on the diagnosis of TBM in pediatrics.[22] (Fig. 10).

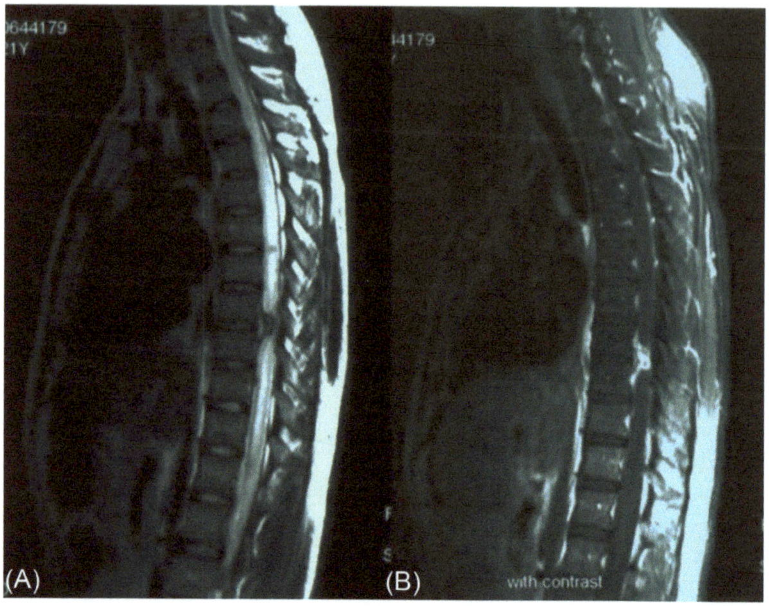

Fig. 8 Spinal thoracic hyperintense signal compatible with tuberculoma.[18]

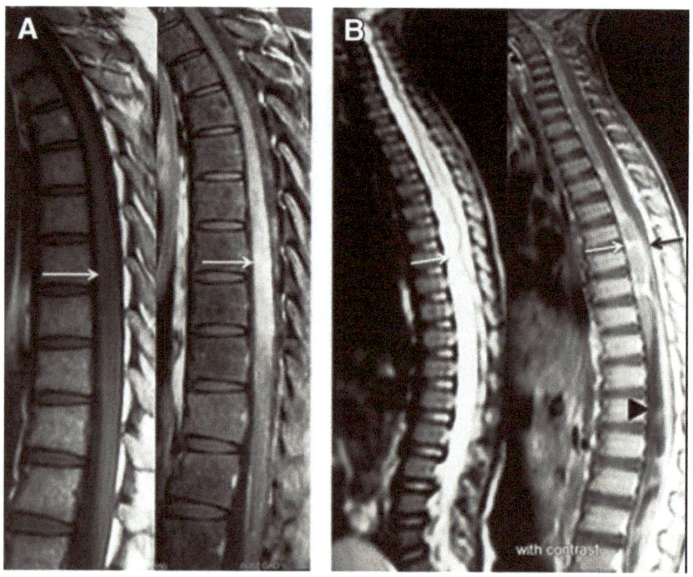

Fig. 9 TB myelitis.[11]

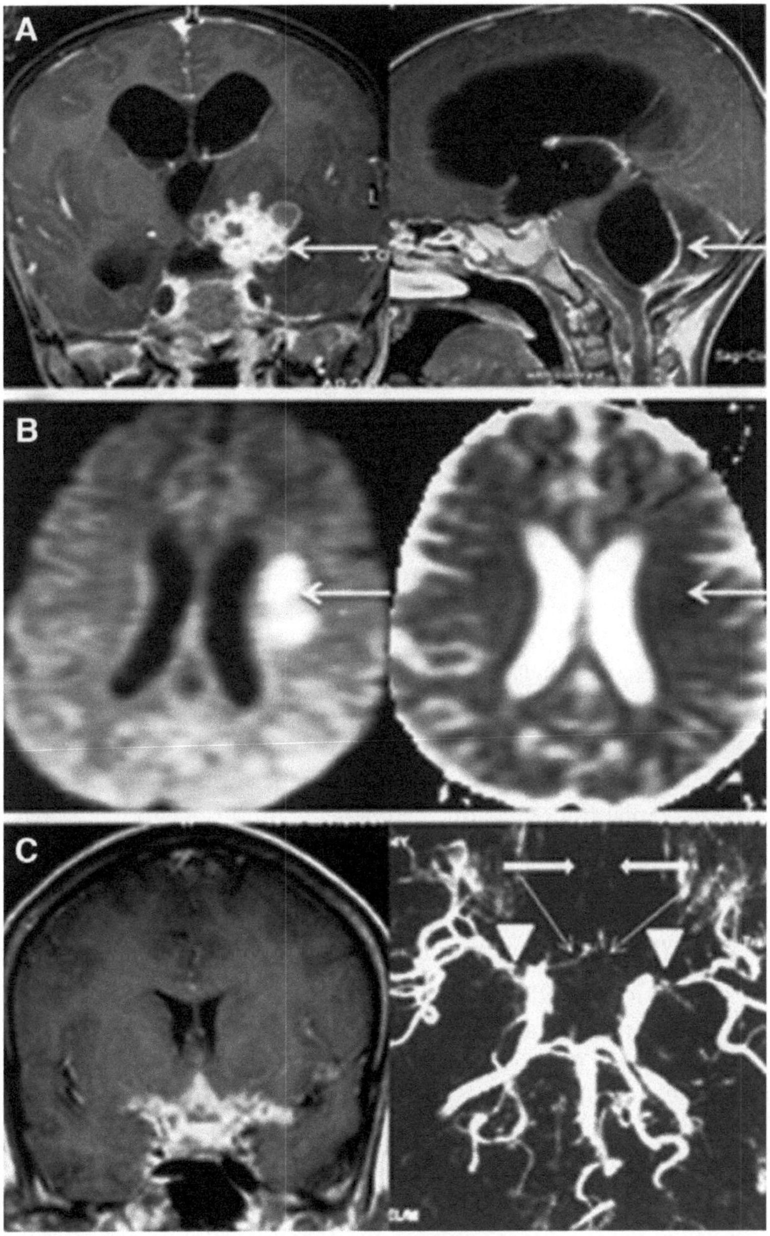

Fig. 10 Complicated TBM with hydrocephalus and infarction.[11]

6.2 Infarct

About 20% of the patients with TBM developed a focal neurological deficit and up to 80% of the patients who die from TBM show vascular involvement in the autopsy. Tuberculosis is one of the causes of strokes in younger adults. Whatever the duration of the disease, vascular involvement becomes more frequent. Overall, the most common affected vessels are related to the middle cerebral artery (MCA) and predominantly occur in the TB-zone, which is fed by thalamus-perforating and lenticulo-triat vessels. A basal ganglia infarction is affected less commonly and is characteristic for TBM.

Pathologic changes in infarction may be seen in three different forms:

(1) *Infiltrative changes*: also called inflammatory or vasculitis-type results after the invasion of microorganisms in the vessel wall and immunological reaction and results vessel inflammation and obstruction. Basilar cerebral vessels and civilian fissures are predominantly affected, with the most tuberculous exudate deposits in this zone.

(2) *Proliferative changes*: It is a common vascular change after TBM, also in the patients who complete treatment. In this type of vascular involvement, vessel medial atrophy and adventitial fibrosis occur.

(3) *Necrotizing changes*: necrotizing or hyaline necrosis usually occur after vascular inflammation, predominantly in arteries and arterioles, not veins.

In many cases, all three vascular involvements occur concomitantly.

Clinical symptoms/signs of infarction are variable and related to the involved vessels. The most common symptom is motor involvement. Other symptoms such as loss of consciousness, seizure, cranial nerve paresis, hypothalamic dysfunction, subarachnoid hemorrhage, intracranial hypertension, papilledema, and syringomyelia have also been reported.

Unlike previous reports, thrombosis is less important in a stroke; the most common cause of thrombotic strokes in tuberculosis is obliterating endarteritis secondary to prolonged disease. Antitubercular drugs have no effect on vasculopathy. The antiinflammatory effects of corticosteroids in the prevention or treatment of vasculitis are controversial. Streptomycin has the most thrombotic drugs among ATT, but it's not recommended to exclude this drug for it's complications.[23] (Fig. 11).

6.3 Tuberculous Ventriculitis

One of the rare and devastating complications of CNS-TB is ventriculitis. Ventriculitis or ependymitis is called ventricular inflammation, which usually occurs after neurosurgical trauma or surgery; the most isolated organisms of this complication are *Staphylococcus* spp. Other pathogens infrequently cause this complication. Very rare reports found throughout the literature have shown that *M. tuberculosis* may be an etiologic of ventriculitis.

Few reports of tuberculous ventriculitis are found throughout the literature. For example, a 3-month child was presented with respiratory distress, seizure, and fever.

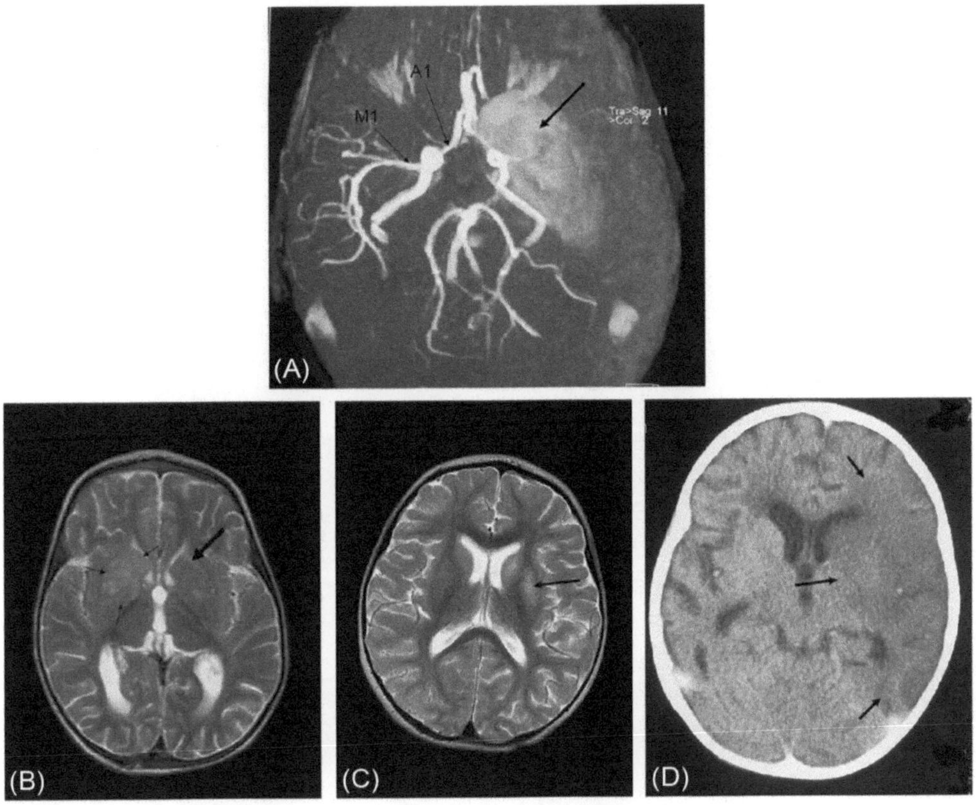

Fig. 11 Narrowing A1 segment of MCA and vasculitis and infarct.[23]

This was diagnosed as ventriculitis, and the child died despite treatment.[24] Another report is about a 74-year-old female who presented with memory loss, vomiting, ataxia, and incontinence. In the brain MRI, ventriculitis was diagnosed. She was treated with ATT and corticosteroid without any sequel.[25]

In a 6-year study on 234 patients with TBM, five patients developed ventriculitis (2.1%). The mean age of the patients was 29 and common symptoms were papilledema, seizure, motor involvement, and visual loss. The most radiological findings were hydrocephalus, ependymal contrast enhancement, and ventricular sludge.[26]

Tuberculous ventriculitis usually occurs after enlargement, extension, and rupture of a tuberculoma in the ventricular system. Clinical findings usually are rigorous, consisting of photophobia, diplopia, deterioration of consciousness, seizure, and papilledema. The most common imaging findings are hydrocephalus, ventricular sludge, and ventricular enhancement. Many authors agree on the use of high doses of steroid concomitant with long-term antituberculosis drugs for the treatment of these patients.[27] (Fig. 12).

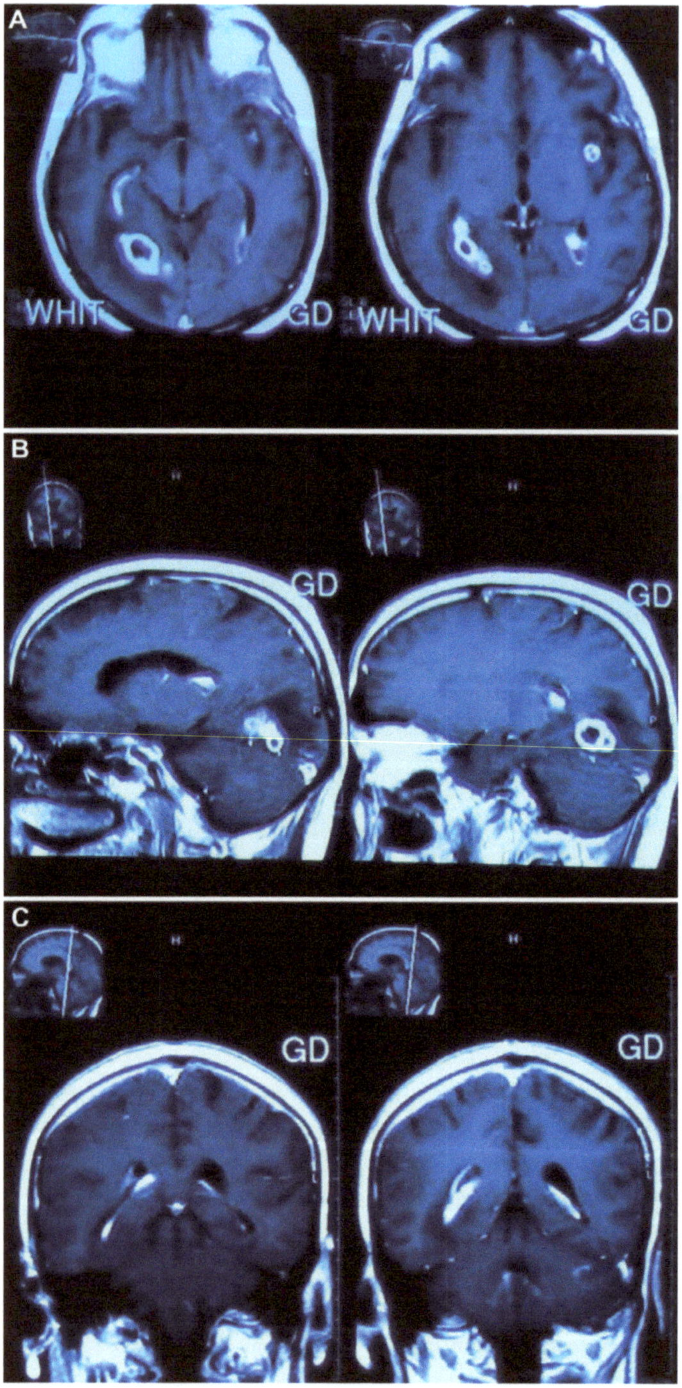

Fig. 12 Tuberculous ventriculitis with ventricular enhancement and sludge.[27]

Interventricular tuberculoma is a form of tuberculoma and has a different presentation from ventriculitis. In a report of interventricular tuberculoma, a 26-year-old male presented with headache, vomiting, and visual loss without systemic or other neurologic signs in 4 months. He was managed by surgical excision of the lesion and 1 year of antituberculosis drugs.[28]

6.4 Tuberculosis Immune Reconstitution Inflammatory Syndrome

Tuberculosis immune reconstitution inflammatory syndrome (TB-IRIS) is a common cause of deterioration in HIV-infected patients who are on ATT and receiving antiretroviral treatment (ART). The prevalence of this phenomenon is about 8%–43% of the patients with HIV/TB coinfection after starting ART. Neurological TB-IRIS is a proportion of these patients, and the most common cause of deterioration of HIV/TB coinfected patients in the first year of ART. In a study of 34 HIV-infected patients with TBM, 16 patients (47%) developed TBM-IRIS after 2 weeks of the initiation of ART. Symptoms consisted of headaches, vomiting, neck stiffness, night sweats, and weight loss. In the laboratory, CSF findings of these patients showed a greater neutrophils count, positive CSF TB-PCR, and also high CSF TNF-α and low IFN-γ concentration. Two patients (6%) died of TB-IRIS complications.[29] In another study on four HIV/TB coinfected children, three of them developed symptoms of TB-IRIS about 10 days after ART, and one patient died related to TB-IRIS.[30]

7. DIAGNOSIS

A physician's judgment based on clinical symptoms plays a crucial role in the diagnosis of CNS-TB. Clinical symptoms of TBM during the initial period of the disease may be seductive. Most patients experience nonspecific symptoms such as fever, headache, vomiting, and loss of appetite. Younger children may present as failure to thrive (FTT) while suffering from weight loss, agitation, sleep disturbance, seizure, focal neurological deficits, abdominal pain, and vomiting. Between 50% and 90% of the children with TBM have had recent close contact. Brain tuberculomas usually are asymptomatic, especially when meningitis is not present or in the initial course of the disease. Some of the clinical symptoms on presentation are headache, fever, weight loss, seizure, and focal neurological deficits. Uncommon symptoms are chorea and brainstem syndromes.

Spinal tuberculosis may present with a variety of clinical symptoms. In the Pott's disease, which is the most common syndrome in spinal tuberculosis, the vertebral body will be the main affected part. Common symptoms are pain, gibbous, and cord pressure findings. Extra dural cord tuberculoma causes nonosseous paraparesis. Tuberculous radiculomyelitis causes subacute paraparesis, radicular pain, and bladder incontinence.[22]

7.1 CSF Analysis

CSF analysis provides valuable clues for the diagnosis of TBM. About two-thirds of patients have lymph-dominant pleocytosis. A high CSF protein concentration above 100 mg/dL and CSF-to-serum glucose lower than 60% are common findings in the majority of the patients.[5] In a study of 306 patients with TBM, CSF analysis showed lymphocytic pleocytosis between 15 and 2500 cells per cubic milliliter, high CSF protein concentration in 87.6%, and a low glucose level in 72.6%. CSF findings of the patients with tuberculoma are high-protein concentrations and pleocytosis (<100 cells per cc) found in more than half the patients.[14,22]

7.2 Bacteriology

At least 10,000 organisms must be present in 1 cc of sputum for a positive smear result.[22] More than 80% of adults and 20% of children with TBM have a positive CSF smear or culture for tuberculosis. However, in brain tuberculoma, the CSF bacteriologic assay has limited value because of its low sensitivity. The most sensitive method in the diagnosis of CNS-tuberculoma is a stereotactic biopsy.[18]

7.3 Tuberculin Skin Test

A tuberculin skin test (TST) will be positive in 10%–50% of patients with CNS-TB. However, it may be more positive after initiation of the treatment. This test may be influenced by age, BCG vaccination, malnutrition, HIV coinfection, and technical error.[5] Thus it has limited value in the diagnosis of CNS-TB.

7.4 Adenosine de Aminase

CSF Adenosine de Aminase has high sensitivity in the diagnosis of CNS-TB, but low specificity due to some diseases such as malaria, lymphoma, brucellosis, and bacterial meningitis, which result in false positives. These limit its use in the diagnosis of CNS-TB. Although some studies excluded the role of the ADA in diagnosing TBM, other reports say that this assay has sensitivity between 44% and 100% and specificity of 100% in diagnosing TBM.

7.5 Immunologic Assay

CSF antibody detection is one of the rapid tests in the diagnosis of TBM, but it cannot diagnosis acute infection from an old infection. The sensitivity and specificity of this method by ELISA are also relatively low, as much as 52%–93% and 58%–100%, respectively. New methods focused on detecting antibodies against the 30 kDa protein of mycobacterium have a sensitivity of 92%. The enzyme-linked immuno spot (ELISPOT) assay on CSF also has high sensitivity and specificity, about 84% and 92%, respectively. Latex agglutination, radio immune assay, and dot-blot assay in CSF are also valuable, but accessibility in poorer countries is the greatest limitation.

7.6 Immune Globulin Release Assay

In comparison to TST, the immune globulin release assay (IGRA) has more diagnostic value for latent tuberculosis. Both tests have limited use in active tuberculosis. One study showed at least 50% of the patients with TBM have negative IGRA.

7.7 Molecular Assay

Based on a recent meta-analysis, commercially available nucleic acid amplification (NAA) has 56% sensitivity and 98% specificity in the diagnosis of TBM. According to the British Infection Association, all patients suspected of CNS-TB must have a CSF-PCR for tuberculosis, but a negative result does not exclude the disease[24]. Another molecular study on 50 CSF samples showed 10% CSF-PCR positives in TBM.[17] Based on a recent meta-analysis, NAA has a sensitivity of 56% and specificity of 98% in the diagnosis of TBM.[5,8,21,31]

7.8 Radiology

In the Philippine cohort study on TBM, common findings on the CT scan were basilar meningeal enhancement, hydrocephalus, transepidermal effusion with or without infarction, and tuberculoma. It has been reported that brain sonography may have the same value in the diagnosis of hydrocephalus, infarct, abscess, and edema in children with TBM. It may show hydrocephalus and rim halo (hyper ventricular echogenicity with normal social). Based on this study, MRI findings are similar to CT scans in TBM.[12] Another study showed that the most radiological picture on a CT scan is hydrocephalus (36.6%), diffuse edema (18.9%), brain tuberculoma (16.3%), basilar enhancement (11.8%), and normal CT found in 18.3%. A brain MRI may be more useful for diagnosis than other clinical findings, which do not appear in a CT. The same study showed basilar enhancement in 31.9%, brain tuberculoma in 57.8%, diffuse edema in 24.9%, infarct in 24.6%, and vasculitis in 9.3%.[15] One study showed that most findings on brain MRI were tuberculoma at 73.1%, arachnoiditis at 48.3%, hydrocephalus and infarct at 20.7%, spondylitis at 17.2%, and myelitis at 6.9%.[32] Ependymitis is another finding in the brain MRI, which is related to extension of inflamed exudates to the ventricular system and resulted border's enhancement. Vasculitis, which is a result of vessel thrombosis, appears as hypodensity areas in CT or a high signal in T2-weighted MRI. Lumpy irregular enhancement areas on CT or cerebral isotone masses in T1-weighted brain MRI are indicative for granulomatosis meningitis.[8]

8. TREATMENT

As for other types of tuberculosis, medical treatment with antituberculosis chemotherapy is the base of management. A combination of other treatments such as steroids, diuretics, and surgical intervention may be needed. Antituberculosis treatment in CNS-TB consists of first-line and second-line drugs such as pulmonary TB. Many experts recommend at least 9–12 months of treatment with the first 2 months utilizing a four-drug regimen consisting of isoniazid (INH), rifampin (RIF), pyrazinamide (PZA), and ethambutol (EMB). That is followed by a two-drug regimen over the next 7–10 months: INH and RIF. This regimen is used for susceptible tuberculosis. In drug-resistant tuberculosis, other drugs may be used and the duration of the treatment may be prolonged. These drugs are prescribed for CNS-TB with the same dose for pulmonary TB, but usually use a maximum dose or above standard dose for some drugs.[31]

According to the British Infection Association, all patients with susceptible CNS-TB must be treated for at least 12 months, and all patients have to receive corticosteroids for 2 months. All patients with CNS-TB must have a brain CT with contrast before treatment, HIV testing, and a cryptococcus antigen assay in HIV-positive patients.[21]

8.1 First-Line Agents

Isoniazid: Isoniazid (INH) is an old anti-TB drug with excellent penetration into the CSF (up to 86%). The usual dose of INH of CNS-TB is 10 mg/kg up to 500 mg daily. Some of the adverse effects of this drug are neuropathy, optic neuritis, epilepsy, psychosis, porphyria, and hepatotoxicity. Pyridoxine supplementation is given to reduce neuropathy. Hepatotoxicity of INH is dose-dependent and more common when in a combination regimen with other hepatotoxic drugs.

Rifampin: Rifampin (RIF) penetrates to 10%–20% of the normal BBB. Recommended dosage for CNS-TB is 20 mg/kg up to 900 mg daily. A common adverse effect is hepatotoxicity, which is not dose-related. Rifabutin may be substituted in HIV-infected patients with rifampin, for drug interaction. A common side effect of the latter drug is anterior uveitis.

Pyrazinamide: Pyrazinamide (PZA) also has excellent penetration to the CSF and is prescribed at a dose of 30–40 mg/kg up to a 2 g maximum daily dose. The most common side effect is hepatotoxicity in as much as 1%.

Ethambutol: Ethambutol (EMB) has poor penetration into the CSF and its role in the treatment of CNS-TB remains questionable. WHO recommended replacing EMB with streptomycin in the treatment of these patients. Retrobulbar neuritis is the most common adverse effect with this drug, which is dose-related. It occurs in 1% with dosage 15 mg/kg and reaches up to 18% in a dosage of 35 mg/kg daily.

8.2 Second-Line Agents

Many experts recommend that drug-resistant tuberculosis be treated in a special center by an expert team. The drugs used in a second-line regimen consist of ethionamide, streptomycin, amikacin, *p*-aminosalicylic acid, and fluoroquinolones.[31]

8.3 Neurosurgery

Hydrocephalus may be the most common complication of CNS-TB, requiring neurosurgical intervention. Two forms of hydrocephalus may occur during the disease course: communicating hydrocephalus, which is more prevalent, and obstructive hydrocephalus. In communicating hydrocephalus, medical treatment with diuretics and corticosteroids may be effective in 70% of patients. Obstructive hydrocephalus and communicating hydrocephalus that doesn't respond to medical treatment require shunt insertion.[21,22]

Other indications for surgery in CNS-TB consist of the brain tubercular abscess, large tuberculoma with mass effect, cord pressure syndrome, and the need for definite diagnosis in brain masses.

8.4 Drug Susceptibility

In the United States, the frequency rate of primary resistance for each of the antituberculosis drugs is 12%; the frequency of MDR is about 1%. A total of 3.7% of new tuberculosis patients and 20% of reactivated former patients are MDR. A history of treatment with antituberculosis drugs is the highest risk factor for drug resistance.[31] In a California TB cohort study, 16 of 114 positive CSF samples (14%) had been resistant to one anti-TB drug, and five samples (4.4%) were MDR. TST was negative in about one-third of the patients.[3]

8.5 Drug-Induced Hepatitis

One of the most adverse effects of ATT is drug-induced hepatitis (DIH), which defines increased ALT more than five times above normal or less with any symptoms of liver dysfunction such as ichter, anorexia, nausea, and vomiting. In a randomized trial of 175 patients with drug-induced hepatitis, the patients were assigned to three groups for reintroduction of ATT. The first group received all hepatotoxic drugs concomitantly with a full dose. The second group received ATT according to the American Thoracic Society (ATS), which consisted of full doses of rifampin on the first day, maximum doses of isoniazid on day 8 and pyrazinamide on day 15. The third group received ATT according to the British Thoracic Society, which consisted of increasing drug dosage and maximum doses of isoniazid, rifampin, and pyrazinamide on days 4, 11, and 15, respectively. The frequency of the second DIH was 10.9%, which was not different in the three groups.[33]

8.6 CNS Tuberculosis in Immunocompromised Patients

HIV infection increases the risk of developing CNS-TB about fivefold. Patients with solid organ transplantation (SOT), except for bone marrow transplants (BMT) also have a greater risk—>20 times—of developing tuberculosis.

In HIV-infected patients, TBM develops more than in HIV-negative patients, especially when the CD4 count is <100 per cubic milliliter. In these patients, extra CNS manifestation of tuberculosis is more common than in HIV-negative patients. Brain tuberculomas in these patients have slower growth and results in headaches, seizures, focal neurological deficits, and papilledema. Brain tubercular abscesses are more common in HIV-infected patients and have acute courses and clinical symptoms, such as seizures, headaches, loss of consciousness, and hemiparesis.

Diagnosis of CNS-TB in immunocompromised patients is similar for patients with normal immune systems. A CSF culture and PCR are the gold standard of tuberculosis in these patients.

CNS-TB IRIS is relatively prevalent in HIV-infected patients with CNS-TB. It usually occurs within 2 weeks after initiation of antiretroviral treatment (ART) and accompanies the worsening of clinical symptoms.

TBM in HIV-infected patients has a high mortality rate, up to 65% in some reports. CD4 counts <50 per cc, advanced disease, and MDR TB are related to a higher mortality rate. In SOT patients, CNS-TB also has a relatively high mortality rate—about 13%–40%—and risk factors of mortality in these patients are diffuse infection, lung transplant, and treatment with muromonab-CD3.[34]

REFERENCES

1. World Health Organization. *Global tuberculosis report 2016*. Geneva: World Health Organization; 2016. Available from: http://www.who.int/tb/publications/global_report/gtbr2016_executive_summary.pdf?ua=1.
2. Phypers M, Harris T, Power C. CNS tuberculosis: a longitudinal analysis of epidemiological and clinical features. *Int J Tuberc Lung Dis* 2006;**10**(1):99–103.
3. Duque-Silva A, Robsky K, Flood J, Barry PM. Risk factors for central nervous system tuberculosis. *Pediatrics* 2015;**136**(5):e1276–84.
4. El Sahly HM, Teeter LD, Pan X, Musser JM, Graviss EA. Mortality associated with central nervous system tuberculosis. *J Infect* 2007;**55**(6):502–9.
5. Cherian A, Thomas SV. Central nervous system tuberculosis. *Afr Health Sci* 2011;**11**(1):116–27.
6. Carrol KC, Butel JS, Morse SA, Mietzner T. *Jawetz medical microbiology*. 27th ed. The McGraw-Hill Companies, Inc.; 2016. p. 309–22 [Chapter 23].
7. Be NA, Sik Kim K, wr B, Jain SK. Pathogenesis of central nervous system tuberculosis. *Curr Mol Med* 2009;**9**(2):94–9.
8. Tai MLS. Tuberculous meningitis: diagnostic and radiological features, pathogenesis and biomarkers. *Neurosci Med* 2013;**4**:101–7.
9. Donald PR, Schaaf HS, Schoeman JF. Tuberculous meningitis and miliary tuberculosis: the Rich focus revisited. *J Infect* 2005;**50**(3):193–5.
10. Yaramiş A, Gurkan F, Elevli M, Söker M, Haspolat K, Kirbaş G, et al. Central nervous system tuberculosis in children: a review of 214 cases. *Pediatrics* 1998;**102**(5).

11. Chaudhary V, Bano SH, Chandra Garga U. Central nervous system tuberculosis: an imaging perspective. *Can Assoc Radiol J* 2017;**68**(2):161–70.
12. Lee LV. Neurotuberculosis among Filipino children: an 11 years' experience at the Philippine Children's Medical Center. *Brain Dev* 2000;**22**(8):469–74.
13. Khoo JLS, Lau KY, Cheung CM, Tsoi TH. Central nervous system tuberculosis. *J HK Coll Radiol* 2003;**6**:217–28.
14. Poalelungi A, Nistor R, Terecoasa E, Gavan C, Tiu C. Tuberculous meningitis mimicking acute polyradiculonevritis. *Rom J Neurol Psychiatry* 2011;**2**(4):194–8.
15. Wasay M, Farooq S, Khowaja ZA, Bawa ZA, Ali SM, Awan S, et al. Cerebral infarction and tuberculoma in central nervous system tuberculosis: frequency and prognostic implications. *J Neurol Neurosurg Psychiatry* 2014;**85**(11):1260–4.
16. Sütlaş PN, Unal A, Forta H, Senol S, Kirbaş D. Tuberculous meningitis in adults: review of 61 cases. *Infection* 2003;**31**(6):387–91.
17. Farinha NJ, Razali KA, Holzel H, Morgan G, Novelli VM. Tuberculosis of the central nervous system in children: a 20-year survey. *J Infect* 2000;**41**(1):61–8.
18. Li H, Liu W, You C. Central nervous system tuberculoma. *J Clin Neurosci* 2012;**19**(5):691–5.
19. Cárdenas G, Soto-Hernández JL, Orozco RV, Silva EG, Revuelta R, Amador JL. Tuberculous brain abscesses in immunocompetent patients: management and outcome. *Neurosurgery* 2010;**67**(4):1081–7 [discussion 1087].
20. Kaushik K, Karade S, Kumar S, Kapila K. Tuberculous brain abscess in a patient with HIV infection. *Indian J Tuberc* 2007;**54**(4):196–8.
21. Thwaites G, Fisher M, Hemingway C, Scott G, Solomon T, Innes J, et al. British Infection Society guidelines for the diagnosis and treatment of tuberculosis of the central nervous system in adults and children. *J Infect* 2009;**59**(3):167–87.
22. Fitzgerald DW, Sterling TR, Haas DW. Mycobacterium tuberculosis. In: Bennet J, Dolin R, Blaser M, editors. *Priciple and practice of infectious diseases*. 8th ed. Canada: Elsevier Publication; 2015. p. 2787–818 [Chapter 251].
23. Lammie GA, Hewlett RH, Schoeman JF, Donald PR. Tuberculous cerebrovascular disease: a review. *J Infect* 2009;**59**(3):156–66.
24. Kumar S, Kumar R, Radotra BD. Tubercular ventriculitis: an uncommon entity. *Indian J Pediatr* 2014;**81**(6):608–10.
25. Guler SK, Güneş N, Cokal BG, Yoldas T. Tuberculous ventriculitis as a relapse following central nervous system tuberculoma. *Neurol Sci* 2015;**36**(1):167–9.
26. Singh P, Paliwal VK, Neyaz Z, Srivastava AK, Verma R, Mohan S. Clinical and magnetic resonance imaging characteristics of tubercular ventriculitis: an under-recognized complication of tubercular meningitis. *J Neurol Sci* 2014;**342**(1–2):137–40.
27. Vaziri S, Soleiman-Meigooni S, Rajabi J, Asgari A. Tuberculous ventriculitis: a rare complication of central nervous system tuberculosis. *Int J Mycobacteriol* 2016;**5**(2):231–4.
28. Coulibaly O, Diawara S, Quenum K, Sogoba Y, Boutarbrouch M, Melhaoul A, et al. Interventricular tuberculoma: a case report. *Case Rep Clin Med* 2013;**2**(1):81–4.
29. Marais S, Meintjes S, Pepper D, Dodd L, Schutz C, Ismail Z. Frequency, severity, and prediction of tuberculous meningitis immune reconstitution inflammatory syndrome. *Clin Infect Dis* 2013;**56** (3):450–60.
30. Toorn R, Rabie H, Dramowski A, Schoeman F. Neurological manifestations of TB-IRIS: a report of 4 children. *Eur J Paediatr Neurol* 2012;**16**(6):676–82.
31. Rock RB, Olin M, Baker CA, Molitor TW, Peterson PK. Central nervous system tuberculosis: pathogenesis and clinical aspects. *Clin Microbiol Rev* 2008;**21**(2):243–61.
32. Abdelmalek R, Kanoun F, Kilani B, Tiouiri H, Zouiten F, Ghoubantini A, et al. Tuberculous meningitis in adults: MRI contribution to the diagnosis in 29 patients. *Int J Infect Dis* 2006;**10**(5):372–7.
33. Sharma SK, Singla R, Sarda P, Mohan A, Makharia G, Jayaswal A, et al. Safety of 3 different reintroduction regimens of antituberculosis drugs after development of antituberculosis treatment-induced hepatotoxicity. *Clin Infect Dis* 2010;**50**(6):833–9.
34. Nelson C, Zunt J. Tuberculosis of the central nervous system in immunocompromised patients: HIV infection and solid organ transplant recipients. *Clin Infect Dis* 2011;**53**(9):915–26.

CHAPTER 10

Tick-Borne Infections of the Central Nervous System

Praveen Sudhindra
Cooper University Hospital, Camden, NJ, United States

1. INTRODUCTION

1.1 The Vector

Ticks are hematophagous ectoparasites, classified into three families—Argasidae (soft ticks), Ixodidae (hard ticks), and the Nutalliellidae (shares features with soft and hard ticks). Ticks lack the capacity to fly or leap, and instead "quest" for suitable hosts by climbing onto vegetation and stretching out their forelegs, awaiting a passing vertebrate they can latch onto. Their lifecycles pass through four stages—egg, larva, nymph, and adult (see Fig. 1). Ticks interact with their environment and sense the presence of potential hosts through specialized organs on their forelegs called the "Haller's organ." They are able to sense body heat, odor, and vibrations caused by hosts' movements.[1]

Ticks are host to a diverse and ever-expanding range of pathogens, including bacteria, viruses and parasites—often transmitting multiple infections during a single blood meal. Hard ticks feed for days at a time, depending on the stage of their lifecycle, while soft ticks feed only for a few hours. Knowledge of this behavior becomes important in determining the risk of acquiring infection following a tick bite and the need for antimicrobial prophylaxis.

2. TICK-BORNE VIRAL INFECTIONS OF THE CNS

The Flaviviridae account for all tick-transmitted viral infections that primarily affect the central nervous system (CNS). These are comprised of several closely related viruses belonging to the genus *Flavivirus*—including the tick-borne encephalitis virus (TBEV), the Powassan virus, the Louping ill virus, and the Langat virus. Key features are summarized in Table 1.

2.1 Tick-Borne Encephalitis Virus (TBEV)

There are three closely related groups of TBEV—Central European tick-borne encephalitis (TBEV-Eu), Russian (Siberian) spring-summer encephalitis (TBEV-Sib), and Far Eastern TBE (TBEV-FE).

The Microbiology of Central Nervous System Infections
https://doi.org/10.1016/B978-0-12-813806-9.00010-X

Fig. 1 *Ixodes ricinus* ticks.

The first detailed description of this entity was published from Austria in 1931 by Schneider,[2] while the virus was first isolated from human blood and the *Ixodes* spp. vector in 1937.

2.1.1 The Virus, Replication, and Pathogenesis

TBEV is an icosahedral, 50 nm enveloped single stranded positive sense ribonucleic acid (RNA) virus. The lipid envelope covers a nucleocapsid layer and the RNA genome encoding a 3414 amino acid polyprotein. This, in turn, is co- and posttranslationally cleaved to yield seven nonstructural proteins.[3] The viral structure, genome, and replication cycle are described in Figs. 2 and 3.

Initial viral replication most likely occurs at the site of inoculation in Langerhan's cells found in the human epidermis. These cells transport the virus via draining lymphatics to the local lymph nodes, following which viremia and systemic infection occur. Invasion of the CNS with neuroinflammation and neuronal death ensue.[3] Although the blood-brain barrier (BBB) is disrupted during the course of TBEV infection, Ruzek and coworkers have demonstrated that this is not a necessary step in neuroinvasion by using rodent models.[4] While the precise mechanisms of CNS invasion remain to be elucidated, a number of pathways including infection of human microvascular endothelial cells, paracellular transfer, or transport within circulating immune cells have been proposed. Infection of neurons and astrocytes likely triggers a robust release of cytokines and chemokines, which play an important role in disrupting the BBB and mediating neuronal injury.[5]

Table 1 Summary of tick-borne flaviviruses causing encephalitis

Virus	Virus subtypes and closely related spp.	Vector	Geographic distribution	Clinical features
TBEV	TBEV-Eu	*Ixodes ricinus*	Europe (except Iberian peninsula)	Biphasic illness, encephalitis, meningitis Residual neurologic deficits common
	TBEV-FE	*I. persulcatus, Hemaphysalis* spp.	Eastern Russia, China, Japan	Highest case fatality rate (up to 20%) Hemorrhagic syndrome reported
	TBEV-Sib (Russian spring-summer encephalitis)	*I. persulcatus, Hemaphysalis* spp.	Russia	Chronic encephalitis more common
Powassan virus	Powassan virus (POWV)	*Ixodes cookei, Hemaphysalis longicornis* (E. Asia)	North America, Far Eastern Russia	Encephalitis, residual neurologic deficits 10% case fatality in patients with neuro-invasive disease
	Deer Tick Virus (DTV)	*I. scapularis, Dermacentor andersoni*	North America	Unclear if distinct clinical manifestations, majority of cases diagnosed by serology which cannot distinguish between DTV and POWV
Louping Ill virus (LIV)[a]	Louping ill virus	*Ixodes ricinus*	Scotland, Britain, Ireland	Human infections are rare, usually associated with occupational exposure (veterinarians, abattoir workers). Presents as a flu-like illness, occasionally followed by neurological signs and symptoms including seizures and flaccid paralysis. One fatal case reported[b]
	Greek goat encephalitis virus	*Ixodes ricinus*	Greece	
	Turkish sheep encephalitis virus	*Ixodes ricinus*	Turkey	
	Spanish sheep encephalitis virus	*Ixodes ricinus*	Spain	
Langat virus	—	*Ixodes* spp.	Malaysia, Thailand, Russia	Encephalitis, hemorrhagic fever

[a]Greek goat, Turkish and Spanish sheep encephalitis viruses are closely related to LIV, but are not subtypes.
[b]Jeffries CL, Mansfield KL, Phipps LP, et al. Louping ill virus: an endemic tick-borne disease of Great Britain. *J Gen Virol* 2014;**95**(Pt 5):1005–1014.

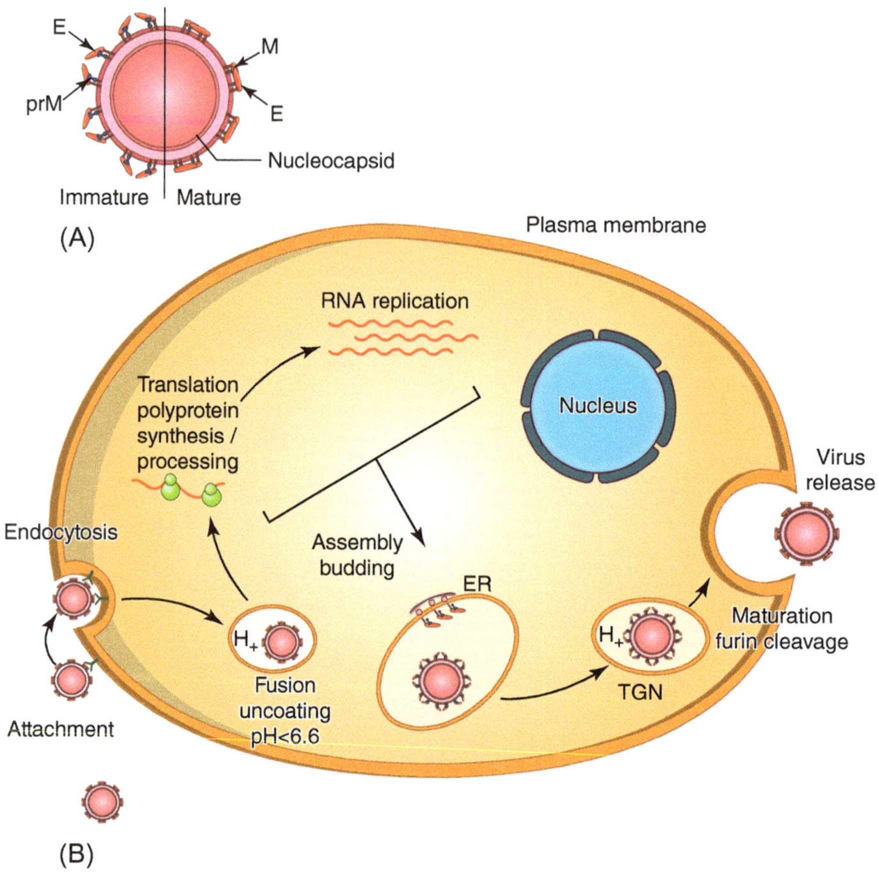

Fig. 2 Schematic diagrams of the structural organization (A) and the replication cycle (B) of TBEV. The viral nucleocapsid, formed by the positive stranded RNA genome and protein C, is surrounded by a lipid envelope in which two surface proteins are carboxy-terminally anchored. In immature virions, proteins prM and E form heterodimers, whereas infectious, mature virus particles carry the small protein M and homodimers of protein E. *ER*, endoplasmic reticulum; *TGN*, trans-Golgi network.

2.1.2 Vectors, Transmission, and Epidemiology

TBEV-Eu is transmitted by *Ixodes ricinus* (sheep tick) while TBEV-Sib and TBEV-FE are transmitted by *Ixodes persulcatus* (taiga tick). The virus is transmitted through the saliva of the infected vector within minutes of the tick bite. Prompt removal of the tick, therefore, does not prevent transmission.[6] Transmission also occurs on occasion by ingesting unpasteurized milk products from or slaughtering viremic livestock. Common source outbreaks, while less frequent in recent times, continue to be reported.[7] Person-to-person transmission through breastfeeding, blood transfusion, and solid-organ transplantation have also been reported.[8]

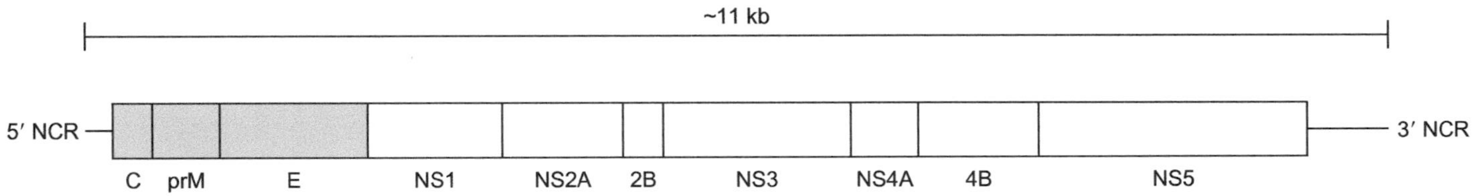

ORF	Function	Immunologic properties
C	Capsid protein—forms homodimeric unit that binds genome into nucleocapsid	
prM	Membrane protein—mediates proper folding of E protein	Humoral immunogen
E	Envelope protein—contributes to virus binding and cellular uptake	Major humoral immunogen; hemmaglutinin; principal target of neutralizing antibody
NS1	Involved in RNA replication	Humoral immunogen
NS2A	Presumptive role in transitioning between viral packaging and replication	May act as interferon antagonist
NS2B	Forms complex with NS3	
NS3	Forms (1) serine protease complex alongside NS2B (amino-terminus); (2) RNA helicase (carboxy-terminus); and (3) RNA triphosphatase	Contains immunodominant peptides presented to CD8+ T cells by MHC class I
NS4A	Interacts with NS1	
NS4B	Interacts with NS3	
NS5	Methyltransferase and RNA-dependent RNA polymerase	Interferon antagonist

C, capsid; *prM*, premembrane; *E*, envelope; *NS*, nonstructural; *NCR*, noncoding region.

Fig. 3 Flavivirus genome.

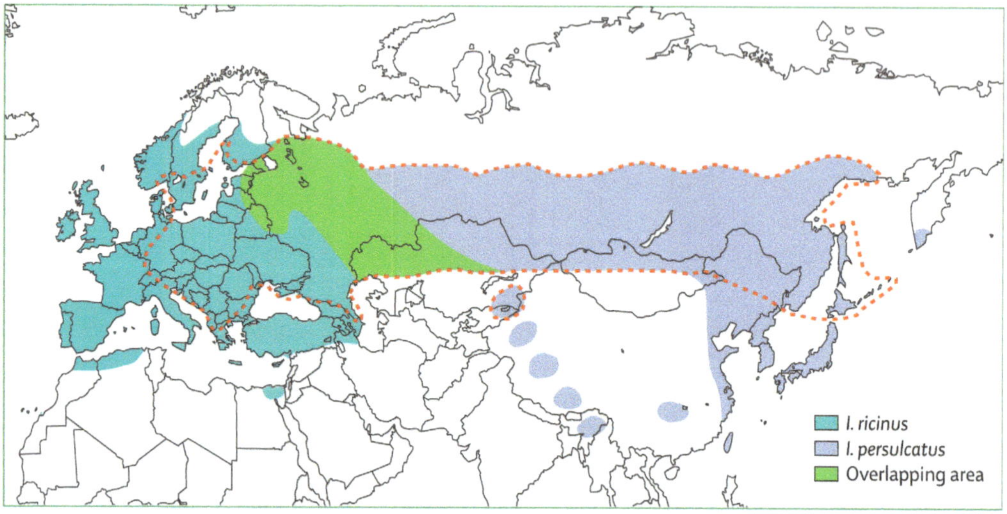

Fig. 4 Geographic distribution of *Ixodes* spp. in Europe and Asia.

More than 10,000 cases of TBE are reported annually worldwide.[9] The epidemiology of TBEV is only partially explained by the range of its vectors. Other factors, such as temperature (>7°C), humidity (>85%), human activity, and vaccine coverage play significant roles.[10] The range of tick vectors is depicted in Fig. 4. Most cases are reported from Russia and the Baltic states of Latvia, Lithuania, and Estonia, with several focal, endemic areas in other countries including the Czech Republic, Slovakia, Slovenia, Poland, Sweden, Switzerland, and Austria. The incidence across Europe ranges from <0.5 to >45 cases/100,000 per year. Over 35 million people live in endemic areas in Russia with incidence ranging from 4 to >30/100,000 per year in areas such as the Tomsk and Kurgan provinces.[11] Among Asian countries, parts of China, Kyrgyzstan, Mongolia, Kazakhstan, Japan, and South Korea are reported to have virus activity.

Risk factors for acquisition of TBEV include residence or travel to endemic areas between April and November, with peak distribution of cases in early and late summer. Socioeconomic factors as well as encroachment into forested regions also play significant roles.

Viral transmission to ticks occurs when different developmental stages cofeed on the same host (Fig. 5). Neither systemic infection nor high levels of viremia are required to transmit infection. Instead, immune cells at the site of an infected tick bite traffic viral particles to other infested sites, probably in response to tick saliva. This results in viral acquisition by hitherto uninfected ticks.[12] Infection rates in ticks range from 0.1% to 5%, generally much lower than the rates of infection with Lyme Borreliosis.[10]

2.1.3 Clinical Manifestations

The incubation period ranges from 4 to 28 days with a median of 8 days. The disease follows a biphasic pattern in more than 70% of patients. The first stage of illness usually

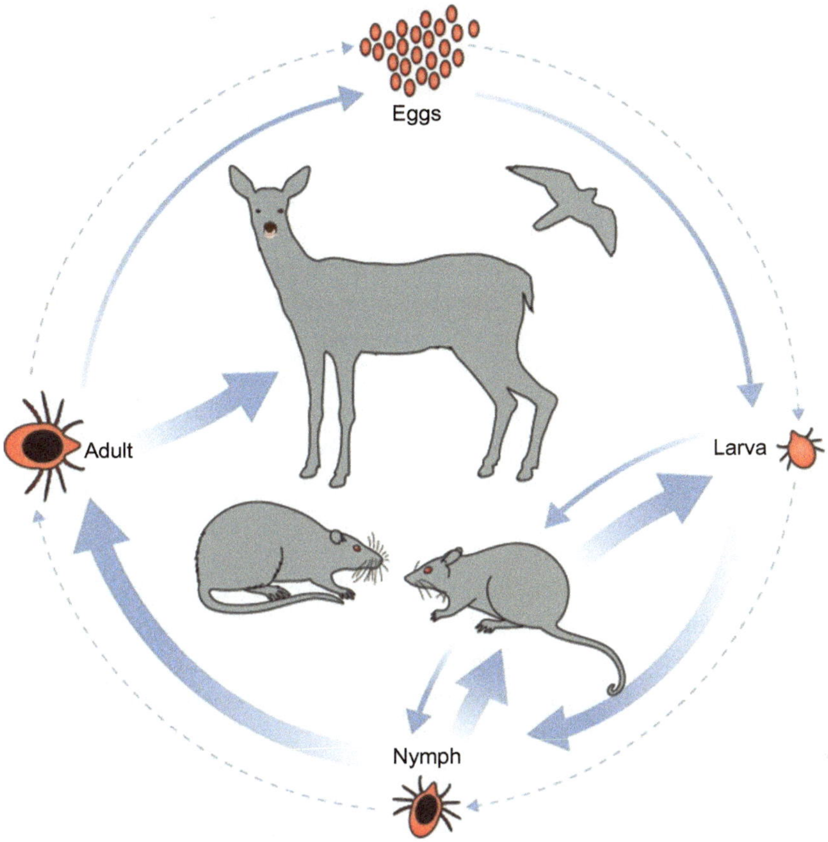

Fig. 5 TBEV transmission cycle.

lasts for 5 days (range 2–10), followed by a 7 day (range 1–21 days) asymptomatic interval. Fever, fatigue, malaise, body pain, and headache are common during the initial phase. Leucopenia, thrombocytopenia, and mildly elevated serum transaminases can be seen. Neurologic manifestations predominate in the second phase of the illness, with several different patterns of involvement including aseptic meningitis, meningoencephalitis, and palsies involving cranial nerves III, VII, IX, X, and XI resulting in dysphagia, gaze abnormalities, and facial paralysis. Some of the more distinctive manifestations are a poliomyelitis-like infection of the anterior horn cells of the cervical spinal cord resulting in acute flaccid paralysis of the shoulder girdle and the upper extremities as well as a self-limiting polyradiculitis that can mimic Guillain-Barré Syndrome.[10,13–15] Seizures are relatively infrequent. In a prospective study of 133 TBE patients in Lithuania, 43.6% had mild meningeal disease, 43.6% had moderate encephalitis, and 12.8% had severe encephalitis. Cranial nerve injury occurred in 5.3%, paralytic disease in 3.8%, and 8.5% had severe disabilities requiring adjustment of daily activities.[16]

Up to two-thirds of patients recover without progression to the second phase. CNS manifestations tend to be more severe with advancing age. The TBEV subtypes differ in their virulence and disease severity, with TBEV-FE associated with a 20% case fatality in hospitalized patients, while 6%–8% of TBEV-Sib and 1%–2% of TBEV-Eu infections are reported to be fatal.[13]

Long-term morbidity is an important sequela of TBEV infections. A range of 26%–46% of patients with TBEV-Eu infections from three studies reported some form of cognitive or neurological symptoms 6–12 months after the acute phase of the illness. Ataxia, headache, memory loss, difficulty concentrating, and tremors are most commonly reported. Epilepsia partialis continua and progressive motor weakness occur less frequently and are probably due to a chronic encephalitic process, more commonly described in TBEV-Sib infections.[13] A fatal hemorrhagic syndrome caused by TBEV-FE was reported from Russia in 1999.[17] Clinical manifestations included a massive gastrointestinal hemorrhage as well as bleeding from mucosa and other viscera.

2.1.4 Diagnosis

A history of tick bites may be absent in about a third of cases, and aseptic meningitis, myeloradiculitis, and cranial nerve palsies can present similarly to Lyme Borreliosis. Dual infections with *Borrelia burgdorferi* and TBEV are not uncommon because they are transmitted by the same ticks. As mentioned earlier, leucopenia, thrombocytopenia, and mild transaminitis occur in the first phase of TBE, while leukocytosis (up to 20,000 cells/mm^3) can be seen in the second phase. CSF analysis reveals a moderate pleocytosis and protein elevation in two-thirds of patients. The cell count is usually less than 100 cells/μL. Polymorphonuclear cells predominate, initially followed by close to 100% mononuclear cells subsequently. MRI abnormalities can be seen in ~18% of patients, with changes usually restricted to the thalamus and basal ganglia. EEG most commonly reveals diffuse slowing.[10]

A real-time reverse transcriptase polymerase chain reaction (RT-PCR) is useful in detecting viremia in the early phase of illness, but is of limited value in detecting virus in the blood or CSF at the beginning of the second phase. Serum TBEV-immunoglobulin M (TBEV-IgM) is usually present by the time neurologic symptoms manifest. TBEV-immunoglobulin G (TBEV-IgG) is also frequently present by this time. IgM and IgG are detectable in the CSF by day 10 of the illness. IgM antibody capture enzyme linked immune assay (ELISA) performed on serum or CSF is the test of choice and is more than 95% sensitive when performed after day 7 of the illness.[10,13] Due to high antigenic cross reactivity among the flaviviruses, diagnostic difficulties may arise where other flaviviral infections such as West Nile virus (WNV), dengue, Japanese encephalitis, etc. cocirculate. In such cases, or when vaccine breakthrough is suspected, one of the following is necessary to help establish the diagnosis: quantitative IgM assays (a value >500 arbitrary units), a fourfold rise in paired acute and convalescent sera, neutralizing antibody assay, or demonstration of intrathecal antibody production.[14,18]

2.1.5 Treatment and Prevention

There are no specific antiviral therapies available, and treatment remains largely supportive. Corticosteroids have not been shown to have any benefits while a small controlled study suggested that tetracycline may hasten recovery via immunomodulatory effects.[19]

Vaccination is the cornerstone of preventative strategies in endemic countries. There are two formaldehyde-inactivated vaccines available in Europe, both based on TBEV-Eu strains: FSME-IMMUN (strain Neudoerfl, Baxter Vaccines, Vienna, Austria) and Encepur (strain K23, Novartis, Basel, Switzerland). In addition, there are two Russian vaccines based on TBEV-FE (EnceVir, Scientific Production Association Microgen) and TBE vaccine Moscow (Federal State Enterprise of Chumakov Institute of Poliomyelitis and Viral Encephalitides, Russian Academy of Medical Sciences). All four vaccines are safe, immunogenic, and provide cross-protection against all TBEV subtypes.[9,10]

The conventional recommended schedule is two intramuscular doses 1–3 months apart prior to the period of transmission, followed by a dose prior to the following tick season. This is believed to result in immunogenic titers in more than 90% of children and adults. People over 60 tend to have poorer responses. Immunity wanes after the third dose, necessitating a booster after 3 years followed by 5 year intervals (3 year intervals are favored for those over 60).[10]

2.2 Powassan Virus

The Powassan virus is the only member of the tick-borne encephalitis group of flaviviruses circulating in North America. There are two closely related "lineages" of the Powassan virus: Lineage I, or the prototype Powassan virus (POWV), and Lineage II, the deer tick virus (DTV). Although serologically indistinguishable, the two lineages represent distinct genotypes of the same virus.

2.2.1 Virus, Vector Transmission, and Epidemiology

POWV/DTV have essentially the same virologic structure as TBEV. The entire single-stranded positive-sense RNA genome was sequenced in 1993.[20] The virus replication cycle is believed to be similar to other members of the TBE group. The two lineages share 84% nucleotide sequence identity and 94% amino acid sequence identity,[21] and are maintained in distinct enzootic cycles. POWV infection occurs in North America and Russia while DTV infection is only known to occur in North America.

POWV is transmitted by *Ixodes cookei*, with the groundhog and the striped skunk acting as reservoirs.[22] DTV is transmitted mainly by *I. scapularis* with the white-footed mouse as the reservoir. Transovarial transmission has been observed.[23] Approximately 6% of adult *I. scapularis* ticks are infected with DTV.[24] Although the range of *I. cookei* is much broader than *I. scapularis*, humans are far more likely to be bitten by the latter. The virus can be transmitted within 15 min of a tick bite.[25]

The epidemiology of the two lineages can be difficult to parse because genetic sequence data have become available only recently. It is possible that a number of earlier infections attributed to POWV were actually caused by DTV. For epidemiological purposes, unless specified, Powassan virus or POWV refers to either lineage in the following discussion.

Powassan virus was first isolated in humans from the brain of a child with fatal encephalitis in Powassan, Ontario, Canada in 1958.[26] About 100 human cases have been reported from North America thus far, with over 80 cases reported since 1997.[27] In North America, the virus has been detected mainly in the North Eastern and North Central United States as well as throughout Ontario, Alberta, Quebec, and British Columbia in Canada.[27] An additional endemic focus exists in the far eastern Primorsky region of Russia.

2.2.2 Pathogenesis

Reports from human pathologic examinations as well as animal models have provided insights into the pathogenesis of the human disease. Ebel reviewed findings from three reported postmortem pathologic examinations, two of whom died of acute encephalitis while the third succumbed a few weeks later.[26,28,29] Histological preparations from the index case demonstrated perivascular and focal parenchymal infiltration predominantly with lymphocytes and monocytes in the frontal, parietal and occipital cortices, basal ganglia, pons, and medulla as well as the cervical and lumbar spinal cord. The other cases revealed denser infiltration in the gray matter compared to white with involvement of spinal anterior horn cells and viral inclusions within neurons. Modeling in primates and rodents has clearly demonstrated viral neurotropism. The white-footed mouse, however, appears to be resistant to clinical infection.[30]

2.2.3 Clinical Manifestations

The Powassan virus is a rare cause of severe neuroinvasive disease in humans. It is unclear what proportion of infections are subclinical. The course of POWV/DTV infection can be characterized as a febrile illness with progressive neurologic dysfunction in a minority with a case fatality of close to 10% and residual neurologic dysfunction in 50% of those that develop neurological disease.[23]

The incubation period is about 1–5 weeks, although most are unable to identify the exact date of the tick bite. This is followed by a prodromal phase characterized by fever, sore throat, headache, and disorientation.[23,27] Those who progress to neurologic involvement develop a variety of manifestations, most commonly meningoencephalitis, aseptic meningitis, or encephalitis. A range of symptoms from nystagmus and poor coordination to spastic or flaccid paralysis can occur. Spastic forms seem to predominate in pediatric cases.[27] In a case series reported by Piantadosi et al. from Massachusetts and New Hampshire, seven out of eight patients with Powassan virus encephalitis had fever, altered sensorium, and focal neurologic deficits; three out of eight had a rash; four out of eight had vomiting; two died; two had residual deficits; and the rest improved.[31]

2.2.4 Diagnosis

The 2011 United States surveillance case definition for POWV/DTV neuroinvasive infection incorporates clinical and laboratory criteria and requires fever ≥38°C with any peripheral or CNS dysfunction documented by a physician, the absence of a more likely explanation, and ≥1 of the following:[22]

Isolation of the POWV/DTV from or detection of a specific viral antigen or nucleic acid in tissue, blood, cerebrospinal fluid (CSF), or another bodily fluid.

A ≥fourfold change in POWV/DTV specific quantitative antibody titers in paired serum samples.

POWV/DTV specific IgM in serum with confirmatory POWV/DTV-specific neutralizing antibodies in the same or a later serum specimen (titer ≥fourfold higher than the corresponding titer to West Nile Virus or St. Louis encephalitis virus).

POWV/DTV-specific IgM antibodies in CSF and a negative result for other IgM antibodies in CSF for arboviruses endemic in the region of exposure.

The majority of patients come to clinical attention in the encephalitic phase. Virus isolation and PCR amplification have a limited role in diagnosis because viremia usually resolves by the end of the first week of infection.

CSF analysis usually reveals a pleocytosis of up to 700 cells/μL with lymphocytic or neutrophilic predominance; the protein level is elevated while glucose levels are normal. In some cases, CSF findings can be nearly normal.[32]

2.2.5 Treatment

There are no currently available specific antiviral therapies or vaccines to treat POWV/DTV infections. High dose corticosteroids have been used to treat severe neuroinvasive disease. All treated patients in these reports survived.[22,33,34] Piantadosi et al. reported the use of intravenous immunoglobulin (IVIG) in two patients with POWV encephalitis, both of whom survived although one sustained significant neurologic deficits at discharge.[31] Pegylated interferon alfa and ribavirin were used unsuccessfully to treat a patient with fatal DTV encephalitis.[22]

Supportive care, including critical support in the form of mechanical ventilation and measures to reduce intracranial hypertension, plays a crucial role in the management of this disease.

3. TICK-BORNE BACTERIAL INFECTIONS OF THE CNS

Unlike the viral infections described thus far, tick-transmitted rickettsial and spirochetal infections do not primarily infect the CNS. However, neurologic involvement occurs with varying frequency as a manifestation of the disseminated infection or as a reflection of an underlying immune compromise. Lyme disease will be discussed in detail below while the salient features of other spirochetal and rickettsial illness that affect the CNS are summarized in Table 2.

Table 2 Summary of spirochetal and rickettsial infections that affect the CNS (excluding Lyme disease)

Disease/organism	Vector	Geographic distribution	Clinical manifestations	Diagnosis	Treatment
Human monocytotropic ehrlichiosis (HME)[a] *Ehrlichia chaffeensis*	*Amblyomma americanum*	South-Central and South Eastern US	Mild-moderate multisystem illness. Fever, headache. Myalgia common. Rash in 25%–30%. ARDS, meningoencephalitis, DIC in severe disease (9%–17%). Fatal illness in immune-compromised	Serology 68% diagnostic in second week. PCR 60%–100% sensitive. Morulae seen in only 7%. Elevated AST/ALT, thrombocytopenia, leukopenia, anemia frequent. Lymphocytic pleocytosis, elevated protein in CSF can be seen	Doxycycline, tetracycline
Rocky mountain spotted fever (RMSF)[b] *Rickettsia rickettsii*	*Dermacentor variabilis*, *D. andersoni*, *Rhipicephalus sanguineus*, *Amblyomma* spp. (Central and S. America)	Widely distributed in small pockets, more common in South Central US. Emerging in Central and South America	Fever, headache, myalgia, Rash (90%), GI symptoms frequent. Meningismus in <20%. Residual neurologic deficits can occur in survivors. Fatal without timely therapy	Early clinical diagnosis essential. Organism can be demonstrated by immunochemistry on skin biopsy specimens. Serology for retrospective confirmation, cross reaction with other Spotted fever rickettsial group. Isolation from blood done in a few labs	Doxycyline is the drug of choice. Chloramphenicol is preferred in pregnancy. Susceptible to rifampin, ciprofloxacin and clarithromycin in vitro

Human granulocytic anaplasmosis (HGA)[a] *Anaplasma phagocytophilum*	*Ixodes scapularis*	Northeastern, Upper mid-western US, Europe, parts of Asia	Similar to HME, life-threatening infection in 3%–7%. Meningo-encephalitis rare. Demyelinating polyneuropathy, brachial plexopathy and facial diplegia reported	Neutropenia, in addition to lab abnormalities described for HME. Morulae in peripheral neutrophils in 20%–80%. PCR 54%–100% sensitive. Serology useful for retrospective diagnosis	Doxycycline, rifampin
Borrelia miyamotoi disease[c]	*Ixodes* spp. (all species that transmit Lyme borreliosis)	US, Russia, Europe, Japan	Flu-like illness common, can be severe with 24% hospitalized in one case series. Meningoencephalitis reported in 3 cases, all elderly with lymphoproliferative disorders. Relapsing fever in 4%–11%	Spirochetes can be stained with Wright or Giemsa stain in peripheral blood smear or CSF. PCR and serologic tests for antibody to GlpQ protein are useful. Cross reactivity seen with Lyme antibody ELISA and C6 peptide ELISA	Doxycycline most commonly used. Ceftriaxone, IV penicillin for CNS disease. Amoxicillin, cefuroxime likely acceptable for non-CNS disease

Continued

Table 2 Summary of spirochetal and rickettsial infections that affect the CNS (excluding Lyme disease)—cont'd

Disease/organism	Vector	Geographic distribution	Clinical manifestations	Diagnosis	Treatment
Q (Query) fever[d] *Coxiella burnetii*	*Dermacentor andersoni* (tick transmission is rare, aerosols from infected animals and ingestion of raw milk are more common modes of transmission)	Worldwide (except New Zealand, the Arctic and Antarctic regions)	Multisystem illness with wide range of manifestations including a self-limited febrile illness, hepatitis, pneumonia and endocarditis. Aseptic meningitis and encephalitis are rare, usually ~1% of cases. Seizures, coma, EP disease, GBS, cranial nerve palsies and demyelinating radiculoneuritis have been reported	CSF studies commonly reveal a mononuclear pleocytosis with normal glucose and elevated protein. PCR and serology are the usual methods of diagnosis. Culture requires Biosafety Level 3 facilities	Tetracyclines are the preferred therapy. Quinolones, rifampin and hydroxychloroquine are also used. Combination therapy is required in chronic Q fever

GlpQ, Glycerophosphodiester phosphodiesterase; *EP*, extrapyramidal; *PCR*, polymerase chain reaction; *ELISA*, enzyme linked immunosorbent assay; *GBS*, Guillain-Barre syndrome.

[a]Dumler JS, Walker DH. Ehrlichia chaffeensis, Anaplasma phagocytophilum and other Anaplasmataceae. In: Bennett JE, Dolin R, Blaser MJ, editors. *Mandell, Douglas, and Bennett's principles and practice of infectious diseases*. 8th ed. Philadelphia: Saunders; 2015. p. 2227–33.

[b]Walker DH, Blanton LS. Rickettsia rickettsii and other spotted fever group rickettsiae. In: Bennett JE, Dolin R, Blaser MJ, editors. *Mandell, Douglas, and Bennett's Principles and practice of infectious diseases*. 8th ed. Philadelphia: Saunders; 2015. p. 2198–205.

[c]Molloy PJ, Telford SR, 3rd, Chowdri HR, Lepore TJ, Gugliotta JL, Weeks KE, et al. *Borrelia miyamotoi* disease in the Northeastern United States: a case series. *Ann Intern Med* 2015;**163**(2):91–8; Krause PJ, Fish D, Narasimhan S, Barbour AG. Borrelia miyamotoi infection in nature and in humans. Clin Microbiol Infect 2015;21(7):631–9.

[d]Marrie TJ, Raoult D. Coxiella burnetii (Q fever). In: Bennett JE, Dolin R, Blaser MJ, editors. *Mandell, Douglas, and Bennett's principles and practice of infectious diseases*. 8th ed. Philadelphia: Saunders; 2015. p. 2208–16.

3.1 Lyme Borreliosis

Lyme disease is the most common tick-borne disease in the United States and Europe with about 300,000 cases estimated annually in the United States. More than 95% of these are reported from 14 states in the Northeast, Mid-Atlantic and Upper Midwest regions.[35]

3.1.1 The Pathogen

Lyme Borreliosis is a multisystem illness mainly caused by three genospecies within the group of spirochetes designated *B. burgdorferi sensu lato* (meaning *B. burgdorferi* in the general sense): *B. burgdorferi sensu stricto* (in the strict sense) in North America and Europe while *B. afzelii* and *B. garinii* are endemic in Europe and parts of Asia.[36]

B. burgdorferi is the longest and narrowest of the *Borrelia* spp., measuring 20–30 μm × 0.2–0.3 μm. The spirochete's genome consists of a linear 950 kb pair of chromosomes with up to 21 linear and circular plasmids comprising ~40% of its DNA. The plasmids encode outer-surface proteins A through F (Osp A-F), which plays a significant role in survival and multiplication in mammalian and tick hosts.[36,37]

3.1.2 Vectors, Transmission, and Pathogenesis

Lyme Borreliosis is transmitted by ticks belonging to the *Ixodes ricinus* complex. In the Northeastern and North Central United States, *Ixodes scapularis* is the main vector while *Ixodes pacificus* transmits disease on the Pacific Coast. In Europe, *I. ricinus* and *I. persulcatus* are the chief vectors. The prevalence of infection in ticks is often more than 25% in endemic regions.[38,39] The tick must feed for at least 36 h to transmit the infection.[40,41]

Ixodes scapularis maintains infection in an efficient enzootic cycle involving white-footed mice, chipmunks (and a few other small rodents), and larval or nymphal stages of the tick. The white-tailed deer is the preferred host of the adult *I. scapularis*. The peak questing periods for the nymphs are between May and July, larval stages from August through September, and adults from spring through fall. On the West coast, the wood rat is often the reservoir while in Europe small rodents and birds play important roles in maintaining *B. afzelii* and *B. garinii* infections, respectively. Although *I. scapularis* is found in a number of areas in the Southeastern United States, the immature forms preferentially feed on lizards, which are not susceptible to *B. burgdorferi* infection; hence, human infections are rare.[36] The transmission cycle is depicted in Fig. 6.

B. burgdorferi adapts to the distinct environments of its tick and mammalian hosts by regulating expression of surface proteins such as Osp and VlsE (vmp-like sequence expression site). OspA is expressed when the spirochete is dormant in the tick midgut while OspC is upregulated during feeding. This protein, in turn, binds Salp15, a tick salivary gland protein that is essential for immune evasion in the mammalian host.[42] Multiplication at the site of the tick bite occurs after an incubation period of 3–32 days. Dissemination to blood, CSF, muscle, and other tissues ensues over the following days

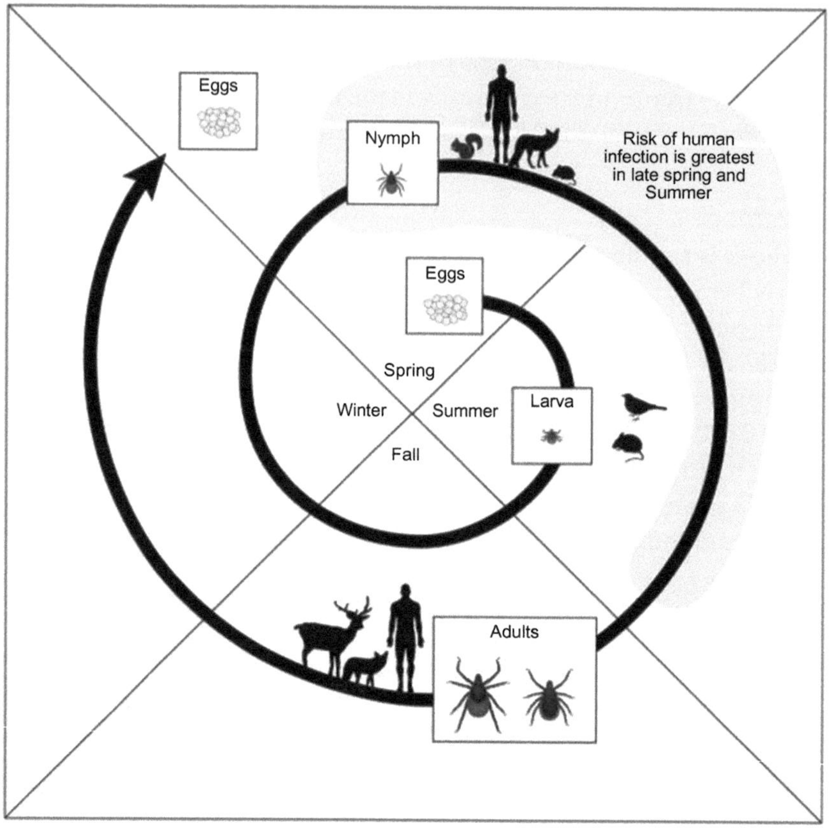

Fig. 6 Lyme Borreliosis transmission cycle.

to weeks. Both innate and adaptive immunity play a role in the host response. Lymphocyte and plasma cell infiltration are seen at all sites of infection. *B. burgdorferi* has not been found in intracellular locations. T-cell primed B cell responses are important in controlling the infection. The IgG response targets more than 100 spirochetal proteins. Opsonization and compliment fixation are the final pathways involved in killing the spirochete.[36,43]

In the absence of appropriate antibiotic treatment, *B. burgdorferi* may persist in certain niches; however, evidence from the natural history of Lyme arthritis indicates that the immune system does succeed in controlling the infection eventually.[44]

3.1.3 Clinical Manifestations

The clinical manifestations are divided into three stages: early localized, early disseminated, and late infection. Up to 10% may have asymptomatic, self-limiting illness.[45]

Early localized infection is characterized by an expanding, oval, erythematous lesion >5 cm at the site of the tick bite called erythema migrans (EM), which occurs

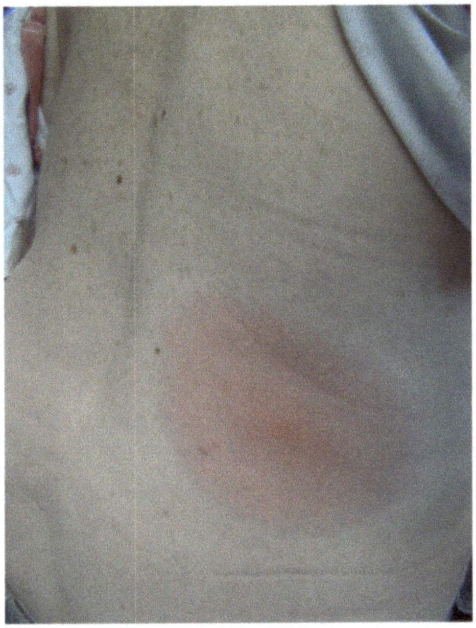

Fig. 7 Erythema migrans.

in 70%–80% of patients diagnosed with Lyme disease.[46] It develops within 1–2 weeks of the tick bite. Central clearing can develop as the lesion expands. Induration, necrotic changes, and vesicle formation have all been described at the center of the lesion. A punctum indicating the site of tick attachment may be seen (see Fig. 7). Local symptoms include itching, burning, and, occasionally, pain. Systemic symptoms are often present, especially in the United States. These include malaise, fatigue, fever, chills, myalgia, arthralgia, and headache. Regional lymphadenopathy has been reported. The axilla, groin, thigh, and popliteal fossa are common sites, although lesions can be found anywhere on the body.[36] EM resolves over days to weeks, even in the absence of treatment.

The spirochete disseminates over the ensuing days to weeks to produce manifestations in the skin, heart, CNS, and peripheral nervous system. Multiple EM lesions occur and are not associated with the site of the tick bite. These lesions fade over 3–4 weeks. Systemic symptoms are often present. Cardiac involvement occurs in 4%–8% of patients and is characterized by fluctuating atrioventricular blocks. Heart blocks resolve with antibiotic therapy, and permanent pacemaker insertion is not required. Rarely, fatal acute pancarditis and dilated cardiomyopathy have been reported in the United States and Europe.[47–49]

Neuroborreliosis occurs in 15% of patients with untreated EM during the early disseminated stage with a median interval of 4 weeks from the onset of EM. The most common manifestation in the United States is facial palsy, which can be bilateral. A concomitant, relatively asymptomatic lymphocytic meningitis is found in about a third of patients. The presence of systemic symptoms, bilateral involvement, and CSF pleocytosis can help differentiate this from Bell's palsy. Bannwarth's syndrome or Garin-Boujadoux-Bannwarth syndrome is the most common manifestation in Europe and is associated with *B. garinii* infection. The syndrome consists of painful neuritis or meningoradiculitis with lymphocytic CSF pleocytosis. Meningitis most commonly manifests as an acute or subacute headache of fluctuating intensity. Neck stiffness is not prominent; Kernig's and Brudzinski's signs are usually absent.[36,50]

Radiculoneuritis is another common manifestation, and can involve motor as well as sensory nerves. It often presents as a deep, severe pain along the nerve distribution that is unresponsive to usual analgesics and is worse at night. This can easily be misdiagnosed as a lumbar disc prolapse or other mechanical causes of neuropathy as well as herpes zoster sine herpete. Mononeuritis multiplex, cerebellar ataxia, and myelitis are other reported manifestations. It is important to note that multiple neurological manifestations can be present concomitantly in a single patient.[50]

In late disease, several months after the onset of infection, a chronic axonal polyneuropathy may develop in rare cases. Radicular pain and distal paresthesias are the most common manifestations of this condition. The existence of this entity in association with *B. burgdorferi sensu stricto* infection has been questioned.[51] *B. garinii* has been linked to a chronic encephalomyelitis characterized by ataxia, spastic paraparesis, bladder dysfunction, cognitive impairment, and cranial neuropathy (VII or VIII). In extremely rare cases, CNS vasculitis and stroke have been reported.[52]

The most common presentation in late infection is arthritis, occurring in about 60% of untreated patients in the United States. This primarily involves the large joint, affecting one or two at a time. Recurrent attacks occur in the absence of treatment; however the number decreases by 10%–20% each year and gradually resolves altogether. A minority experience intermittent or persistent arthritis even after appropriate antibiotic therapy.[36,44]

3.1.4 Diagnosis

The diagnosis in early localized infection is made clinically by evaluating the EM rash and considering the epidemiologic risk factors. Serologic testing has a low sensitivity at this stage (<40%).[46] Southern tick-associated rash illness (STARI) caused by the bite of the Lone Star tick (*Amblyomma americanum*) manifests a rash that is indistinguishable from EM but is uncommon in Lyme-endemic areas. *B. burgdorferi* has been cultured in a Barbour-Stoenner-Kelly (BSK) medium from biopsies of EM lesions as well as from plasma in the early stages.[36]

Serologic testing is the cornerstone of establishing the diagnosis in the disseminated and late stages of infection. The Centers for Disease Control (CDC) in the United States currently recommends a two-tier testing approach. An ELISA is first performed, and those with equivocal or positive results are tested by Western Blotting. In the United States, a whole cell sonicate (WCS) of *B. burgdorferi* strain B31 is most commonly used. As per criteria, an IgM blot is considered positive if two of the following three bands are present: 23, 39, and 41 kDa while the IgG blot is considered positive if five of the following 10 bands are present: 18, 23, 28, 30, 39, 41, 45, 58, 66, and 93 kDa. Two caveats with regard to the Western blot are the presence of the 41 kDa IgG band in about half the normal population and the possibility of a false-positive result when only 23 and 41 kDa IgM bands are present.[36]

IgM testing alone should not be used to support the diagnosis in patients after the first 2 months of illness. In patients with meningitis, CSF analysis usually reveals a lymphocytic pleocytosis of around a 100 cells/cu mm, a mild to moderate protein elevation, and a normal glucose level. Intrathecal antibody production (IgM, IgG, and IgA) can be demonstrated and supports the diagnosis. PCR testing for *B. burgdorgeri* in the CSF has a low sensitivity (10%–30%). In late neurologic disease (encephalopathy, axonal neuropathy), CSF analysis is often normal; however. intrathecal antibody production is usually demonstrable. Testing for intrathecal antibodies can be performed using an antibody capture assay or by diluting serum and CSF specimens to a similar final immunoglobulin concentration and simultaneously performing a *B. burgdorferi* specific ELISA. A CSF index of 1.3 (CSF antibody titer exceeds that in the serum) is indicative of intrathecal antibody production. These steps are necessary to exclude passive transfer of antibodies into the CSF from serum. This test is positive in about 50% of patients with early Lyme neuroborreliosis in the United States.[36,46,50]

The C6 peptide ELISA employing a 26-mer peptide of the sixth invariant region of the VlsE lipoprotein of *B. burgdorferi* is a promising second-generation test. The main advantage is the early detection of IgG, eliminating the need for IgM testing. A two-tier test with the traditional WCS ELISA as step 1 followed by C6 ELISA as step 2 has comparable specificity with the conventional testing method. Disease acquired in Europe can be diagnosed with this test while the traditional B31 strain WCS testing has lower sensitivity.[36]

The diagnosis of Lyme disease has been fraught with ambiguity and an exaggerated perception of its protean manifestations. This is further compounded by limited opportunities during the illness to successfully culture *B. burgdorferi* from human specimens as well as the interpretation of serologic tests without considering the clinical context. One study found that the seroprevalence of *B. burgdorferi* antibodies in healthy subjects in the Northeastern United States was 9.4%.[53] Serologic tests, including those for C6 peptide, IgM, and intrathecal antibodies, can remain positive for years after treatment.[54] Therefore inappropriate use of serologic testing in endemic areas can lead to misattribution of symptoms to Lyme disease and inappropriate therapy.

Lyme disease should be approached as an illness with discrete, recognizable manifestations. The patient should have a history of exposure or risk of exposure to Ixodes ticks in an endemic area during the appropriate season. Serologic testing should be employed in the right context, recognizing the limitations of using an antibody response to diagnose an active infection.

3.1.5 Treatment

The most commonly used regimen to treat Lyme Neuroborreliosis is 2–4 weeks of IV ceftriaxone, 2 g daily.[55] Oral doxycycline (200–400 mg total daily dose) has been shown to be equivalent to ceftriaxone in European studies.[56,57] Oral regimens are recommended to treat facial palsy.[46] Because doxycycline is contraindicated in children <8 years, pregnant women, and lactating women, alternative agents must be considered in these populations.

Amoxicillin and cefuroxime have good efficacy while the macrolides (azithromycin) have been associated with treatment failure.[46] Recovery in patients with meningitis and radiculitis occurs fairly quickly after initiating antibiotic therapy. Facial palsy also usually resolves, with or without antibiotic therapy. On occasion, with severe nerve injuries, recovery may be only partial. In late Lyme neuroborreliosis recovery is more gradual and can continue for 1–2 years after treatment.[50]

4. PREVENTION OF TICK-BORNE DISEASES

General measures to prevent tick-transmitted infections include frequent body checks for ticks, wearing long-sleeved shirts, long pants tucked into socks, use of DEET containing insect repellants, wearing permethrin-impregnated clothing, and bathing or showering within 2 h of tick exposure.[46]

Vaccines are available to prevent TBEV. There is virtually no risk of acquiring Lyme disease if the tick is removed within 24 h of attachment.[40] A single 200 mg dose of oral doxycycline was found to be 87% effective in preventing Lyme disease if administered within 72 h of the tick bite.[58]

5. CONCLUSION

Newly recognized tick-borne diseases are being added to the list of emerging infections on a regular basis. Climate change is driving an expansion in the distribution of both tick vectors and the infectious agents they transmit. Awareness of these infections can facilitate appropriate diagnosis and therapy, which can be life-saving.

REFERENCES

1. Diaz JH. Ticks, including tick paralysis. In: Bennett JE, Dolin R, Blaser MJ, editors. *Mandell, Douglas, and Bennett's principles and practice of infectious diseases*. 8th ed. Philadelphia: Saunders; 2015. p. 3266–79.
2. Schneider H. Uber epidemische akute Meningitis serosa. *Wien Klin Wochenschr* 1931;**44**:350–2.
3. Mandl CW. Steps of the tick-borne encephalitis virus replication cycle that affect neuropathogenesis. *Virus Res* 2005;**111**(2):161–74.

4. Palus M, Vancova M, Sirmarova J, Elsterova J, Perner J, Ruzek D. Tick-borne encephalitis virus infects human brain microvascular endothelial cells without compromising blood-brain barrier integrity. *Virology* 2017;**507**:110–22.
5. Zhang X, Zheng Z, Liu X, Shu B, Mao P, Bai B, et al. Tick-borne encephalitis virus induces chemokine RANTES expression via activation of IRF-3 pathway. *J Neuroinflammation* 2016;**13**(1):209.
6. Alekseev AN, Chunikhin SP. The experimental transmission of the tick-borne encephalitis virus by ixodid ticks (the mechanisms, time periods, species and sex differences). *Parazitologiia* 1990;**24**(3): 177–85.
7. Kohl I, Kozuch O, Eleckova E, Labuda M, Zaludko J. Family outbreak of alimentary tick-borne encephalitis in Slovakia associated with a natural focus of infection. *Eur J Epidemiol* 1996;**12**(4): 373–5.
8. Lipowski D, Popiel M, Perlejewski K, Nakamura S, Bukowska-Osko I, Rzadkiewicz E, et al. A cluster of fatal tick-borne encephalitis virus infection in organ transplant setting. *J Infect Dis* 2017;**215**(6): 896–901.
9. Fisher M, Rabe IB, Rollin PE. *Tickborne encephalitis*. In: *CDC health information for international travel 2018 [Internet]*. New York: Oxford University Press; 2018. Available from: https://wwwnc.cdc.gov/travel/yellowbook/2016/infectious-diseases-related-to-travel/tickborne-encephalitis.
10. Lindquist L, Vapalahti O. Tick-borne encephalitis. *Lancet* 2008;**371**(9627):1861–71.
11. Kunze U. Tick-borne encephalitis as a notifiable disease—Status quo and the way forward. Report of the 17th annual meeting of the International Scientific Working Group on Tick-Borne Encephalitis (ISW-TBE). *Ticks Tick Borne Dis* 2015;**6**(5):545–8.
12. Randolph SE. Transmission of tick-borne pathogens between co-feeding ticks: Milan Labuda's enduring paradigm. *Ticks Tick Borne Dis* 2011;**2**(4):179–82.
13. Thomas SJ, Endy TP, Rothman AL, Barrett AD. Flaviviruses. In: Bennett JE, Dolin R, Blaser MJ, editors. *Mandell, Douglas, and Bennett's principles and practice of infectious diseases*. 8th ed. Philadelphia: Saunders; 2015. p. 1881–903.
14. Bogovic P, Strle F. Tick-borne encephalitis: a review of epidemiology, clinical characteristics, and management. *World J Clin Cases* 2015;**3**(5):430–41.
15. Sips GJ, Wilschut J, Smit JM. Neuroinvasive flavivirus infections. *Rev Med Virol* 2012;**22**(2):69–87.
16. Mickiene A, Laiskonis A, Gunther G, Vene S, Lundkvist A, Lindquist L. Tickborne encephalitis in an area of high endemicity in lithuania: disease severity and long-term prognosis. *Clin Infect Dis* 2002;**35**(6):650–8.
17. Ternovoi VA, Kurzhukov GP, Sokolov YV, Ivanov GY, Ivanisenko VA, Loktev AV, et al. Tick-borne encephalitis with hemorrhagic syndrome, Novosibirsk Region, Russia, 1999. *Emerg Infect Dis* 2003;**9**(6):743–6.
18. Stiasny K, Aberle JH, Chmelik V, Karrer U, Holzmann H, Heinz FX. Quantitative determination of IgM antibodies reduces the pitfalls in the serodiagnosis of tick-borne encephalitis. *J Clin Virol* 2012;**54**(2):115–20.
19. Atrasheuskaya AV, Fredeking TM, Ignatyev GM. Changes in immune parameters and their correction in human cases of tick-borne encephalitis. *Clin Exp Immunol* 2003;**131**(1):148–54.
20. Mandl CW, Holzmann H, Kunz C, Heinz FX. Complete genomic sequence of Powassan virus: evaluation of genetic elements in tick-borne versus mosquito-borne flaviviruses. *Virology* 1993;**194**(1): 173–84.
21. Kuno G, Artsob H, Karabatsos N, Tsuchiya KR, Chang GJ. Genomic sequencing of deer tick virus and phylogeny of powassan-related viruses of North America. *Am J Trop Med Hyg* 2001;**65**(5):671–6.
22. El Khoury MY, Camargo JF, White JL, Backenson BP, Dupuis 2nd AP, Escuyer KL, et al. Potential role of deer tick virus in Powassan encephalitis cases in Lyme disease-endemic areas of New York, U.S.A. *Emerg Infect Dis* 2013;**19**(12):1926–33.
23. Ebel GD. Update on Powassan virus: emergence of a North American tick-borne flavivirus. *Annu Rev Entomol* 2010;**55**:95–110.
24. Dupuis 2nd AP, Peters RJ, Prusinski MA, Falco RC, Ostfeld RS, Kramer LD. Isolation of deer tick virus (Powassan virus, lineage II) from Ixodes scapularis and detection of antibody in vertebrate hosts sampled in the Hudson Valley, New York State. *Parasit Vectors* 2013;**6**:185.

25. Ebel GD, Kramer LD. Short report: duration of tick attachment required for transmission of powassan virus by deer ticks. *Am J Trop Med Hyg* 2004;**71**(3):268–71.
26. McLean DM, Donohue WL. Powassan virus: isolation of virus from a fatal case of encephalitis. *Can Med Assoc J* 1959;**80**(9):708–11.
27. Hermance ME, Thangamani S. Powassan virus: an emerging arbovirus of public health concern in North America. *Vector Borne Zoonotic Dis* 2017;**17**(7):453–62.
28. Gholam BI, Puksa S, Provias JP. Powassan encephalitis: a case report with neuropathology and literature review. *CMAJ* 1999;**161**(11):1419–22.
29. Tavakoli NP, Wang H, Dupuis M, Hull R, Ebel GD, Gilmore EJ, et al. Fatal case of deer tick virus encephalitis. *N Engl J Med* 2009;**360**(20):2099–107.
30. Mlera L, Meade-White K, Saturday G, Scott D, Bloom ME. Modeling Powassan virus infection in *Peromyscus leucopus*, a natural host. *PLoS Negl Trop Dis* 2017;**11**(1).
31. Piantadosi A, Rubin DB, McQuillen DP, Hsu L, Lederer PA, Ashbaugh CD, et al. Emerging cases of Powassan virus encephalitis in New England: clinical presentation, imaging, and review of the literature. *Clin Infect Dis* 2016;**62**(6):707–13.
32. Raval M, Singhal M, Guerrero D, Alonto A. Powassan virus infection: case series and literature review from a single institution. *BMC Res Notes* 2012;**5**:594.
33. Sung S, Wurcel AG, Whittier S, Kulas K, Kramer LD, Flam R, et al. Powassan meningoencephalitis, New York, New York, USA. *Emerg Infect Dis* 2013;**19**(9).
34. Hicar MD, Edwards K, Bloch K. Powassan virus infection presenting as acute disseminated encephalomyelitis in Tennessee. *Pediatr Infect Dis J* 2011;**30**(1):86–8.
35. CDC. Lyme disease. Available from: https://www.cdc.gov/lyme/stats/.
36. Steere AC. Lyme disease (Lyme Borreliosis) due to *Borrelia burgdorferi*. In: Bennett JE, Dolin R, Blaser MJ, editors. *Mandell, Douglas, and Bennett's principles and practice of infectious diseases*. 8th ed. Philadelphia: Saunders; 2015. p. 2725–35.
37. Fraser CM, Casjens S, Huang WM, Sutton GG, Clayton R, Lathigra R, et al. Genomic sequence of a Lyme disease spirochaete, *Borrelia burgdorferi*. *Nature* 1997;**390**(6660):580–6.
38. Wroblewski D, Gebhardt L, Prusinski MA, Meehan LJ, Halse TA, Musser KA. Detection of *Borrelia miyamotoi* and other tick-borne pathogens in human clinical specimens and Ixodes scapularis ticks in New York State, 2012–2015. *Ticks Tick Borne Dis* 2017;**8**(3):407–11.
39. Waindok P, Schicht S, Fingerle V, Strube C. Lyme borreliae prevalence and genospecies distribution in ticks removed from humans. *Ticks Tick Borne Dis* 2017;**8**(5):709–14.
40. des Vignes F, Piesman J, Heffernan R, Schulze TL, Stafford 3rd KC, Fish D. Effect of tick removal on transmission of *Borrelia burgdorferi* and Ehrlichia phagocytophila by Ixodes scapularis nymphs. *J Infect Dis* 2001;**183**(5):773–8.
41. Sood SK, Salzman MB, Johnson BJ, Happ CM, Feig K, Carmody L, et al. Duration of tick attachment as a predictor of the risk of Lyme disease in an area in which Lyme disease is endemic. *J Infect Dis* 1997;**175**(4):996–9.
42. Fikrig E, Narasimhan S. *Borrelia burgdorferi*—traveling incognito? *Microbes Infect* 2006;**8**(5):1390–9.
43. Montgomery RR, Lusitani D, de Boisfleury Chevance A, Malawista SE. Human phagocytic cells in the early innate immune response to *Borrelia burgdorferi*. *J Infect Dis* 2002;**185**(12):1773–9.
44. Steere AC, Schoen RT, Taylor E. The clinical evolution of Lyme arthritis. *Ann Intern Med* 1987;**107**(5):725–31.
45. Steere AC, Sikand VK, Schoen RT, Nowakowski J. Asymptomatic infection with *Borrelia burgdorferi*. *Clin Infect Dis* 2003;**37**(4):528–32.
46. Sanchez E, Vannier E, Wormser GP, Hu LT. Diagnosis, treatment, and prevention of lyme disease, human granulocytic anaplasmosis, and babesiosis: a review. *JAMA* 2016;**315**(16):1767–77.
47. Marcus LC, Steere AC, Duray PH, Anderson AE, Mahoney EB. Fatal pancarditis in a patient with coexistent Lyme disease and babesiosis. Demonstration of spirochetes in the myocardium. *Ann Intern Med* 1985;**103**(3):374–6.
48. Lardieri G, Salvi A, Camerini F, Cinco M, Trevisan G. Isolation of *Borrelia burgdorferi* from myocardium. *Lancet* 1993;**342**(8869):490.

49. Centers for Disease Control and Prevention (CDC), Three sudden cardiac deaths associated with Lyme carditis – United States, November 2012–July 2013. *MMWR Morb Mortal Wkly Rep* 2013;**62**(49):993–6.
50. Marques AR. Lyme neuroborreliosis. *Continuum* 2015;**21**(6 Neuroinfectious Disease):1729–44.
51. Wormser GP, Strle F, Shapiro ED, Dattwyler RJ, Auwaerter PG. A critical appraisal of the mild axonal peripheral neuropathy of late neurologic Lyme disease. *Diagn Microbiol Infect Dis* 2017;**87**(2):163–7.
52. Garkowski A, Zajkowska J, Zajkowska A, Kulakowska A, Zajkowska O, Kubas B, et al. Cerebrovascular manifestations of Lyme Neuroborreliosis—a systematic review of published cases. *Front Neurol* 2017;**8**:146.
53. Krause PJ, Narasimhan S, Wormser GP, Barbour AG, Platonov AE, Brancato J, et al. *Borrelia miyamotoi* sensu lato seroreactivity and seroprevalence in the northeastern United States. *Emerg Infect Dis* 2014;**20**(7):1183–90.
54. Steere AC, McHugh G, Damle N, Sikand VK. Prospective study of serologic tests for lyme disease. *Clin Infect Dis* 2008;**47**(2):188–95.
55. Wormser GP, Dattwyler RJ, Shapiro ED, Halperin JJ, Steere AC, Klempner MS, et al. The clinical assessment, treatment, and prevention of lyme disease, human granulocytic anaplasmosis, and babesiosis: clinical practice guidelines by the Infectious Diseases Society of America. *Clin Infect Dis* 2006;**43**(9):1089–134.
56. Ljostad U, Skogvoll E, Eikeland R, Midgard R, Skarpaas T, Berg A, et al. Oral doxycycline versus intravenous ceftriaxone for European Lyme neuroborreliosis: a multicentre, non-inferiority, double-blind, randomised trial. *Lancet Neurol* 2008;**7**(8):690–5.
57. Bremell D, Dotevall L. Oral doxycycline for Lyme neuroborreliosis with symptoms of encephalitis, myelitis, vasculitis or intracranial hypertension. *Eur J Neurol* 2014;**21**(9):1162–7.
58. Nadelman RB, Nowakowski J, Fish D, Falco RC, Freeman K, McKenna D, et al. Prophylaxis with single-dose doxycycline for the prevention of Lyme disease after an Ixodes scapularis tick bite. *N Engl J Med* 2001;**345**(2):79–84.

CHAPTER 11

Interactions of Human Retroviruses With the Blood-Brain Barrier

Céline Curis*,†,‡, Philippe V. Afonso*,†
*Institut Pasteur, Paris, France
†Centre National de la Recherche Scientifique (CNRS), Paris, France
‡Université Paris Diderot, Sorbonne Paris Cité, Paris, France

1. INTRODUCTION

The internal milieu is delimited by barriers, such as the skin and mucosa, such as the intestinal barrier. They provide physical and immunological protections. Additionally, a physiological barrier surrounds the central nervous system (CNS), which is composed of the brain and spinal cord, and delineates a protected niche within the host.[1]

P. Ehrlich first demonstrated the existence of an impermeable structure between the blood and the CNS in 1880. When injecting trypan blue in blood, the dye was found in all organs but the brain and the spinal cord; conversely, the injection of trypan blue in the cerebrospinal fluid (CSF) colored the neural tissue but did not diffuse into the blood. There are two specific interfaces separating the CNS from the blood: the blood cerebrospinal fluid barrier (BCSFB), which is established by choroid plexus epithelial cells, and the blood-brain barrier (BBB).

The physical barrier of the BBB is made of a continuous endothelium of cells sealed together through protein complexes called tight junctions (TJs) (Fig. 1A). TJs are composed of transmembrane proteins [claudins, occludins, and junction adhesion molecules (JAMs)] linked to the actin cytoskeleton through cytoplasmic accessory proteins (Zonula Occludens (ZO)-1 to 3, and cingulin). TJs limit the paracellular diffusion of molecules through the BBB.[2] Moreover, pinocytosis through endothelial cells is rare. Transport of water, nutrients, amino acids, or hormones into the CNS, and efflux of metabolites and xenobiotics from the CNS, are regulated by active transporters. Thus the BBB forms a selective barrier responsible for maintaining strict CNS fluid homeostasis.

The vascular barrier is surrounded by astrocytes (through their endfeet), pericytes, and perivascular antigen-presenting cells.[3] These cells interact with each other and with the acellular layer (the basal membrane), forming the neurovascular unit of the BBB. Astrocytes participate in BBB maturation and stability. Pericytes support BBB integrity and regulate blood flow though their contractile properties.

The Microbiology of Central Nervous System Infections
https://doi.org/10.1016/B978-0-12-813806-9.00011-1

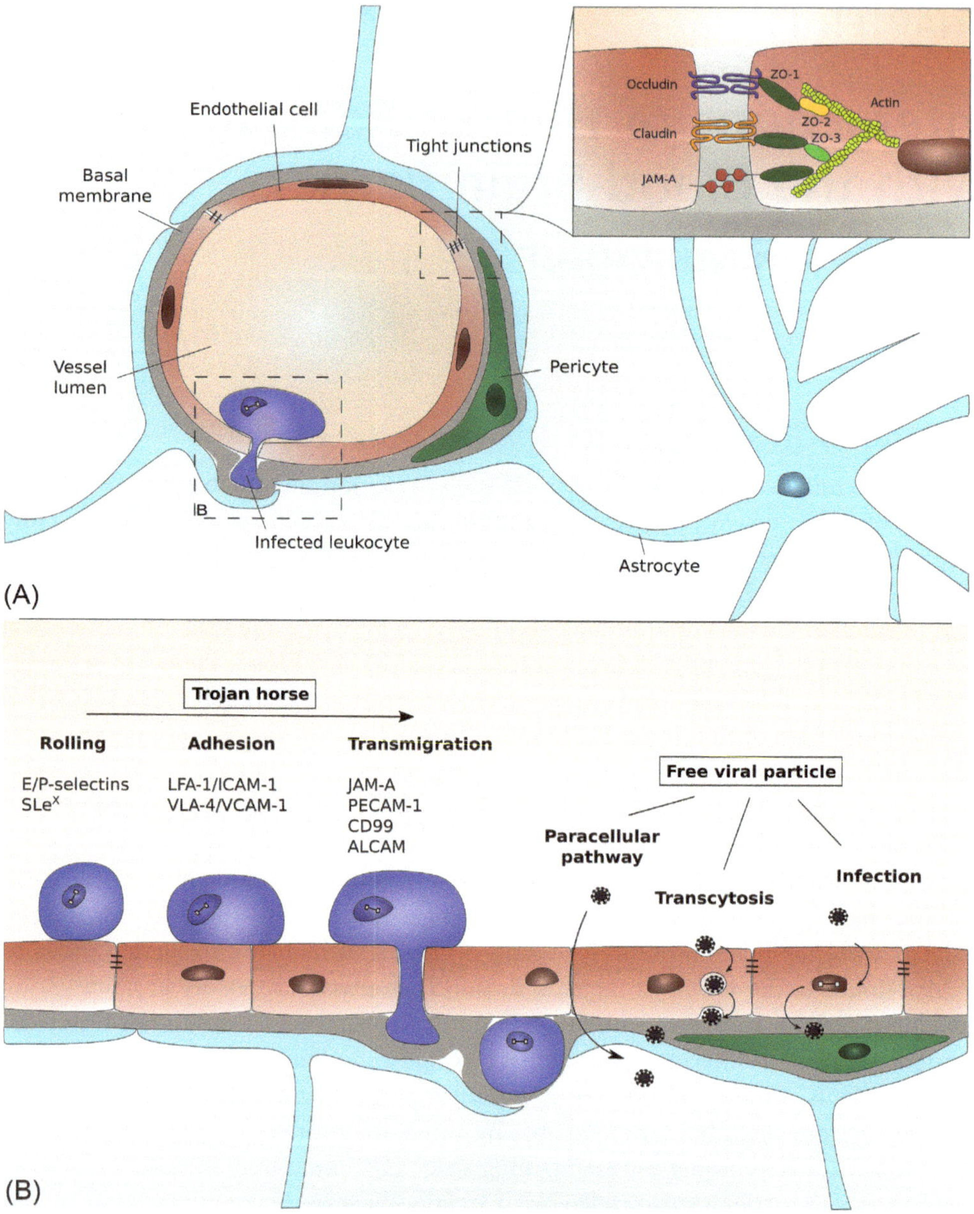

Fig. 1 The BBB can be crossed by human retroviruses targeting the CNS. (A) Schematic representation of the BBB. Endothelial cells are sealed together with tight junctions. Surrounded by pericytes and astrocytes, and together with the basal membrane, they form the neurovascular unit of the BBB. Retroviral-infected leukocytes can overcome this barrier. (B) Retroviruses cross the BBB via two main routes. The migration of infected cells across the BBB, also known as a Trojan horse, is a multiple-step mechanism. It involves rolling and attachment of the leukocyte to the endothelium, tight junctions disruption, and finally extravasation. Free viral particles can also directly access the CNS by paracellular transport, transcytosis, or through productive infection of endothelial cells.

Multiple CNS pathologies are associated with BBB dysfunction. BBB leakage may allow fluid to accumulate and generate edema within the CNS; inflammatory cells may cross the barriers and participate in neuroinflammation, or autoimmune diseases.[4] Some infectious agents have developed CNS invasion strategies,[5] among them viruses and in particular retroviruses.

2. RETROVIRUS-ASSOCIATED NEUROLOGICAL DISORDERS

Retroviruses are enveloped RNA viruses that transcribe RNA into DNA prior to integration into the host chromosomal DNA. Two human retroviruses have been associated with CNS diseases: human T-lymphotropic virus type 1 (HTLV-1) and human immunodeficiency virus type 1 (HIV-1).

HTLV-1 was the first human retrovirus identified. It is a member of the *deltaretrovirus* genus, and infects at least 5–10 million people worldwide.[6] While most HTLV-1-infected individuals remain asymptomatic, HTLV-1 is the causal agent of a severe lymphoproliferation, the adult T-cell leukemia/lymphoma (ATL), and a wide range of inflammatory diseases. In particular, HTLV-1 has been linked to a neuroinflammatory disease, the tropical spastic paraparesis or HTLV-1-associated myelopathy (HAM/TSP).[7] HAM/TSP is a progressive neurological disorder characterized by limb weakness, spasticity, and autonomic dysfunctions, which develop in 0.1%–2% of infected individuals. Over the course of the disease, patients evolve from difficulties in walking to being bedridden. The progression occurs mainly during the first year after the onset of the disease. HAM/TSP is characterized by spinal cord atrophy and demyelinization, in particular in the thoracic region of the spinal cord.[8] Demyelinization is associated with the presence of perivascular infiltrates in the CNS. Among the infiltrated cells, infected CD4 T cells are present, and HAM/TSP diagnosis can be confirmed by detection of the provirus in the CSF of patients.

Some rare cases of pathologies resembling HAM/TSP have been associated with HTLV-2 infection.[9]

HIV-1 is a member of the *lentivirus* genus. It was isolated in 1983 by Montagnier and colleagues, although the associated pathology, the acquired immunodeficiency syndrome (AIDS), had been known since the 1970s. HIV-1 infection is often associated with neurological complications of AIDS, called neuroAIDS or HIV-associated neurocognitive disorders (HAND).[10] Without any treatment, these neurological impairments affect 60% of HIV-infected patients and are directly associated with HIV infection, as they can occur in the absence of opportunistic infections. The most common syndromes are HIV-associated dementia (HAD), HIV-encephalitis (HIVE), or minor cognitive and motor disorder (MCMD). HAD is characterized by motor, cognitive, and behavioral changes such as impaired short-term memory, reduced mental concentration, muscle weakness, and depression. A progressive cortical atrophy can be observed by magnetic resonance imaging (MRI). This is linked with infiltration of HIV-1-infected macrophages into the CNS, leading to axonal and myelin damage and neuronal loss.

Of note, HIV-2 was also reported to cause encephalitis in a few cases.[11]

Finally, foamy viruses are retroviruses of the *spumavirus* genus, which infect simians, apes, horses, cats, and cows, among other species. Sporadically, they can be transmitted from simian to human through severe bites.[12] The foamy provirus was detected in the CNS in African Green Monkeys.[13] Moreover, in transgenic mice, the expression of foamy virus regulatory proteins generates neurotoxicity.[14] However, foamy virus infection has not been clearly associated with CNS pathologies.

3. RETROVIRUS-ASSOCIATED BBB DISRUPTION

BBB disruption during retrovirus-associated neurological diseases was first proven by studying the diffusion of serum markers within the brain parenchyma. Albumin was detected in the brain parenchyma of patients with HIV-1-related neurologic disease.[15] Immunoglobulin and fibrinogen deposits were also found in the brain parenchyma of HIV-infected patients[16] and of a HAM/TSP patient.[17] In HAM/TSP patients, spinal cord edemas, which are evidenced by MRI, indicate active inflammatory processes and BBB disruption.[18] Such a disruption occurs early during disease onset, as spinal cord edemas can be detected in some asymptomatic carriers.

The presence of cellular infiltrates in the CNS is another marker of a dysfunctional BBB. Perivascular accumulations of HIV-1-infected macrophages are detected in HIVE patients.[19] Perivascular lymphocyte infiltration is also characteristic in spinal cord lesions of HAM/TSP patients.[20]

The loss of function of the BBB is associated with the disruption of the TJs. Immunohistological studies showed fragmentation or absence of immunoreactivity for occludin and ZO-1 in the vicinity of macrophage perivascular accumulation in CNS tissues from patients with HIVE.[19] The level of ZO-1 disorganization correlates with the extent of dementia in HAD patients.[21] Similarly, ZO-1 staining on spinal cord sections of HAM/TSP patients revealed that TJs are disorganized on the BBB endothelium, especially close to leukocyte infiltrates.[22]

In the context of a retroviral infection, BBB disruption can be induced by several mechanisms: infection of BBB cells, proinflammatory cytokine production, and viral protein secretion.

3.1 BBB Disruption by Retroviral Infection

Cerebral microvascular cells express HIV receptors CCR5, CCR3, and CCR1 but not CD4. This is consistent with in vitro studies demonstrating that primary human brain capillaries could be infected with HIV-1 in a CD4-independent manner.[23] Yet infection of microvascular endothelial cells in vivo is still debated. In situ hybridization early demonstrated HIV infection around the BBB vasculature; however these results could correspond to viral particles adsorbed at the surface of the endothelium, or to the presence of perivascular infected macrophages.[24]

BBB endothelial cells express different receptors for HTLV-1 (GLUT-1, Neuropilin-1, and heparan sulfate proteoglycans).[25] In vitro, the human cerebral microvascular endothelial cell line hCMEC/D3 can be productively infected with HTLV-1. Upon infection, these cells can no longer form efficient TJs nor an impermeable and functional barrier.[25] In situ hybridization suggested that BBB endothelial cells could be infected with HTLV-1 in vivo, although this seems to be a rare event.[25]

Astrocytes can be infected with HTLV-1 in vivo.[26] Upon infection, astrocytes may display an altered glutamate metabolism and imbalanced matrix metalloproteinase (MMP)/MMP inhibitor expression, which can eventually lead to altered functionality of the BBB.[27] Moreover, infected astrocytes could be killed by anti-Tax cytotoxic T lymphocytes. Astrocytes are also efficiently infected by HIV via cell-to-cell contact, although astrocytes were reported as resistant to infection via cell-free HIV in vitro.[28]

Pericytes can be productively infected with HIV in vitro.[29] Microglia cells are also susceptible to HTLV-1 infection,[30] and may lead to viral persistence in tissues.

Upon infection of BBB elements, the barrier may be disrupted. This could occur through infection-induced cell death, alterations of cellular pathways and TJ regulation, or through the secretion of many factors (e.g., cytokines, viral proteins, MMPs) that could alter the barrier physiology. The BBB may also be disrupted by the secretion of similar factors by the infected cells (HIV-infected monocytes or HTLV-1-infected lymphocytes).

3.2 BBB Disruption by Soluble Cellular Factors

3.2.1 Proinflammatory Cytokines

Neurological diseases of retroviral origin are associated with an increase in the concentration of cytokines in the serum and the CSF in vivo. Upon HIV-1 infection, even during the asymptomatic phase, a chronic inflammation takes place that is related to elevated levels of cytokines in plasma and in the CSF.[31] Of note, even during highly active antiretroviral therapy (HAART), the levels of inflammatory cytokines are still elevated.[32] In HTLV-1 infected patients, CXCL10, LTB-4, IL-6, TNF-α, and IFN-γ are present at high levels in plasma.[33] CXCL10, CXCL9, and neopterin levels are elevated in the CSF of HAM/TSP patients and are prognostic biomarkers of HAM/TSP.[34]

These different mediators of inflammation participate in BBB disruption. In vitro, treatment of endothelial cells monolayers with inflammatory cytokines leads to endothelial monolayer disruption.[22,35]

These cytokines can be produced directly by infected cells. HIV-1-infected monocytes overproduce inflammatory cytokines such as TNF-α and IL-1ß.[36] HTLV-1-infected CD4 T cells secrete a wide range of inflammatory mediators. In vitro, secretion of IL-1α and TNF-α by HTLV-1-infected lymphocytes is necessary and sufficient to induce an increase in brain endothelial monolayer permeability.[22]

Cytokines can also be produced by uninfected cells in the presence of the virus or infected cells. For example, TNF-α and IL-1ß, secreted by HIV-infected cells, activate microglial cells and astrocytes, which become the major contributors to neuroinflammation and neural cell death.[37]

3.2.2 MMPs

Matrix metalloproteinases (MMPs) are a family of zinc-dependent endopeptidases targeting proteins of the extracellular matrix. MMPs are known to increase the permeability of the BBB.[38] These enzymes alter the basal membrane of the neurovascular unit. They also trigger occludin degradation, leading to TJ alteration and apoptosis of endothelial cells.

In HIV-infected patients, CSF levels of MMPs are elevated.[39] MMP-9 upregulation correlates with the number of infiltrated cells, suggesting a causal relationship between MMP activity and BBB leakage. The downregulation of MMP regulators TIMPs observed in brain and CSF samples from HAD patients might also contribute to the pathogenesis of neural tissue damage.

Similarly, the MMP/TIMP balance is modified in astrocytes upon HTLV-1 infection.[27] Inhibiting MMPs was shown to reduce the migration of HTLV-1-infected CD4 T cells through the BBB in vitro.[40]

3.3 BBB Disruption by Viral Proteins

3.3.1 Soluble Viral Transactivators: HIV-1 Tat and HTLV-1 Tax

In vitro, the HIV-1 transactivator Tat is found in the supernatant of HIV-infected cells, and is actively secreted.[41] While this protein is essential for efficient viral replication, an estimated 65% of the produced Tat is secreted into the extracellular space.[42] In vivo, soluble Tat can be detected in the CSF of infected individuals.

Soluble Tat has been found to disrupt cerebral endothelial monolayers in vitro. On brain microvascular endothelial cells, soluble Tat induces a decrease in claudin-1, claudin-5, and ZO-2 expressions and a disorganization of TJs.[43] Tat modulation of TJ proteins was confirmed in vivo. Injection of Tat into the hippocampus of C57BL/6 mice leads to disruption of ZO-1 and occludin distribution on BBB cells.[44] Interestingly, HIV-1 Tat could be associated with the severity of neurological disorders. HIV-1 in clade B is more neuropathic than HIV-1 clade C, and in vitro, Tat clade B disrupts the BBB to a greater extent than Tat from clade C.[45]

The HTLV-1 transactivator Tax could display similar functions. First, Tax can be secreted by HTLV-1 infected cells, either in a soluble form[46] or in exosomes.[47] Cell-free Tax has been found in the plasma of infected patients[48] and in the CSF of HAM/TSP patients.[49] It has been shown that both soluble Tax and exosomal Tax are able to activate the NF-κB pathway in recipient cells.[47,50] Thus, extracellular Tax could participate in endothelial activation and BBB disruption. Moreover, soluble Tax can alter the functions of astrocytes[51] and induce cytokine secretion by neuronal NT2-N cells;[52] these could disrupt the BBB in vivo.

3.3.2 HIV-1 Envelope Protein gp120

The HIV-1 envelope glycoprotein is composed of two subunits: the transmembrane gp41 and the surface gp120. While gp120 participates in receptor recognition, gp41 is implicated in membrane fusion. Gp120 can be shed from infected cells and is found circulating in the serum of HIV-infected individuals.[53]

In vitro, gp120 can disrupt rat brain endothelial cell monolayers, as assessed by an increase in permeability of the cultured cells.[54] In vivo, the BBB is permeable in gp120 transgenic mice, as demonstrated by diffusion of albumin into the CNS parenchyma.[55]

Several mechanisms for gp120-induced BBB permeabilization have been proposed. Gp120 could trigger degradation of TJ proteins[56] or cytotoxicity.[57] Moreover, a gp120-induced increase in permeability is likely associated with MMP upregulation and basal lamina degradation.[58]

3.3.3 HIV-1 Regulatory Proteins

HIV-1 negative regulatory factor (Nef) is a regulatory protein that is essential for viral persistence. It is anchored to cell membranes, and can function as a scaffold protein to activate and alter many cellular pathways. It is devoid of enzymatic activity. Nef can be detected in the plasma of HIV-1-infected patients at concentrations ranging from 5 to 10 ng/mL. It is, however, unclear if Nef is present on a soluble form or sequestered in circulating extracellular vesicles such as exosomes.[59,60] First, Nef can disrupt TJs. Exosomes containing Nef that are shed by HIV-infected microglial cells downregulate ZO-1 expression on HBMEC cells.[61] Nef may also be toxic to endothelial cells as it induces apoptosis via caspase activation in HBMECs.[62] Finally, BBB disruption by Nef seems to depend on MMP-9 activation.[63]

Similarly, the soluble HIV-1 regulatory viral protein Vpr is detected in the CSF and serum of HIV-1-infected patients with neurological disorders.[64] Vpr can alter astrocyte metabolism[65] and induce an inflammatory microenvironment.[66]

In a nutshell, retroviral proteins are capable of activating the vascular compartment of the BBB and triggering TJ disruption. The question of whether free viral protein concentration is high enough in vivo to induce significant changes in the BBB is not solved. The accumulation of infected leukocytes could locally generate a high concentration in secreted viral proteins.

4. CROSSING THE BBB

The CNS has long been considered to be immune to pathogens. However, retroviruses can be detected in the nervous tissue in many cases.

4.1 Detection of the Virus in the CNS

HIV antigens can be detected in the CSF of HIV-infected patients[67] and HIV-1 envelope subunit gp41 is detected on infiltrated cells by immunoperoxidase staining.[68] Similarly, HIV-1 nucleic acid can be amplified by PCR on several brain areas.[69] Finally, HIV particles can be isolated from the CSF of patients with neurological complications and in asymptomatic patients, though less frequently.[70]

In the CNS, HIV can infect a wide variety of cell types.[71] Productive infection is mostly studied in perivascular macrophages and microglia (resident brain macrophages). HIV is also detected in endothelial cells, brain monocytes, astrocytes, and possibly in oligodendrocytes and neurons, though this is still controversial.

Similarly, the presence of HTLV-1 in the CNS is reported in HAM/TSP patients. Though the virus is mostly detected in infiltrated lymphocytes or other hematopoietic cells, the nervous tissue can also be directly infected. In situ hybridization in postmortem brains of HAM/TSP patients revealed that astrocytes can be productively infected by HTLV-1;[26,72] microglia cells are also susceptible to HTLV-1 infection.[30] These could contribute to viral persistence in the tissue.

Retroviruses enter the CNS by crossing the BBB. Different routes can occur: either a passage as free viral particles, or as infected cells (Fig. 1B).

4.2 Crossing the BBB as a Free Viral Particle

Viral particles can cross the endothelial layer by transcytosis and infect the nervous tissue without altering the structure of the BBB.

For HIV-1, adsorptive endocytosis followed by transcytosis seems to be mediated by the gp120 envelope protein.[73] Transcytosis is amplified by cytokine production. LPS-induced transcytosis is dependent on IL-6 and granulocyte-macrophage colony-stimulating factor secretion by endothelial cells.[74] It is unclear whether cytokines enhance transcytosis in an autocrine/paracrine manner or through a crosstalk of the endothelial cells with pericytes.[75]

Transcytosis has also been suggested for HTLV-1 virions by ultrastructural analysis on rat brain capillary endothelial cells.[76] HTLV-1 virions are uptaken by endothelial cells and released on the basolateral side of the BBB.

Passage of viral particles could also occur through the altered BBB in a paracellular way. Fiala et al. showed that HIV-1 p24 antigen is detected in the CNS only after TNF-α- and IL-6- induced BBB disruption.[77] They concluded that HIV invades the brain by a paracellular route. It is, however, difficult to distinguish between transcytosis through the endothelial cells (even after the BBB is disrupted) and paracellular diffusion.

4.3 Crossing the BBB as an Infected Cell

Infected cells appear to be able to traffic directly into the CNS. A study by Cavrois et al. has demonstrated that HTLV-1-infected cells present in the CNS originate from clones present in the periphery,[78] demonstrating the capacity of HTLV-1 infected cells to cross the BBB in vivo. It is called the "Trojan horse mechanism." Leukocyte migration into the CNS is complex. Infected cells roll and adhere to the endothelium before they extravasate and penetrate the nervous tissue.

4.3.1 Increased Rolling of Infected Cells on the BBB Endothelium

Leukocyte rolling can be mediated by weak interactions between proteins of the selectin family and their ligands. Selectins are commonly expressed on endothelial cells. On one hand, in HIV patients, selectins can be overexpressed on endothelial cells. Endothelial E-selectin can be activated directly in the presence of HIV-infected circulating monocytes,[79] or upon infection of endothelial cells.[80] On the other hand, selectin ligands such as the sialyl Lewis^{x} antigen are overexpressed on CD4 T cells in HAM/TSP patients.[81]

4.3.2 Increased Adhesion of Infected Cells to the BBB Endothelium

After rolling, leukocytes tightly adhere to the endothelium. Ex vivo experiments indicated that mononuclear cells from HAM/TSP patients adhere more to primary activated endothelial monolayers than cells from asymptomatic carriers or healthy subjects.[82] Similarly, the adhesion of HIV-infected monocytes to human brain capillary endothelial cell monolayers is increased compared to uninfected monocytes.[83] Leukocyte firm adhesion to endothelial cells is mainly mediated through integrins, such as LFA-1 and VLA-4, with their endothelial receptors, ICAM-1 and VCAM-1, respectively.

Upon Tax expression, HTLV-1-infected lymphocytes overexpress integrins and immunoglobulins such as LFA-1, VCAM-1, and ICAM-1,[84,85] which may favor adhesion. Moreover, adhesion can be increased via integrin activation: the HTLV-1 accessory protein p8 enhances the clustering of LFA-1 and thus, adhesion.[86] Similarly, LFA-1 and LFA-3 are upregulated in HIV-infected monocytes.[83] In inflammatory conditions, adhesion molecules can also be overexpressed on endothelial cells. ICAM-1 and VCAM-1 expressions are enhanced on the endothelium of the spinal cord in HAM/TSP patients as well as on the endothelium of the brain in AIDS patients.[87] Interestingly, endothelial surface levels of ICAM-1 may influence the route of cellular transmigration through the BBB. High levels of ICAM-1 are correlated with a rapid transcellular diapedesis whereas intermediate levels of ICAM-1 promote the paracellular pathway.[88]

4.3.3 Increased Transmigration of Infected Cells Through the BBB Endothelium

Leukocyte attachment to the endothelium is ultimately followed by active crawling on the BBB and transmigration. Some studies suggest that infected cells could cross the BBB without disruption via emperipolesis. It is a sort of cellular transcytosis when the endothelial cell engulfs the invading leukocytes in so-called transmigratory cups. They are then released in the CNS without having caused any structural or functional abnormalities to the BBB. Though leukemic cells were shown to cross endothelial cells by this process at the level of the leptomeningeal vein in the meninges,[89] it has not been reported for HTLV-1 or HIV-1 infection yet.

As the BBB is disrupted in retrovirus-associated neurological disorders, it is more likely that infected cells (lymphocytes and monocytes/macrophages) gain access to the CNS by paracellular migration. This process may account for 90% of transmigration events,[90] and it is the most-studied mechanism for retroviral transmigration through the BBB. Extended interactions between leukocytes and endothelial cells induce deformation of the cells and cytoskeleton reorganization. These are regulated by a series of cell surface molecules.

Junctional adhesion molecules (JAMs) are transmembrane molecules of the TJs involved in transmigration.[91] In normal conditions, JAM-A is present at the TJs. In vitro, the culture of HIV-infected monocytes with human brain microvascular endothelial cells results in decreased expression of endothelial JAM-A.[92] Similarly, JAM-B is downregulated in the presence of gp120 or Tat.[93,94] These downregulations weaken the TJs, leading to BBB rupture. JAM-A is, however, critical for transmigration of $CD14^+ CD16^+$ monocytes across the BBB.[95] The remaining endothelial JAM-A is redistributed from the TJs to the endothelial surface, and can then interact with leukocytic JAM-A. This homophilic interaction participates in leukocyte transmigration.

Platelet/endothelial cell adhesion molecule 1 PECAM-1 (or CD31) was the first adhesion protein reported to be required for leukocyte transmigration through endothelial cells.[96] It was thought that homophilic PECAM-1 interactions between endothelial cells and leukocytes mediate transmigration. However, PECAM-1 function may be indirect. HIV-infected cells secrete soluble PECAM-1, which may interact with endothelial PECAM-1 and enhance BBB disruption.[97]

CD99 is a transmembrane protein present at endothelial cell junctions, and overexpressed by HIV-infected monocytes.[98] Blocking CD99 arrests cellular transmigration;[99] of note, it seems that CD99 acts downstream of PECAM-1.

Finally, ALCAM is a member of the immunoglobulin superfamily, which regulates actin cytoskeleton dynamics and participates in tissue development and maintenance. It is expressed by endothelial cells and leukocytes. Endothelial ALCAM mediates homophilic interactions with leukocytic ALCAM, and heterophilic interactions with leukocytic CD6. In vitro, ALCAM downregulation on primary human umbilical vein endothelial cells reduces monocyte transmigration.[100] Moreover, ALCAM is overexpressed by

HTLV-1-infected lymphocytes and is implicated in lymphocyte transmigration through the BBB.[101] Similarly, ALCAM expression is enhanced in HIV-infected monocytes, facilitating the transmigration of CD14+ CD16+ monocytes into the CNS of HIV-seropositive individuals.[95]

4.3.4 Amplification of the Infiltration Into the CNS via Chemotaxis

Once leukocytes have reached the CNS, they may amplify the recruitment of other leukocytes through chemotaxis in a positive feedback loop.

For example, secretion of HIV Tat by infiltrated monocytes can amplify monocyte recruitment. Indeed, Tat displays a chemotactic capacity for monocytes.[102] Moreover, Tat stimulates the production of the chemoattractant CCL2/MCP-1 by astrocytes and endothelial cells.[103,104] HIV-infected monocytes, which overexpress CCR2, are then preferentially recruited to the CNS.[98,105]

In the case of HTLV-1, infected lymphocytes secrete IFN-γ, which activates astrocytes that in turn secrete chemokine CXCL10.[106] Thus, leukocytes expressing CXCR3, especially activated Th1 cells, will be recruited to the central compartment.

5. CONCLUSION

HIV-1 and HTLV-1 are two human retroviruses that are linked to neurological disorders. The passage of the virus through the brain barriers into the CNS involves a wide range of adhesion proteins, cytokines, and chemokines. Once in the CNS, it is difficult to control either HTLV-1 or HIV-1 replication or subsequent neuroinflammation in the brain parenchyma. HAM/TSP outcomes such as neurodegeneration are irreversible. Moreover, although the BBB is disrupted, current retroviral treatments hardly penetrate the CNS. Though HAART targeting HIV efficiently decreases the proviral load, the viral reservoir persists within the CNS of treated patients.[71] One major difficulty is achieving efficient drug concentration in the CNS. Indeed, drugs may be extruded through efflux pumps present on the BBB. Identifying alternative approaches in reducing CNS neurodisorders remains a major challenge.

REFERENCES

1. Doran KS, Banerjee A, Disson O, Lecuit M. Concepts and mechanisms: crossing host barriers. *Cold Spring Harb Perspect Med* 2013;**3**.
2. Lossinsky AS, Shivers RR. Structural pathways for macromolecular and cellular transport across the blood-brain barrier during inflammatory conditions. *Histol Histopathol* 2004;**19**:535–64 [Review].
3. Engelhardt B, Liebner S. Novel insights into the development and maintenance of the blood-brain barrier. *Cell Tissue Res* 2014;**355**:687–99.
4. Weiss N, Miller F, Cazaubon S, Couraud PO. The blood-brain barrier in brain homeostasis and neurological diseases. *Biochim Biophys Acta* 2009;**1788**:842–57.
5. Pulzova L, Bhide MR, Andrej K. Pathogen translocation across the blood-brain barrier. *FEMS Immunol Med Microbiol* 2009;**57**:203–13.

6. Gessain A, Cassar O. Epidemiological aspects and world distribution of HTLV-1 infection. *Front Microbiol* 2012;**3**:388.
7. Gessain A, Barin F, Vernant JC, Gout O, Maurs L, Calender A, et al. Antibodies to human T-lymphotropic virus type-I in patients with tropical spastic paraparesis. *Lancet* 1985;**2**:407–10.
8. Osame M. Pathological mechanisms of human T-cell lymphotropic virus type I-associated myelopathy (HAM/TSP). *J Neurovirol* 2002;**8**:359–64.
9. Araujo A, Hall WW. Human T-lymphotropic virus type II and neurological disease. *Ann Neurol* 2004;**56**:10–9.
10. Gonzalez-Scarano F, Martin-Garcia J. The neuropathogenesis of AIDS. *Nat Rev Immunol* 2005;**5**: 69–81.
11. Wood BR, Klein JP, Lyons JL, Milner DA, Phillips RE, Schutten M, et al. HIV-2 encephalitis: case report and literature review. *AIDS Patient Care STDs* 2012;**26**:383–7.
12. Betsem E, Rua R, Tortevoye P, Froment A, Gessain A. Frequent and recent human acquisition of simian foamy viruses through apes' bites in central Africa. *PLoS Pathog* 2011;**7**.
13. Falcone V, Leupold J, Clotten J, Urbanyi E, Herchenroder O, Spatz W, et al. Sites of simian foamy virus persistence in naturally infected African green monkeys: latent provirus is ubiquitous, whereas viral replication is restricted to the oral mucosa. *Virology* 1999;**257**:7–14.
14. Aguzzi A, Bothe K, Anhauser I, Horak I, Rethwilm A, Wagner EF. Expression of human foamy virus is differentially regulated during development in transgenic mice. *New Biol* 1992;**4**:225–37.
15. Singer EJ, Syndulko K, Fahy-Chandon B, Schmid P, Conrad A, Tourtellotte WW. Intrathecal IgG synthesis and albumin leakage are increased in subjects with HIV-1 neurologic disease. *J Acquir Immune Defic Syndr* 1994;**7**:265–71.
16. Petito CK, Cash KS. Blood-brain barrier abnormalities in the acquired immunodeficiency syndrome: immunohistochemical localization of serum proteins in postmortem brain. *Ann Neurol* 1992;**32**: 658–66.
17. Ozden S, Seilhean D, Gessain A, Hauw JJ, Gout O. Severe demyelinating myelopathy with low human T cell lymphotropic virus type 1 expression after transfusion in an immunosuppressed patient. *Clin Infect Dis* 2002;**34**:855–60.
18. Silva MT, Araujo A. Spinal cord swelling in human T-lymphotropic virus 1-associated myelopathy/ tropical spastic paraparesis: magnetic resonance indication for early anti-inflammatory treatment? *Arch Neurol* 2004;**61**:1134–5.
19. Dallasta LM, Pisarov LA, Esplen JE, Werley JV, Moses AV, Nelson JA, et al. Blood-brain barrier tight junction disruption in human immunodeficiency virus-1 encephalitis. *Am J Pathol* 1999;**155**:1915–27.
20. Kubota R, Umehara F, Izumo S, Ijichi S, Matsumuro K, Yashiki S, et al. HTLV-I proviral DNA amount correlates with infiltrating CD4+ lymphocytes in the spinal cord from patients with HTLV-I-associated myelopathy. *J Neuroimmunol* 1994;**53**:23–9.
21. Boven LA, Middel J, Verhoef J, De Groot CJ, Nottet HS. Monocyte infiltration is highly associated with loss of the tight junction protein zonula occludens in HIV-1-associated dementia. *Neuropathol Appl Neurobiol* 2000;**26**:356–60.
22. Afonso PV, Ozden S, Prevost MC, Schmitt C, Seilhean D, Weksler B, et al. Human blood-brain barrier disruption by retroviral-infected lymphocytes: role of myosin light chain kinase in endothelial tight-junction disorganization. *J Immunol* 2007;**179**:2576–83.
23. Moses AV, Bloom FE, Pauza CD, Nelson JA. Human immunodeficiency virus infection of human brain capillary endothelial cells occurs via a CD4/galactosylceramide-independent mechanism. *Proc Natl Acad Sci U S A* 1993;**90**:10474–8.
24. Wiley CA, Schrier RD, Nelson JA, Lampert PW, Oldstone MB. Cellular localization of human immunodeficiency virus infection within the brains of acquired immune deficiency syndrome patients. *Proc Natl Acad Sci U S A* 1986;**83**:7089–93.
25. Afonso PV, Ozden S, Cumont MC, Seilhean D, Cartier L, Rezaie P, et al. Alteration of blood-brain barrier integrity by retroviral infection. *PLoS Pathog* 2008;**4**.
26. Lehky TJ, Fox CH, Koenig S, Levin MC, Flerlage N, Izumo S, et al. Detection of human T-lymphotropic virus type I (HTLV-I) tax RNA in the central nervous system of HTLV-I-associated myelopathy/tropical spastic paraparesis patients by in situ hybridization. *Ann Neurol* 1995;**37**:167–75.
27. Giraudon P, Szymocha R, Buart S, Bernard A, Cartier L, Belin MF, et al. T lymphocytes activated by persistent viral infection differentially modify the expression of metalloproteinases and their

endogenous inhibitors, TIMPs, in human astrocytes: relevance to HTLV-I-induced neurological disease. *J Immunol* 2000;**164**:2718–27.
28. Li GH, Henderson L, Nath A. Astrocytes as an HIV reservoir: mechanism of HIV infection. *Curr HIV Res* 2016;**14**:373–81.
29. Nakagawa S, Castro V, Toborek M. Infection of human pericytes by HIV-1 disrupts the integrity of the blood-brain barrier. *J Cell Mol Med* 2012;**16**:2950–7.
30. Hoffman PM, Dhib-Jalbut S, Mikovits JA, Robbins DS, Wolf AL, Bergey GK, et al. Human T-cell leukemia virus type I infection of monocytes and microglial cells in primary human cultures. *Proc Natl Acad Sci U S A* 1992;**89**:11784–8.
31. Nixon DE, Landay AL. Biomarkers of immune dysfunction in HIV. *Curr Opin HIV AIDS* 2010;**5**:498–503.
32. Anthony IC, Ramage SN, Carnie FW, Simmonds P, Bell JE. Influence of HAART on HIV-related CNS disease and neuroinflammation. *J Neuropathol Exp Neurol* 2005;**64**:529–36.
33. Starling AL, Coelho-Dos-Reis JG, Peruhype-Magalhaes V, Pascoal-Xavier MA, Goncalves DU, Bela SR, et al. Immunological signature of the different clinical stages of the HTLV-1 infection: establishing serum biomarkers for HTLV-1-associated disease morbidity. *Biomarkers* 2015;**20**:502–12.
34. Sato T, Coler-Reilly A, Utsunomiya A, Araya N, Yagishita N, Ando H, et al. CSF CXCL10, CXCL9, and neopterin as candidate prognostic biomarkers for HTLV-1-associated myelopathy/tropical spastic paraparesis. *PLoS Negl Trop Dis* 2013;**7**.
35. de Vries HE, Blom-Roosemalen MC, van Oosten M, de Boer AG, van Berkel TJ, Breimer DD, et al. The influence of cytokines on the integrity of the blood-brain barrier in vitro. *J Neuroimmunol* 1996;**64**:37–43.
36. Roux-Lombard P, Modoux C, Cruchaud A, Dayer JM. Purified blood monocytes from HIV 1-infected patients produce high levels of TNF alpha and IL-1. *Clin Immunol Immunopathol* 1989;**50**:374–84.
37. Kaul M, Garden GA, Lipton SA. Pathways to neuronal injury and apoptosis in HIV-associated dementia. *Nature* 2001;**410**:988–94.
38. Candelario-Jalil E, Yang Y, Rosenberg GA. Diverse roles of matrix metalloproteinases and tissue inhibitors of metalloproteinases in neuroinflammation and cerebral ischemia. *Neuroscience* 2009;**158**: 983–94.
39. Conant K, McArthur JC, Griffin DE, Sjulson L, Wahl LM, Irani DN. Cerebrospinal fluid levels of MMP-2, 7, and 9 are elevated in association with human immunodeficiency virus dementia. *Ann Neurol* 1999;**46**:391–8.
40. Ikegami M, Umehara F, Ikegami N, Maekawa R, Osame M. Selective matrix metalloproteinase inhibitor, N-biphenyl sulfonyl phenylalanine hydroxamic acid, inhibits the migration of CD4+ T lymphocytes in patients with HTLV-I-associated myelopathy. *J Neuroimmunol* 2002;**127**:134–8.
41. Ensoli B, Buonaguro L, Barillari G, Fiorelli V, Gendelman R, Morgan RA, et al. Release, uptake, and effects of extracellular human immunodeficiency virus type 1 Tat protein on cell growth and viral transactivation. *J Virol* 1993;**67**:277–87.
42. Rayne F, Debaisieux S, Bonhoure A, Beaumelle B. HIV-1 Tat is unconventionally secreted through the plasma membrane. *Cell Biol Int* 2010;**34**:409–13.
43. Andras IE, Pu H, Deli MA, Nath A, Hennig B, Toborek M. HIV-1 Tat protein alters tight junction protein expression and distribution in cultured brain endothelial cells. *J Neurosci Res* 2003;**74**:255–65.
44. Pu H, Tian J, Andras IE, Hayashi K, Flora G, Hennig B, et al. HIV-1 Tat protein-induced alterations of ZO-1 expression are mediated by redox-regulated ERK 1/2 activation. *J Cereb Blood Flow Metab* 2005;**25**:1325–35.
45. Gandhi N, Saiyed ZM, Napuri J, Samikkannu T, Reddy PV, Agudelo M, et al. Interactive role of human immunodeficiency virus type 1 (HIV-1) clade-specific Tat protein and cocaine in blood-brain barrier dysfunction: implications for HIV-1-associated neurocognitive disorder. *J Neurovirol* 2010;**16**: 294–305.
46. Alefantis T, Mostoller K, Jain P, Harhaj E, Grant C, Wigdahl B. Secretion of the human T cell leukemia virus type I transactivator protein tax. *J Biol Chem* 2005;**280**:17353–62.
47. Jaworski E, Narayanan A, Van Duyne R, Shabbeer-Meyering S, Iordanskiy S, Saifuddin M, et al. Human T-lymphotropic virus type 1-infected cells secrete exosomes that contain Tax protein. *J Biol Chem* 2014;**289**:22284–305.

48. Medina F, Quintremil S, Alberti C, Godoy F, Pando ME, Bustamante A, et al. Tax secretion from peripheral blood mononuclear cells and Tax detection in plasma of patients with human T-lymphotropic virus-type 1-associated myelopathy/tropical spastic paraparesis and asymptomatic carriers. *J Med Virol* 2016;**88**:521–31.
49. Cartier L, Ramirez E. Presence of HTLV-I Tax protein in cerebrospinal fluid from HAM/TSP patients. *Arch Virol* 2005;**150**:743–53.
50. Lindholm PF, Marriott SJ, Gitlin SD, Bohan CA, Brady JN. Induction of nuclear NF-kappa B DNA binding activity after exposure of lymphoid cells to soluble tax1 protein. *New Biol* 1990;**2**:1034–43.
51. Szymocha R, Akaoka H, Dutuit M, Malcus C, Didier-Bazes M, Belin MF, et al. Human T-cell lymphotropic virus type 1-infected T lymphocytes impair catabolism and uptake of glutamate by astrocytes via Tax-1 and tumor necrosis factor alpha. *J Virol* 2000;**74**:6433–41.
52. Cowan EP, Alexander RK, Daniel S, Kashanchi F, Brady JN. Induction of tumor necrosis factor alpha in human neuronal cells by extracellular human T-cell lymphotropic virus type 1 Tax. *J Virol* 1997;**71**:6982–9.
53. SK O, Cruikshank WW, Raina J, Blanchard GC, Adler WH, Walker J, et al. Identification of HIV-1 envelope glycoprotein in the serum of AIDS and ARC patients. *J Acquir Immune Defic Syndr* 1992;**5**:251–6.
54. Annunziata P, Cioni C, Toneatto S, Paccagnini E. HIV-1 gp120 increases the permeability of rat brain endothelium cultures by a mechanism involving substance P. *AIDS* 1998;**12**:2377–85.
55. Toneatto S, Finco O, van der Putten H, Abrignani S, Annunziata P. Evidence of blood-brain barrier alteration and activation in HIV-1 gp120 transgenic mice. *AIDS* 1999;**13**:2343–8.
56. Kanmogne GD, Primeaux C, Grammas P. HIV-1 gp120 proteins alter tight junction protein expression and brain endothelial cell permeability: implications for the pathogenesis of HIV-associated dementia. *J Neuropathol Exp Neurol* 2005;**64**:498–505.
57. Khan NA, Di Cello F, Stins M, Kim KS. Gp120-mediated cytotoxicity of human brain microvascular endothelial cells is dependent on p38 mitogen-activated protein kinase activation. *J Neurovirol* 2007;**13**:242–51.
58. Louboutin JP, Strayer DS. Blood-brain barrier abnormalities caused by HIV-1 gp120: mechanistic and therapeutic implications. *Sci World J* 2012;**2012**.
59. Luo X, Fan Y, Park IW, He JJ. Exosomes are unlikely involved in intercellular Nef transfer. *PLoS One* 2015;**10**.
60. Lenassi M, Cagney G, Liao M, Vaupotic T, Bartholomeeusen K, Cheng Y, et al. HIV Nef is secreted in exosomes and triggers apoptosis in bystander CD4+ T cells. *Traffic* 2010;**11**:110–22.
61. Raymond AD, Diaz P, Chevelon S, Agudelo M, Yndart-Arias A, Ding H, et al. Microglia-derived HIV Nef+ exosome impairment of the blood-brain barrier is treatable by nanomedicine-based delivery of Nef peptides. *J Neurovirol* 2016;**22**:129–39.
62. Acheampong EA, Parveen Z, Muthoga LW, Kalayeh M, Mukhtar M, Pomerantz RJ. Human Immunodeficiency virus type 1 Nef potently induces apoptosis in primary human brain microvascular endothelial cells via the activation of caspases. *J Virol* 2005;**79**:4257–69.
63. Sporer B, Koedel U, Paul R, Kohleisen B, Erfle V, Fontana A, et al. Human immunodeficiency virus type-1 Nef protein induces blood-brain barrier disruption in the rat: role of matrix metalloproteinase-9. *J Neuroimmunol* 2000;**102**:125–30.
64. Levy DN, Refaeli Y, MacGregor RR, Weiner DB. Serum Vpr regulates productive infection and latency of human immunodeficiency virus type 1. *Proc Natl Acad Sci U S A* 1994;**91**:10873–7.
65. Ferrucci A, Nonnemacher MR, Wigdahl B. Extracellular HIV-1 viral protein R affects astrocytic glyceraldehyde 3-phosphate dehydrogenase activity and neuronal survival. *J Neurovirol* 2013;**19**: 239–53.
66. Gangwani MR, Noel Jr. RJ, Shah A, Rivera-Amill V, Kumar A. Human immunodeficiency virus type 1 viral protein R (Vpr) induces CCL5 expression in astrocytes via PI3K and MAPK signaling pathways. *J Neuroinflammation* 2013;**10**:136.
67. Goudsmit J, de Wolf F, Paul DA, Epstein LG, Lange JM, Krone WJ, et al. Expression of human immunodeficiency virus antigen (HIV-Ag) in serum and cerebrospinal fluid during acute and chronic infection. *Lancet* 1986;**2**:177–80.

68. Davis LE, Hjelle BL, Miller VE, Palmer DL, Llewellyn AL, Merlin TL, et al. Early viral brain invasion in iatrogenic human immunodeficiency virus infection. *Neurology* 1992;**42**(9):1736.
69. Torres-Munoz J, Stockton P, Tacoronte N, Roberts B, Maronpot RR, Petito CK. Detection of HIV-1 gene sequences in hippocampal neurons isolated from postmortem AIDS brains by laser capture microdissection. *J Neuropathol Exp Neurol* 2001;**60**:885–92.
70. Chiodi F, Albert J, Olausson E, Norkrans G, Hagberg L, Sonnerborg A, et al. Isolation frequency of human immunodeficiency virus from cerebrospinal fluid and blood of patients with varying severity of HIV infection. *AIDS Res Hum Retrovir* 1988;**4**:351–8.
71. Kramer-Hammerle S, Rothenaigner I, Wolff H, Bell JE, Brack-Werner R. Cells of the central nervous system as targets and reservoirs of the human immunodeficiency virus. *Virus Res* 2005;**111**:194–213.
72. Levin MC, Rosenblum MK, Fox CH, Jacobson S. Localization of retrovirus in the central nervous system of a patient co-infected with HTLV-1 and HIV with HAM/TSP and HIV-associated dementia. *J Neurovirol* 2001;**7**:61–5.
73. Banks WA, Freed EO, Wolf KM, Robinson SM, Franko M, Kumar VB. Transport of human immunodeficiency virus type 1 pseudoviruses across the blood-brain barrier: role of envelope proteins and adsorptive endocytosis. *J Virol* 2001;**75**:4681–91.
74. Dohgu S, Fleegal-DeMotta MA, Banks WA. Lipopolysaccharide-enhanced transcellular transport of HIV-1 across the blood-brain barrier is mediated by luminal microvessel IL-6 and GM-CSF. *J Neuroinflammation* 2011;**8**:167.
75. Dohgu S, Banks WA. Brain pericytes increase the lipopolysaccharide-enhanced transcytosis of HIV-1 free virus across the in vitro blood-brain barrier: evidence for cytokine-mediated pericyte-endothelial cell crosstalk. *Fluids Barriers CNS* 2013;**10**:23.
76. Romero IA, Prevost MC, Perret E, Adamson P, Greenwood J, Couraud PO, et al. Interactions between brain endothelial cells and human T-cell leukemia virus type 1-infected lymphocytes: mechanisms of viral entry into the central nervous system. *J Virol* 2000;**74**:6021–30.
77. Fiala M, Looney DJ, Stins M, Way DD, Zhang L, Gan X, et al. TNF-alpha opens a paracellular route for HIV-1 invasion across the blood-brain barrier. *Mol Med* 1997;**3**:553–64.
78. Cavrois M, Gessain A, Gout O, Wain-Hobson S, Wattel E. Common human T cell leukemia virus type 1 (HTLV-1) integration sites in cerebrospinal fluid and blood lymphocytes of patients with HTLV-1-associated myelopathy/tropical spastic paraparesis indicate that HTLV-1 crosses the blood-brain barrier via clonal HTLV-1-infected cells. *J Infect Dis* 2000;**182**:1044–50.
79. Nottet HS, Persidsky Y, Sasseville VG, Nukuna AN, Bock P, Zhai QH, et al. Mechanisms for the transendothelial migration of HIV-1-infected monocytes into brain. *J Immunol* 1996;**156**:1284–95.
80. Cota-Gomez A, Flores NC, Cruz C, Casullo A, Aw TY, Ichikawa H, et al. The human immunodeficiency virus-1 Tat protein activates human umbilical vein endothelial cell E-selectin expression via an NF-kappa B-dependent mechanism. *J Biol Chem* 2002;**277**:14390–9.
81. Kambara C, Nakamura T, Furuya T, Nishiura Y, Kawakami A, Ichinose K, et al. Increased sialyl Lewis(x) antigen-positive cells mediated by HTLV-1 infection in peripheral blood CD4(+) T lymphocytes in patients with HTLV-1-associated myelopathy. *J Neuroimmunol* 2002;**125**:179–84.
82. Al-Fahim A, Cabre P, Kastrukoff L, Dorovini-Zis K, Oger J. Blood mononuclear cells in patients with HTLV-I-associated myelopathy: lymphocytes are highly activated and adhesion to endothelial cells is increased. *Cell Immunol* 1999;**198**:1–10.
83. Dhawan S, Weeks BS, Soderland C, Schnaper HW, Toro LA, Asthana SP, et al. HIV-1 infection alters monocyte interactions with human microvascular endothelial cells. *J Immunol* 1995;**154**:422–32.
84. Tanaka Y, Fukudome K, Hayashi M, Takagi S, Yoshie O. Induction of ICAM-1 and LFA-3 by Tax1 of human T-cell leukemia virus type 1 and mechanism of down-regulation of ICAM-1 or LFA-1 in adult-T-cell-leukemia cell lines. *Int J Cancer* 1995;**60**:554–61.
85. Valentin H, Lemasson I, Hamaia S, Casse H, Konig S, Devaux C, et al. Transcriptional activation of the vascular cell adhesion molecule-1 gene in T lymphocytes expressing human T-cell leukemia virus type 1 Tax protein. *J Virol* 1997;**71**:8522–30.
86. Van Prooyen N, Gold H, Andresen V, Schwartz O, Jones K, Ruscetti F, et al. Human T-cell leukemia virus type 1 p8 protein increases cellular conduits and virus transmission. *Proc Natl Acad Sci U S A* 2010;**107**:20738–43.

87. Umehara F, Izumo S, Takeya M, Takahashi K, Sato E, Osame M. Expression of adhesion molecules and monocyte chemoattractant protein-1 (MCP-1) in the spinal cord lesions in HTLV-I-associated myelopathy. *Acta Neuropathol* 1996;**91**:343–50.
88. Abadier M, Haghayegh Jahromi N, Cardoso Alves L, Boscacci R, Vestweber D, Barnum S, et al. Cell surface levels of endothelial ICAM-1 influence the transcellular or paracellular T-cell diapedesis across the blood-brain barrier. *Eur J Immunol* 2015;**45**:1043–58.
89. Azzarelli B, Mirkin LD, Goheen M, Muller J, Crockett C. The leptomeningeal vein. A site of re-entry of leukemic cells into the systemic circulation. *Cancer* 1984;**54**:1333–43.
90. Carman CV, Springer TA. A transmigratory cup in leukocyte diapedesis both through individual vascular endothelial cells and between them. *J Cell Biol* 2004;**167**:377–88.
91. Weber C, Fraemohs L, Dejana E. The role of junctional adhesion molecules in vascular inflammation. *Nat Rev Immunol* 2007;**7**:467–77.
92. Huang W, Eum SY, Andras IE, Hennig B, Toborek M. PPARalpha and PPARgamma attenuate HIV-induced dysregulation of tight junction proteins by modulations of matrix metalloproteinase and proteasome activities. *FASEB J* 2009;**23**:1596–606.
93. Mahajan SD, Aalinkeel R, Sykes DE, Reynolds JL, Bindukumar B, Adal A, et al. Methamphetamine alters blood brain barrier permeability via the modulation of tight junction expression: implication for HIV-1 neuropathogenesis in the context of drug abuse. *Brain Res* 2008;**1203**:133–48.
94. Mahajan SD, Aalinkeel R, Sykes DE, Reynolds JL, Bindukumar B, Fernandez SF, et al. Tight junction regulation by morphine and HIV-1 tat modulates blood-brain barrier permeability. *J Clin Immunol* 2008;**28**:528–41.
95. Williams DW, Anastos K, Morgello S, Berman JW. JAM-A and ALCAM are therapeutic targets to inhibit diapedesis across the BBB of CD14+CD16+ monocytes in HIV-infected individuals. *J Leukoc Biol* 2015;**97**:401–12.
96. Muller WA, Weigl SA, Deng X, Phillips DM. PECAM-1 is required for transendothelial migration of leukocytes. *J Exp Med* 1993;**178**:449–60.
97. Eugenin EA, Gamss R, Buckner C, Buono D, Klein RS, Schoenbaum EE, et al. Shedding of PECAM-1 during HIV infection: a potential role for soluble PECAM-1 in the pathogenesis of NeuroAIDS. *J Leukoc Biol* 2006;**79**:444–52.
98. Williams DW, Calderon TM, Lopez L, Carvallo-Torres L, Gaskill PJ, Eugenin EA, et al. Mechanisms of HIV entry into the CNS: increased sensitivity of HIV infected CD14+CD16+ monocytes to CCL2 and key roles of CCR2, JAM-A, and ALCAM in diapedesis. *PLoS One* 2013;**8**.
99. Lou O, Alcaide P, Luscinskas FW, Muller WA. CD99 is a key mediator of the transendothelial migration of neutrophils. *J Immunol* 2007;**178**:1136–43.
100. Poissonnier L, Villain G, Soncin F, Mattot V. miR126-5p repression of ALCAM and SetD5 in endothelial cells regulates leucocyte adhesion and transmigration. *Cardiovasc Res* 2014;**102**:436–47.
101. Curis C, Percher F, Jeannin P, Montange T, Chevalier SA, Seilhean D, et al. Human T-lymphotropic virus type 1-induced overexpression of activated leukocyte cell adhesion molecule (ALCAM) facilitates trafficking of infected lymphocytes through the blood-brain barrier. *J Virol* 2016;**90**:7303–12.
102. Lafrenie RM, Wahl LM, Epstein JS, Hewlett IK, Yamada KM, Dhawan S. HIV-1-Tat protein promotes chemotaxis and invasive behavior by monocytes. *J Immunol* 1996;**157**:974–7.
103. Weiss JM, Nath A, Major EO, Berman JW. HIV-1 Tat induces monocyte chemoattractant protein-1-mediated monocyte transmigration across a model of the human blood-brain barrier and up-regulates CCR5 expression on human monocytes. *J Immunol* 1999;**163**:2953–9.
104. Park IW, Wang JF, Groopman JE. HIV-1 Tat promotes monocyte chemoattractant protein-1 secretion followed by transmigration of monocytes. *Blood* 2001;**97**:352–8.
105. Eugenin EA, Osiecki K, Lopez L, Goldstein H, Calderon TM, Berman JW. CCL2/monocyte chemoattractant protein-1 mediates enhanced transmigration of human immunodeficiency virus (HIV)-infected leukocytes across the blood-brain barrier: a potential mechanism of HIV-CNS invasion and NeuroAIDS. *J Neurosci* 2006;**26**:1098–106.
106. Ando H, Sato T, Tomaru U, Yoshida M, Utsunomiya A, Yamauchi J, et al. Positive feedback loop via astrocytes causes chronic inflammation in virus-associated myelopathy. *Brain* 2013;**136**:2876–87.

SECTION III

Fungal and Protozoal Infections of the CNS

CHAPTER 12

Recent Advances in Fungal Infections of the Central Nervous System: From Etiology to Diagnosis and Management

Mohammadhassan Gholami-Shabani*, Samira Zamani*, Hoda Moosa*, Masoomeh Shams-Ghahfarokhi†, Fatemehsadat Jamzivar*, Mehdi Razzaghi-Abyaneh*

*Pasteur Institute of Iran, Tehran, Iran
†Tarbiat Modares University, Tehran, Iran

1. INTRODUCTION

Fungal infections of the central nervous system (CNS) are relatively uncommon neuroinfections that are increasingly being recognized these days.[1,2] The infections of the CNS differ from infections of other organ systems. Various infections of the CNS are able to progress rapidly and cause substantial damage or even death if they are not recognized and treated promptly and aggressively.[3] The brain is endowed with a blood brain barrier (BBB) that prevents entry of pathogens and inflammatory cells, providing an effective protection against infections. In a healthy individual, organisms can only reach the brain through hematogenous spread, extension from the juxtacranial structures (middle ear cavity, paranasal sinus), or through neural pathways (e.g., olfactory nerves). However, iatrogenic factors can increase the risk for CNS infections by either causing a breach in the barrier (e.g., following neurosurgery), the presence of a foreign body, (e.g., ventriculoperitoneal shunt), or by medications that affect immunity or bacterial colonization (e.g., immunosuppressive drugs).[4] There are a variety of organisms that can cause CNS infections, including bacterial, viral, fungal, and protozoal agents. Bacterial infections can be caused by pyogenic organisms or may be due to mycobacteria or spirochetes. Bacterial infections, particularly those due to pyogenic organisms, may lead to meningitis, brain abscess and epidural or subdural abscesses. Viral infections may also lead to meningitis or cause encephalitis or myelitis. Protozoal infections that may affect the brain include toxoplasmosis, malaria, and amoebiasis.[5] While the majority of infections in the CNS are caused by bacteria and viruses, fungi are increasingly being recognized as important pathogens. This is mainly due to the increased awareness of clinicians about these conditions, widespread use of antibacterial agents, a rapid increase in the number of the

The Microbiology of Central Nervous System Infections
https://doi.org/10.1016/B978-0-12-813806-9.00012-3

immunocompromised population, advances in imaging, and the availability of microbiological techniques to confirm the diagnosis from bodily fluids and specimens.[6,7] Fungal infections often follow the use of antibiotics, which kill nonpathogenic as well as pathogenic bacteria, thereby providing a free field in the body for fungal invasion. Modern therapeutic modalities such as cancer chemotherapy and organ transplantation have greatly increased the immunocompromised population who are also at risk.[8] With improved therapies, patients with cancer survive longer. Both the acute complications of intensive therapies and the risks of chronic immunosuppression have led to an increased incidence of CNS infections. The presentation and course of common infections may be different from those in patients without cancer, and new syndromes related both to the underlying diseases and to their treatment have complicated the differential diagnosis.[9] New fungal pathogens are being discovered and new syndromes are being described for both known and emerging fungal pathogens. Opportunistic fungal infection occurs when a fungus that is normally nonpathogenic in normal hosts enters a compromised host, as in patients with acquired immunodeficiency disease (AIDS), resulting in a disease process.[10] Fungal infections of the CNS are life-threatening and almost always a clinical surprise. Their presentation is subtle, often without any diagnostic characteristics, and they are frequently mistaken for tuberculous meningitis, pyogenic abscess, or brain tumors. Granulocytopenia, cellular, and humoral-mediated immune dysfunctions are predisposing factors to the development of CNS infections in immunosuppressed patients. Environmental factors in tropical countries such as India play a significant role in the pathogenesis of CNS fungal infections. There are several published large series and reviews from India about various fungal organisms involving the CNS, including aspergillosis, zygomycosis, and phaeohyphomycosis.[7] Table 1 shows fungi that cause infections of the CNS. Important risk factors for fungal infections of the CNS are summarized in Table 2.

Table 1 Fungal pathogens that cause infections of the central nervous system

Outbreak	Meningitis	Brain abscess
Common	*Cryptococcus neoformans*	*Aspergillus* spp.
	Histoplasma capsulatum	*Candida* spp.
	Coccidioides immitis	*Zygomycetes*
Uncommon	*Aspergillus* spp.	*Blastomyces dermatitidis*
	Blastomyces dermatitidis	*Coccidioides immitis*
	Candida spp.	*Fusarium* spp.
	Paracoccidioides brasiliensis	*Histoplasma capsulatum*
	Sporothrix schenckei	*Paracoccidioides brasiliensis*
	Zygomycetes	*Penicillium* spp.
		Pseudallescheria boydii
		Sporothrix schenckei
		Stilago

Table 1 Fungal pathogens that cause infections of the central nervous system—cont'd

Outbreak	Meningitis	Brain abscess
Rare	*Blastoschizomyces capitatus* *Rhodotorula rubra*	*Acrophialophora fusispora* *Bipolaris* spp. *Blastoschizomyces capitatus* *Chaetomiumper lucidum* *Chaetomium strumarium* *Cladophialophora bantiana* *Paecilomyces* spp. *Ramichloridium* spp. *Trichoderma brachiatum* *Trichophyton* spp. *Trichosporon* spp.

Table 2 Important risk factors for fungal infections of the central nervous system

Risk factors	Immunodeficiency	Important pathogens
AIDS	T cells	*Cryptococcus neoformans* *Coccidioides immitis* *Histoplasma capsulatum* *Penicillium marneffei* *Aspergillus* spp.
Leukemia/neutropenia	Granulocytes	*Aspergillus* spp.
Hematopoietic stem cell transplant (HSCT)	Granulocytes and T, B cells Breaking blood-brain barrier Intravascular catheters Drugs immunosuppressive	*Aspergillus* spp.
Taking corticosteroids	Monocytes and macrophages Complement receptors Immunoglobulin	*Cryptococcus neoformans* *Candida* spp.
TNF inhibitors	Cytokines, B, T, cells	*Cryptococcus neoformans* *Histoplasma capsulatum* *Aspergillus* spp. *Zygomycetes* *Fusarium* spp. *Pseudallescheria boydii*

2. HISTORICAL ASPECTS

Fungal infections have been recognized since early times, but CNS fungal infections have mainly been recognized since the 19th century.[11] Generally, Hippocrates is credited with the first description of candidiasis in his book "Epidemic," in which he described white patches in the oral cavity of a debilitated patient. The fungal etiology of these thrushes was established in the 1840s by Berg and Bennett. Zenkar described the first patient with intracerebral candidiasis, who died in 1861. Smith and Sano were the first to report a case of *Candida* meningitis in 1933. Various postmortem studies have established that candidiasis is a more common CNS fungal infection than aspergillosis, zygomycosis, or cryptococcosis.[12–15] In 1792, Micheli (a priest and botanist) described *Aspergillus* to refer to the nine species of fungi that resembled aspergillum, a perforated globe frequently used to sprinkle holy water during religious ceremonies. In 1856, Virchow published the first complete microscopic description of *Aspergillus*. However, it was Oppe in 1897 who reported the first case of rhino-orbitocerebral aspergillosis and of cerebral aspergillosis in a man due to an extension of fungal infection from sphenoid sinusitis. The first case of human zygomycetes infection was described by Kurchenmeister in 1855; that isolated nonseptate hypha from a cancerous lung. In 1943, Gregory described in detail three cases of rhinocerebral zygomycosis.[7] *Cryptococcus neoformans* is encapsulated yeast first identified as a human pathogen in 1894, when it was isolated from the tibia of a patient in Germany by Buese and Buschke. Von Hansemann (1905) described the first cases of cryptococcal meningitis with lesions that mimicked tuberculous meningitis and showed gelatinous cysts.[7,16] Coccidioidomycosis was first reported by Alejandro Posadas and Robert Wernicke in 1892. In 1905 Ophuls described the first case of coccidioidal meningitis.[17] Gilchrist described blastomycosis in 1894 and successfully grew the organism in culture. Gonyea reported on three patients with blastomycosis meningitis in 1978 without extracranial infection. In 1952, Binford published a case of cerebral abscess due to *Cladosporium*. Histoplasmosis was first described by Darling in 1906 in a patient with disseminated granulomatous infection. Histiocytes were studded with *Histoplasma capsulatum*.[18–20] Paracoccidioidomycosis was first described in 1908 by Lutz in Brazil.[21] In 1888, Nocard described acid-fast aerobic actinomycetes in cattle, which were called *Nocardia farcinica* by Trevisan in 1889. Eppinger in 1891 reported the first case of metastatic (from the lung) cerebral nocardiosis.[22]

3. CLASSIFICATION AND EPIDEMIOLOGY

Fungal infections are not notifiable diseases and precise information on their prevalence throughout the world is not available. In general, fungi are cited as ubiquitous; however, some forms have a more restricted geographical distribution than others. More than 100,000 fungal species are recognized by now and only a couple hundred are found tobe pathogenic to humans. Fortunately, only about 10%–15% of pathological fungi usually produce systemic/CNS mycosis.[7] True pathogenic fungi (having a restricted geographic distribution, mostly in the United States) are *Blastomyces*, *Coccidioides*, *Paracoccidioides*, *Histoplasma*, and *Sporothrix*.[23] They produce clinical lesions in normal individuals and then provide long-term immunity to patients who have recovered from the active infections. Whereas the opportunistic fungi (having ubiquitous distribution), including *Aspergillus fumigatus*, *Candida albicans*, *Cryptococcus neoformans*, and *Rhizopus arrhizus*, provide no long-term immunity to patients; hence relapses are noted.[7] Fungi present mainly in three forms: molds (colonies of branching hyphae-mycelium), yeasts (colonies of single cells) and intermediate forms. Some fungi are thermodimorphic, changing their forms from mold to yeast under different environmental conditions. Clinically important true pathogenic dimorphic fungi such as *Coccidioides, Histoplasma, Sporotrichum, Paracoccidioides,* and *Blastomyces* grow in mycelial form at 25°C (filamentous in nature), and convert to a yeast form (spherules) at normal body temperature (37°C) in humans. Encapsulated yeast, *C. neoformans,* maintains its morphology in normal human tissues and in nature, similar to various septate and nonseptate mycetes. However, *Nocardia* and *Actinomyces* are not true fungi but are fungi-like bacteria, producing CNS lesions mimicking fungal infections; they are therefore included in this chapter. Fungi are classified in three major groups: pseudomycetes/yeasts *(Blastomyces, Candida, Coccidioides, Cryptococcus, Histoplasma, Paracoccidioides,* and *Sporotrichum),* septate mycetes *(Aspergillus, Cephalosporium, Cladosporium, Diplorhinotrichum, Hormodendrum, Paecilomyces,* and *Penicillium),* and nonseptate mycetes *(Absidia, Basidiobolus, Cunninghamella, Mortierella, Mucor,* and *Rhizopus).*[7,24]

4. PATHOGENESIS

Fungi grow profoundly in environments with an abundance of organic matter and water. Enzymes produced by the fungal cell wall act on the organic matter and break down their proteins, carbohydrates, and other macromolecules into micromolecules, which are then easily used by the fungi to maintain their life processes.[25] Generally, the brain and spinal cord are regarded as immunologically privileged sites. The brain has a specialized, relatively impermeable BBB, and its surrounding subarachnoid spaces have effective meningeal barriers. These highly vascular structures provide effective resistance to fungal infections. For immune surveillance, mainly activated T-lymphocytes are usually permitted across the BBB in normal individuals. However, in immunocompromised states, these anatomic and functional barriers are easily overcome by common opportunistic and/or pathogenic fungi to produce clinical manifestations.[7] Fungal infections of the CNS also evoke humoral and cellular responses like those in bacterial infections with the scope to enable the host to eliminate the pathogen. Activation of the resident brain cells by fungi combined with relative expression of immune-enhancing and immune-suppressing cytokines and chemokines may play a determinant role and partially explain the immunopathogenesis of CNS fungal infections. Activated resident brain cells such as microglia, astrocytes, and endothelial cells express major histocompatibility complex class I and II molecules and therefore act as antigen-presenting cells. In addition, they express complement (C) receptors and produce cytokines, chemokines, and molecules with antifungal activity such as nitric oxide (NO). They are also capable of phagocytosis. Microglia acting as antigen-presenting cells stimulate T-cell proliferation and cytokine secretion, which in turn stimulates these microglial cells to ingest and more effectively kill invading fungi. The C system is a key component of the innate immune system, playing a central role in host defenses against pathogens. It is also a powerful drive to initiate inflammation and, if unregulated, can result in pathologic changes that lead to severe tissue damage. It is generally accepted that the C system is essential to mediate cytolysis of fungi. Furthermore, it is well known that C receptors expressed by activated microglial cells are important to mediate phagocytosis. Immunopathogenesis of CNS fungal infections in humans is not completely understood. The exploration of the genomic sequence of most fungal pathogens can help better understand the pathogenesis, virulence, and immune response of host defenses against these pathogens. Further studies in these areas will advance our understanding of CNS mycoses. Fungi produce diseases due to their allergenicity, toxigenicity, pathogenicity, and neurotoxicity, but CNS mycoses are essentially due to serious infective processes.[26,27] Usually, the inhaled aerosolized fungi initiate a primary mycotic infection in the lungs and or paranasal sinuses that is usually self-limiting but may spread to other organs. Frequently, hematogenous dissemination (from lungs, intestines, or prosthetic heart valves) results in the systemic mycosis; less frequently fungal infections appear in the CNS.[28] Direct contiguous spread to the CNS occurs from the paranasal sinuses, orbits, petromastoid region, and retropharyngeal area in some patients. Direct implantation may result during the trauma period, intensive care procedures, and intracranial operations. Identifying the etiological agent as fungal and not bacterial is vital since antibacterial therapy is not effective against fungi and CNS mycoses lead to high morbidity and mortality. Actinomycetes are bacteria with narrow branched vegetative cells that superficially resemble fungi. The common ones belong to *Nocardia* (aerobic) and *Actinomyces* (anaerobic) species; therefore, these are also considered in this write-up.[29,30]

5. CLINICOPATHOLOGIC SYNDROMES

The CNS fungal infections usually result from pulmonary, intestinal, cardiac, or craniofacial mycoses; therefore intracranial seedlings occur either during hematogenous dissemination or by direct extension from the juxtacranial sites.[7,31] The CNS mycosis may be disseminated (cryptococcosis, coccidioidomycosis, etc.), focal (aspergillosis, zygomycosis, etc.), or multifocal (candidiasis). Fungi produce diseases due to their allergenicity, toxigenicity, pathogenicity, neurotoxicity, and/or virulence.[32] Fungi producing distinctive clinical syndromes are found in three major morphological forms:

1. *Small pseudomycetes (yeasts) causing leptomeningitis*: Because of their small size (cryptococcosis, blastomycosis, histoplasmosis, coccidioidomycosis, etc.) these fungi gain access to the cerebral microcirculation. From there they seed and infect the CSF and its contained leptomeninges. Fungi may reach the brain parenchyma along the Virchow Robin spaces around the penetrating and perforating small cortical/cerebral

vessels arising from the major vessels in the subarachnoid spaces. Therefore, these fungi may result in meningitis and or meningoencephalitis.[7,33–35]

2. *Large pseudomycetes (candidiasis) producing cerebral abscesses and granulomas*: These fungi are larger than 20 μm and can occlude cerebral arterioles. Such occlusions lead to focal cerebral ischemia and infarctions. Depending upon the virulence of the fungi and host resistance, a variety of the cerebral fungal lesions appear, such as ischemic areas, focal infarctions, cerebritis, abscesses, granulomas and combinations of these lesions. The tissue necrosis and highly virulent fungal infection rapidly convert infected cerebral areas into microabscesses; whereas a good host resistance but persistence of infection causes granulomatous inflammatory reactions in adjacent leptomeninges, neural parenchyma or in both sites. This may result in hydrocephalus that needs to be treated promptly.[36–38]
3. *Septate (aspergillosis) and nonseptate (zygomycosis) mycetes* are very large in size and normally grow with large branched hyphae. Usually they infect juxtacranial sites (paranasal sinuses, orbits, oral cavity, etc.) for a considerable period of time and then are capable of invading contiguous cranial bones, meningeal tissues, basal cerebral venous sinuses, etc., as well as intermediate and large-sized intracranial arteries. This results in arterial thrombosis and occlusions, which, in turn, causes extensive cerebral infarctions. These patients clinically present as a case of cerebral stroke. The evolving hemorrhagic cerebral infarct is then converted to septic infarcts with associated cerebritis and abscesses, whereas a good host defense results in granuloma formation.[39–41]

6. TRUE PATHOGENIC FUNGI CAUSING INFECTIONS OF THE CNS

Blastomycetes, Coccidioides, Paracoccidioides, Histoplasma, and *Sporothrix* cause CNS infections. These fungi can invade tissues without an underlying predisposition (true pathogenic fungi).

6.1 Blastomycosis (North American Blastomycosis)

Blastomycosis is the systemic pyogranulomatous (blastomycoma) disease caused by the inhalation of conidia (spores) of *Blastomyces dermatitidis*. This fungus is a spherical thick-walled yeast cell 8–15 μm in diameter with a broad-based budding daughter cell. It is found in the soil, may have a natural reservoir in dogs, and has worldwide distribution. Hyphae to yeast transition is believed to be a consequence of heat shock protein (HSP)-induced uncoupling of oxidative phosphorylation. Growth is observed as a fluffy white mold on appropriate media at room temperature and as brown, wrinkled colonies of yeast at 37°C.[42] Blastomycosis is an uncommon mycosis endemic in the southeastern and central United States and widely reported in Africa. It begins as an initial subclinical pulmonary lesion and then hematogenously spreads to other organ systems. The yeasts are phagocytized by pulmonary macrophages, which may disseminate to produce secondary lesions in the skin, bone, and genitourinary system; however, the CNS rarely gets involved. CNS blastomycosis usually presents with meningitis, subdural focal leptomeningeal abscess, or intraparenchymal abscess (Fig. 1). However, it is a less common mycotic infection when involving the CNS compared to histoplasmosis and coccidioidomycosis.[44]

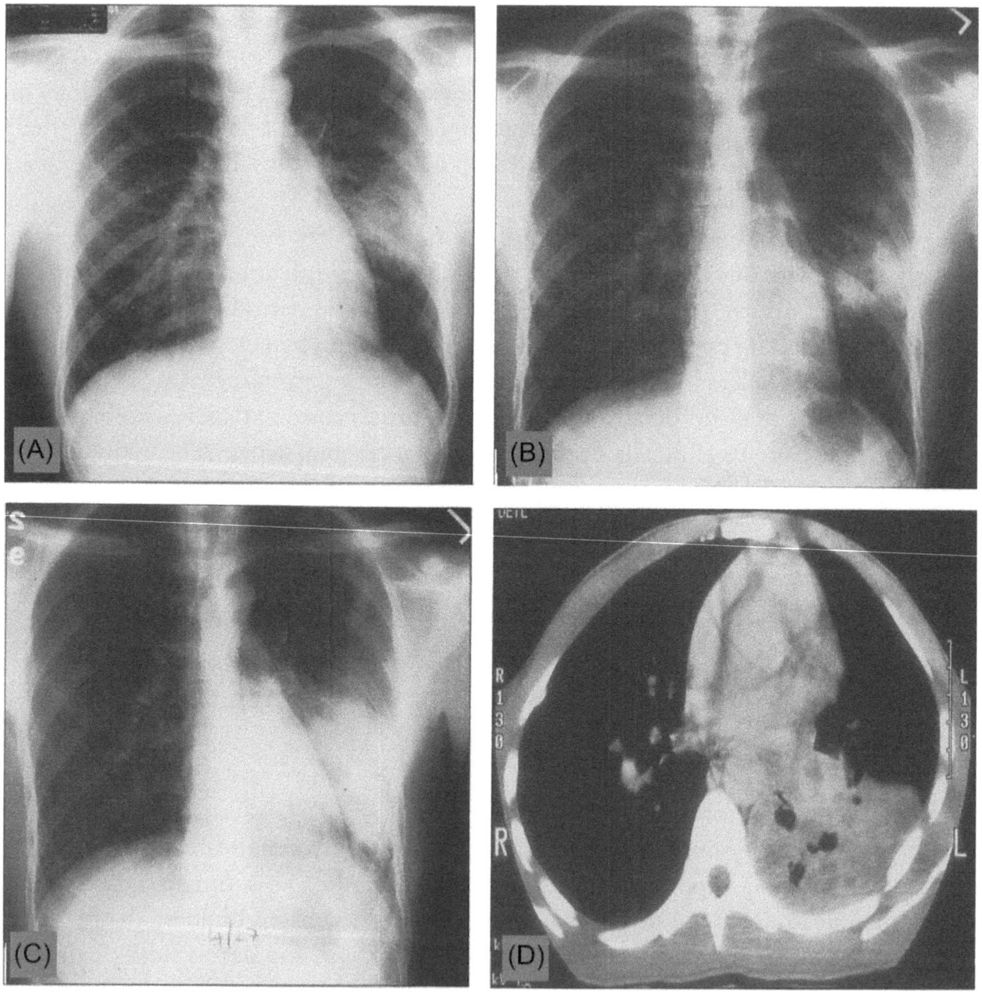

Fig. 1 Series of chest radiographs and chest CT scan showing pulmonary blastomycosis. A 23-year-old man with chronic schizophrenia who rarely left his home developed a left lower lobe pneumonia, which was believed to be a result of aspiration but was unresponsive to conventional oral antimicrobial therapy. After several weeks of unsuccessful therapy characterized by persistent fever, nonproductive cough, and progressive infiltrate on a chest radiograph, he was admitted for evaluation and bronchoscopy. Yeast forms consistent with *Blastomyces dermatitidis* were seen on Gomori's methenamine silver stain, and cultures confirmed a diagnosis of blastomycosis. (A) Chest radiograph from March 1993 reveals a superior segment left upper lobe infiltrate, (B and C) studies from April 1993 reveal progression of the infiltrate. (D) CT scan of the chest from May 1993 confirms a dense left lower lobe infiltrate with multiple areas of necrosis and a small effusion.[43]

6.1.1 Neuropathology

Clinical involvement of the CNS accounts for 5%–10% of extrapulmonary blastomycosis in clinical reviews, but has been observed in up to 33% of patients in autopsy series. The last major review of CNS blastomycosis was published 30 years ago. Most subsequent reports of CNS blastomycosis have been limited to single case reports or small series that were reported prior to the widespread use of modern imaging modalities and expanded-spectrum azole therapy.[42,45] Typically, cerebral blastomycosis produces leptomeningitis, meningoencephalitis, brain abscesses, paradural abscesses, and adjacent granulomas. Grossly, two-fifths of cases present with chronic leptomeningitis, one-third with granulomas and abscesses, and one-fourth with spinal epidural granulomas and abscesses. Epidural abscesses occur rarely, secondary to craniovertebral osteomyelitis. Fibrosis in subarachnoid spaces can cause hydrocephalus. Abscesses may be extradural, subdural, or intraparenchymatous in location. No region in the CNS or peripheral nervous system is immune. It may be either focal or disseminated basal leptomeningitis with fibrinopurulent exudate. This can block subarachnoid spaces, causing hydrocephalus. Pachymeningitis may result. Histologic features elicit a mixed granulomatous and suppurative reaction. The center of an abscess contains caseous necrotic material with cells (lymphocytes, neutrophils, plasma cells, macrophages, and Langhans giant cells) and organisms. Hematoxylin and eosin (H&E) could not stain wet preparations, while periodic acid-Schiff (PAS) and methenamine silver stain demonstrate the organism in tissues. Bone and vertebral disc destruction with paraspinal abscess closely simulates tuberculous disease of the spine.[46]

6.1.2 Clinical Presentation

Patients with CNS blastomycotic lesions may have varying degrees of symptoms, which usually relate to the location of their lesion. Symptoms and signs of presentation may include headache, seizures, memory loss, confusion, coma, paraparesis, hemiparesis, and aphasia.[46] CNS blastomycosis presents with headaches and neck stiffness; intracranial blastomycotic abscesses or granulomas result in increased intracranial pressure (ICP) with or without localized signs. Eventually, these patients develop convulsions, mental deterioration, confusion, and lethargy.[7,46] The skull X-ray findings of calvarial lesions in granulomatous diseases such as blastomycosis are nonspecific. The lytic lesions can be single or multiple and with smooth or blurred borders. Clinically, these lesions are usually associated with a soft tissue mass of varying size that may demonstrate drainage. The X-ray views of spinal lesions will have findings similar to those seen in tuberculosis spondylitis. Tuberculosis and blastomycosis begin in the body of the vertebra. Both have a tendency to spare the disc and usually involve more than one vertebra. A collapsed vertebra due to granulomatous disease may be difficult to differentiate from metastatic disease. MRI and CT scans are a valuable means of determining the presence and extent of granulomatous infections of the CNS. The advantage of CT is its ability to demonstrate bone architecture better than MRI; however, MRI is superior in demonstrating early pathologic changes and spinal cord abnormalities that may be missed by CT. Plain

CT scans may show iso- or hyperattenuation lesions. However, small solitary lesions show homogeneous enhancement on CT scanning with surrounding edema. Slightly bigger lesions may show ring-enhancing lesions with surrounding edema. Motor impairment, cranial nerve palsies, and visual symptoms were also encountered in varying degrees.[7,46] Solitary blastomycotic abscesses of the CNS can displace neural structures, resulting in neurologic deficits. These masses are especially threatening when found in the posterior fossa or spinal canal and must be evacuated in a timely fashion to avoid progressive neurologic deterioration. Lesions can be found in the subdural, epidural, and subgaleal spaces or in the spinal canal. The lesions may be hard and tumorlike or purely suppurative. Surgery is required to remove the mass lesion and drain the purulent material. Debridement of the granulomatous tissue and necrotic bone, if present, should be done, which also facilitates antifungal therapy given postoperatively. Blastomycosis in African countries commonly affects the thoracic spine, ribs, and sternum by direct extension from the lungs. Osteomyelitis and discitis are associated with paraspinal abscess formation. Clinical and radiologic features are similar to tuberculosis, with which blastomycosis sometimes coexists.[7,46]

6.1.3 Diagnosis

Diagnosis of blastomycosis is established by biopsy, cultures, CSF examination, and serologic tests (the C fixation test and the double immunodiffusion test). CSF from patients with blastomycotic meningitis may be clear or cloudy, show pleocytosis, with lymphocytes or polymorphonuclear leukocytes predominant, and have a total number of leukocytes up to 15,000/mm^3. The protein concentration is often elevated and may be greater than 300 mg/dL, whereas the glucose concentration is either normal or decreased. The CSF pressure is increased. Vertebral biopsy and culture establish the diagnosis of spinal blastomycosis.[47,48]

6.1.4 Treatment

Although there are few published reports of the use of lipid formulations of AmB to treat CNS blastomycosis, these agents are preferred because of the prolonged therapy that must be given for this form of the disease. There are experimental animal data showing superior CNS penetration by liposomal AmB when compared with AmB lipid complex and AmB deoxycholate. Whether this might translate to better efficacy of this formulation in patients with CNS blastomycosis is not clear.[49] Azoles should not be used as primary therapy for CNS blastomycosis but should be used as follow-up therapy after an initial response to AmB. The most appropriate azole to use is not clear. Fluconazole has excellent CSF penetration but less activity against *B. dermatitidis* than other azoles. There are a few anecdotal reports of success. The dosage should be 800 mg/day. Itraconazole achieves minimal levels in the CSF but has greater intrinsic activity against *B. dermatitidis* than fluconazole. There are few case reports of CNS blastomycosis treated with itraconazole. Voriconazole has good CSF penetration and excellent in vitro activity against *B. dermatitidis* and has been used to treat CNS blastomycosis.[50] There is very little experience with this agent for other forms of blastomycosis. To date, there are no data for posaconazole as an agent for treatment of CNS blastomycosis. Neurosurgical intervention for cerebral abscesses, granuloma, and hydrocephalus as well as spinal decompressive and stabilization procedures may be required if significant neurological compromise is present. Surgical management of mass lesions is necessary to prevent neurologic deterioration as well as to establish the diagnosis. In conjunction with antifungal drugs, surgery helps improve the overall prognosis in these patients. Prognosis is poor if patients are untreated but much better with appropriate management. For selected patients, surgery may also be important in the management of CNS blastomycosis that causes focal neurological dysfunction. Surgical drainage of an epidural abscess or other critical lesions may be important in limiting morbidity and mortality from this disorder.[48]

6.2 Coccidioidomycosis (Modeling Valley Fever)

Coccidioidomycosis was first described in 1892 by Alejandro Posadas (a medical student from Argentina), when he evaluated Domingo Ezcurra, a soldier who presented with a verrucous papule on the right cheek. Many others arose from careful epidemiologic and clinical investigations in the San Joaquin Valley during the 1930s, when people migrated there from the "Dust Bowl" of the Midwest, and during the 1940s, when the events of World War II brought military recruits, prisoners of war, and persons of Japanese descent to camps and other areas of endemicity. Especially impressive were the contributions of Charles E. Smith, who tirelessly studied this disease throughout his professional career.[51,52] Coccidioidomycosis, also known as "San Joaquin Valley Fever" or "Desert Rheumatism," is the oldest of the major systemic mycoses, considered to be one of the most infectious fungal diseases. It is a geographically restricted mycosis, including regions of semiarid climate. Its distribution corresponds to areas where warm temperatures and dry conditions exist. It is endemic in the southwestern United States (especially in the San Joaquin Valley, California, and Arizona, where 85% of the population is skin-test positive), Mexico, and South America, particularly Argentina and Paraguay. However, due to migration and travel, cases have been reported across the world and with increasing frequency.[53] Coccidioidomycosis is caused by *Coccidioides immitis,* which is probably the most virulent fungi, causing human mycoses accounting for about 100,000 cases per year in the United States, with 70–80 deaths annually.[7,53] Both mycelia and spores are carried a considerable distance by wind or rodents. The causative organism *C. immitis* is a nonbudding spherical structure measuring 20–70 μm in diameter with a double refractile capsule. Mature forms are filled with numerous small (2–5) endospores. *C. immitis* is a dimorphic fungus producing hypha and arthroconidia (arthrospores) in its saprophytic (soil) environment and spherules (sporangia) in infected tissues. Pulmonary infection is contracted by inhalation of the airborne arthroconidia in immunocompetent or immunocompromised individuals. Most infections are self-limited. About two-thirds of cases of pulmonary infections are asymptomatic while one-third of cases develops mild through severe pulmonary disorders. However, the majority of patients recover and develop strong, specific immunity against infection. The fungus is therefore considered both a pathogen and an opportunist. Hematogenous spread to the CNS occurs in about 50% of cases as a terminal event.[7,54]

6.2.1 Neuropathology

With coccidioidomycosis, meningeal inflammation results in accumulation of exudate, opacification of leptomeninges, and obliteration of sulci with caseous granulomatous nodules at the base of the brain and in the cervical region. Extensive fibrosis causes obstructive hydrocephalus. Invasion of blood vessels leads to multiple aneurysms. Unusually large granulomatous lesions and frank abscesses can occur in the brain or spinal cord parenchyma. The microscopic picture mimics tuberculous meningitis. Meningeal fibrosis, granulomas (organisms surrounded by epithelioid cells, giant cells, lymphocytes, and plasma cells), small abscesses with caseous necrosis, and vascular invasion are seen. On H&E stains, *Coccidioides* are basophilic, and they are well demonstrated with methenamine silver stains.[7,54]

6.2.2 Clinical Presentation

Fig. 2 shows radiographic and CT scans of patients with coccidioidomycosis. Approximately 50% of patients with disseminated coccidioidomycosis acquire CNS disease. It can occur acutely with primary infection or later with dissemination. The meninges can be the only site of dissemination, in which case the patient is at increased risk of complications and death. Coccidioidal meningitis can present as an acute process but is usually chronic with insidious onset, in contrast to meningitis from bacterial causes. Persistent headaches should be evaluated thoroughly upon worsening, especially in cases of unusual severity, associated nausea and vomiting, blurry vision, or a change in mental status (e.g., drowsiness and confusion). Other common manifestations include nuchal rigidity and photophobia.[56] CNS coccidioidomycosis presents as acute, subacute, or chronic symptomatic meningitis; transient focal deficits (aphasia and hemiparesis); confusion; restlessness; and depression. Multiple cranial nerve palsies, raised ICP, and hydrocephalus occur as complications of basal meningitis and/or mass lesion. Symptoms related to increased intracranial pressure (e.g., nausea, vomiting, altered mental status) are relatively common. Less-common presentations include focal neurologic deficits, cranial nerve palsies, tremulousness, intention tremors, papilledema, gait abnormalities, seizure, and coma.[7,51] Typically a granulomatous and suppurative basilar process, coccidioidal meningitis can also involve the brain parenchyma and spinal cord with granulomas and abscesses. Hydrocephalus is a common sequela and is often present at initial diagnosis in children.[57] Donor-related coccidioidomycosis in organ transplant recipients is also described. In our

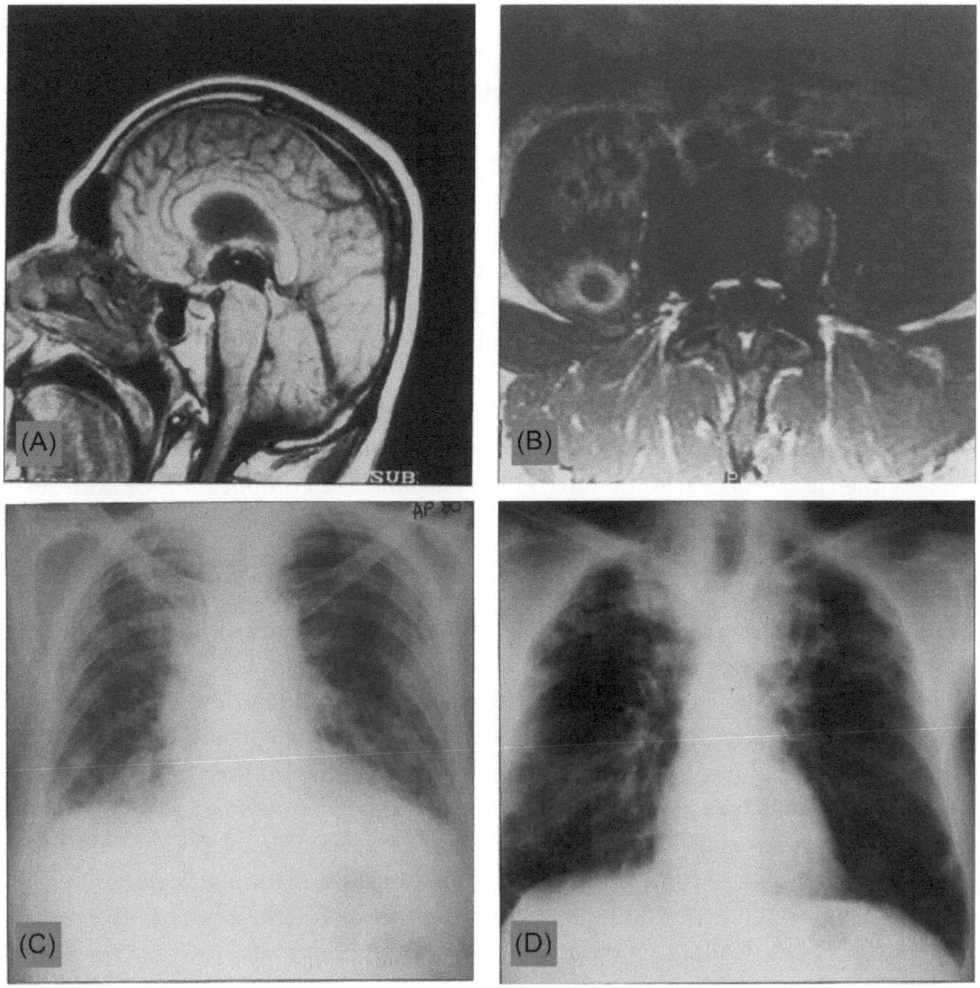

Fig. 2 (A) Coccidioidomycosis of the central nervous system. Coccidioidomycosis may present as a chronic lymphocytic meningitis with extensive inflammation of the basilar meninges. Both communicating and noncommunicating hydrocephalus may occur. In this MRI image, the Sylvian aqueduct is patent whereas the third ventricle is enlarged, which is consistent with a communicating hydrocephalus. (B) Coccidioidal abscess of the paraspinous muscles with vertebral involvement. Disseminated coccidioidomycosis may occur in multiple sites. In the MRI image shown in the figure, involvement of the fourth vertebra as well as paraspinous muscles abscesses can be seen. Surgical therapy in addition to antifungal treatment usually is required. This complication is particularly common among black men. (C) Radiograph showing diffuse pneumonia in a patient with an HIV infection. Although bilateral pneumonia is sometimes the result of inhalation of a large inoculum of *Coccidioides immitis*, more often it is a reflection of fungemia and overwhelming dissemination. In that form, diffuse coccidioidal pneumonia usually occurs in hosts with underlying cell-mediated immune deficiency, such as in patients with HIV infection, patients who have undergone allogeneic organ transplantations, or patients receiving chronic corticosteroid therapy. Sometimes called reticulonodular pneumonia, the pattern is similar to that seen in miliary tuberculosis. (D) Radiograph showing chronic pneumonia. Chronic coccidioidal pneumonia is characterized by persistent symptoms and progressive pulmonary involvement. Patients have chronic cough, weight loss, and fever. Radiographically, both fibrosis and alveolar infiltrates are present, which often worsen in the apical regions.[55]

series, one fatal case of CNS coccidioidomycosis presented with signs of raised ICP, multiple cranial nerve palsies, and rapidly progressive sensory motor impairment in the limbs.[7] The most common complication of coccidioidal meningitis is hydrocephalus, as noted above. Most—but not all—individuals with late-onset hydrocephalus continue to have CSF-based evidence of active disease, despite receipt of therapy. Cerebral infarction may be a presenting manifestation of coccidioidal meningitis or may present later in the course of the recognized illness. As with hydrocephalus, it appears that an active inflammatory process in the CSF is a risk factor for this complication. Vasculitic infarctions have usually been ascribed to inflammation of small- to middle-sized blood vessels. On rare occasions, venous and dural thrombosis may occur. Spinal arachnoiditis may occur as a complication, but unlike infarctions and hydrocephalus, arachnoiditis tends to occur after presentation.[51] Cerebral abscesses secondary to *Coccidioides* infection have been rarely reported. There are also reports of mass lesions. Both these complications—especially the latter—appear to be rare. Hyponatremia as a complication of both severe coccidioida pneumonia and meningitis has been observed on many occasions. Apparently, the association between hyponatremia and the syndrome of inappropriate antidiuretic hormone secretions has only recently been reported.[51] Coccidioidal infection is an infrequent but potentially devastating complication of pregnancy. This is particularly true if meningitis develops in a pregnant female or if a patient with coccidioidal meningitis becomes pregnant.[54] MRI studies showed evidence of leptomeningitis, intraparenchymatous brain stem granulomatous lesions, and hydrocephalus. A nonfatal case with cerebral granulomas and leptomeningitis was biopsied and then treated with antifungal medications. Myelopathy may result from adhesive meningitis, a mass lesion, or an extradural abscess associated with a vertebral osteomyelitis. Spinal coccidioidomycosis may present as acute or chronic spondylitis in the thoracic and lumbar regions, especially in cases of disseminated coccidioidomycosis. Apart from vertebral bodies, the pedicles, transverse processes, laminae, spinous processes, and contiguous ribs may be affected; however, intervertebral discs are relatively spared. Coccidioidomycosis usually presents as progressive spinal cord compression. Paraspinal masses and sinuses are commonly seen, and meningitis rarely occurs, with fatal outcome.[7]

6.2.3 Diagnosis

With coccidioidomycosis, CSF is usually under increased pressure but occasionally shows evidence of a spinal block. Low-to-moderate elevation of proteins, moderate decline in glucose, and a persistent pleocytosis may be present; the organisms can be recognized in wet preparations of the CSF.[58] The definitive diagnosis of coccidioidomycosis rests on the careful histopathologic identification of endosporulating spherules, a positive culture result with confirmation of *Coccidioides* species, and identification by serologic techniques. Thus far, there is no antigen detection method for *Coccidioides* infection. Histopathologic examination is not routinely applicable for coccidioidal meningitis, although if there is a pulmonary focus or other focus of disseminated infection, histopathologic examination of these lesions may be useful. Culture of CSF specimens on bacteriologic or fungal media is diagnostic. Unfortunately, culture results are positive for only a small percentage of meningitis cases. The foregoing notwithstanding, on occasion, *Coccidioides* organisms can be observed on direct microscopic examination of the CSF specimen. This rare event suggests an unusually high meningeal fungal burden. Even more perverse is the visualization of the mycelial or arthrocondial forms of the fungus in CSF samples. The majority of these reports have involved individuals with ventriculoperitoneal or other shunt devices in place. Less often, individuals without such foreign material are reported to have hyphal elements visualized.[59] Serologic examination is the mainstay for confirming the diagnosis. Although a negative result of a serologic test cannot exclude the diagnosis of a *Coccidioides* infection, a negative serologic test result for a patient with untreated disseminated disease, if found in an experienced laboratory, is quite rare. An exception to this may be the patient with an HIV infection or another severe immunocompromising illness. The measurement of coccidioidal antibodies in the CSF is considerably less sensitive but of greater diagnostic specificity. If a patient has a high titer of IgG antibody in the serum, some low-titer "spill over" can occasionally be seen in the CSF. In cases of meningitis, the finding of coccidioidal IgG antibody in CSF samples is virtually diagnostic.[60] A wide variety of serologic studies are available. The most reliable are the immunodiffusion tests for IgM and IgG and the complement fixation test for IgG, performed in a laboratory with considerable experience. ELISAs for IgM and IgG and latex agglutination tests are less reliable and should not be used to include or exclude the diagnosis of coccidioidal meningitis without subsequent confirmatory testing. It should be noted that monitoring of the serum coccidioidal IgG level should not be used as a guide for the course of the meningeal illness, as it is with primary and nonmeningeal disseminated disease.[51] Radiographic evaluation of the patient with coccidioidal meningitis is useful, as suggested above. A high index of suspicion for raised erythrocyte sedimentation rate, calcified nodes on chest X-rays, and craniospinal bone lesions on CT scans/MRI of the cerebrospinal axis (mass lesion and hydrocephalus) form the mainstays of investigations. Osteomyelitis of the skull or vertebrae may be visible on the plain X-ray film. Osteomyelitis may also present as causing punched-out lesions on skull films and an underlying abscess on CT/MRI studies. Radiolucent vertebral lesions devoid of surrounding sclerosis in early cases and dense sclerotic vertebrae with normal disc spaces in late states are features seen on plain X-ray films. However, bone changes and paraspinal lesions are well delineated on CT/MRI studies. Neuroimaging studies will reveal hydrocephalus in 30%–50% of patients at some time in the course of disease. MRI with contrast enhancement is more sensitive at identifying the typical basilar cisternal enhancement. About 15%–20% of patients will have evidence of vasculitic infarction. MRI is also useful at identifying and evaluating spinal arachnoiditis. A CT or MRI will rarely reveal an abscess or other focal brain complication of a *Coccidioides* infection.[61] Differential diagnosis includes multiple myeloma, sarcoidosis, histiocytosis, and metastases. It is unusual for coccidioidomycosis to present as a solitary bone lesion, which may mimic a primary bone tumor. A fine-needle aspiration biopsy helps the clinician differentiate between inflammatory and neoplastic processes involving bone by acquiring material for cytologic studies and cultures.[51]

6.2.4 Treatment

Antifungal drugs effective against coccidioidomycosis include amphotericin-B and azoles (ketoconazole, fluconazole, and itraconazole). Intravenous amphotericin-B is the most promising drug, with intrathecal administration via lumbar puncture or a subcutaneous reservoir in seriously ill or deteriorating cases. In cases of severe pulmonary compromise where an early response is highly desired, amphotericin-B is used for its rapid onset. Azoles are used for the chronic processes. When a localized mass lesion is causing compression of the brain/spinal cord, appropriate neurosurgical intervention is carried out and intrathecal azoles have been used. Coccidioidomycosis osteomyelitis remains a rare but difficult disease to treat, with a lifelong risk of recurrence. Spinal decompression and stabilization where indicated may be required, along with antifungal drugs (amphotericin-B/ketoconazole). A combined medical and surgical approach has been shown to be effective, but medical therapy alone with intravenous amphotericin-B followed by suppressive azole therapy may be effective in only selected patients. The overall prognosis in coccidioidomycosis is improving with aggressive management.[62]

6.3 Paracoccidioidomycosis (South American Blastomycosis)

Paracoccidioidomycosis is a chronic granulomatous infectious disease. The etiologic agent is the thermally dimorphic fungus *Paracoccidioides brasiliensis*. At 19–28°C the fungus develops as a mold and produces slow-growing colonies; the microscopic structures are thin septet hyphae with chlamydospores, which is the form found in nature. Chlamydospores are the infectious form of the fungus. In cultures at 37°C, as well as in tissues and exudates, the fungus appears as an oval-to-round yeast cell that reproduces by multiple budding; the "pilot wheel" cell is characteristic.[63] Paracoccidioidomycosis is the most prevalent systemic mycosis in many countries in Latin America. Its distribution is limited to subtropical regions of Central and South America, where it is endemic. Nowadays, inhalation is considered the most important route of infection. After inhalation, the chlamydospores, which are the infecting particles, transform to the yeast form in the patient's body. The primary site of infection, which is often not apparent, is the lungs. After being inhaled, *P. brasiliensis* causes a benign and transient pulmonary or oral infection in normal individuals. The primary infection is almost always subclinical in individuals with a normal immune system. If the host becomes immune depressed, a later subsequent reactivation results in chronic infection of the lungs or other organs, especially mucous and cutaneous tissue, lymph nodes, adrenals, and the CNS. Paracoccidioidomycosis is not contagious from person to person because at body temperature, the fungus is in the yeast form.[45,64]

6.3.1 Neuropathology

Fig. 3 shows clinical and paraclinical findings of the disease. CNS involvement occurs in about one-eighth of cases of systemic paracoccidioidomycosis.[66] It is far more common in males than in females. Systemic symptoms predominate in one-half of cases, neurologic manifestations predominate in one-fifth of cases, and the remaining cases present with combined symptomatology. Epilepsy is the most frequent neurologic presentation (50% of the cases). Three main types of pathologic CNS lesions are seen: granulomas-South American blastomycomas/paracoccidioidomycomas (most common); meningitis with or without hydrocephalus (uncommon), and an abscess (rare).[66] The pseudotumors or granulomas may be intraparenchymatous (common) or meningeal (uncommon). They may occur in the dura mater with clinical characteristics of meningiomas.[67] Leptomeningitis is typically of the basal granulomatous type. Spread to the brain parenchyma occurs through the Virchow-Robin spaces, especially at the level of the hypothalamus and the lateral cerebral fissures. Granulomas are formed by epithelioid cells; Langhans giant cells, and lymphohistiocytic inflammatory infiltrate. Nodules may resemble tubercles.[66]

6.3.2 Diagnosis

The diagnosis of systemic paracoccidioidomycosis is not difficult; it is established by visualization of *P. brasiliensis* upon light microscopic examination, isolation of the fungus by culture from biological specimens (sputum or biopsy samples) and by immunological tests of serum such as double immunodiffusion (DID) or ELISA with specific antigens. On the other hand, the diagnosis of neuroparacoccidioidomycosis is often difficult, and a strong clinical suspicion is necessary. This frequently occurs when a patient from an endemic area or with a history of traveling to an endemic region, or a patient who has systemic paracoccidioidomycosis, has neurological signs or symptoms.[68] A definitive diagnosis is usually made by visualization or isolation of the fungus *P. brasiliensis* from a CNS biopsy or necropsy material. However, finding the fungus in the CSF is time consuming; it is also inefficient and rarely occurs. Biopsy of brain lesions is a definitive but invasive procedure, and not infrequently the lesions are located in regions where a biopsy cannot be safely performed.[63] Neuroradiological methods (brain CT or MRI) are useful for diagnosis but are not specific; the features of this disease are similar to those of all granulomatous diseases of the CNS.[56,69] CT scan images are usually hypodense, with annular or nodular enhancing surrounded by mild edema after contrast injection. In 65% of the patients, there were multiple mass lesions, and 35% had a single mass lesion. The granuloma was most frequently located in brain hemispheres (67%); in 25% the location was

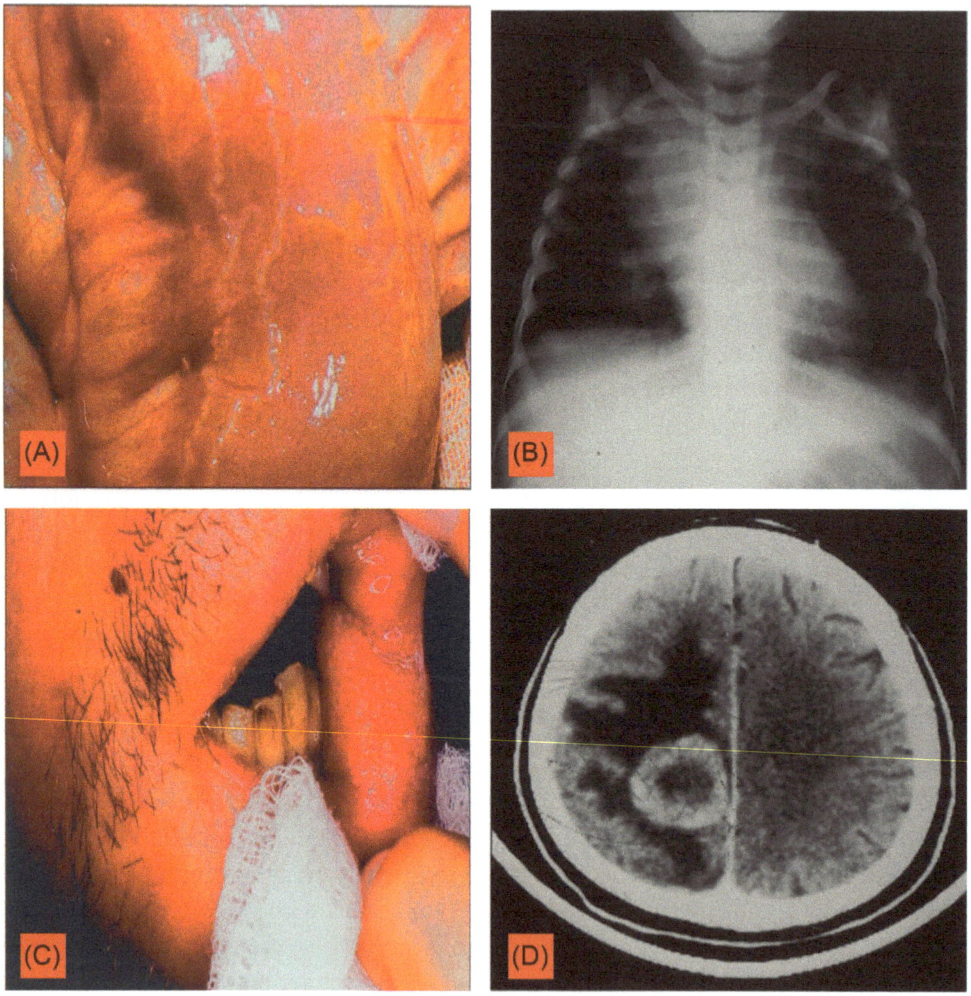

Fig. 3 (A) Intestinal lymphangiectasia. This patient, a 32-year-old man, had a clinical history of fever, weight loss, abdominal pain, vomiting, jaundice, and diarrhea for 3 months. This case was unusual because the patient's jaundice was related to a stenosis of the biliary tract caused by paracoccidioidomycosis (PCM). Jaundice in PCM is usually a consequence of biliary tract obstruction caused by enlarged surrounding lymph nodes. The patient's diarrhea (note the intense intestinal lymphangiectasia) was due to obstruction of the mesenteric lymphatic circulation. Laboratory investigation revealed hypoalbuminemia and reduced immunoglobulin and lymphocyte levels in the blood. (B) Chest radiograph from a 5-year-old patient with paracoccidioidomycosis (PCM). This patient showed striking mediastinal lymphadenopathy with no pulmonary infiltrates. In contrast to the chronic form of PCM, patients with the acute form of the disease rarely have clinical or radiographic evidence of lung involvement. (C) Mulberry-like stomatitis. Oropharyngeallesions from hematogenous spreading of the fungus are very common in paracoccidioidomycosis (PCM). They were found in 50% of our 150 cases studied between 1986 and 1996. It is believed higher local pO_2 and lower temperature of the oropharyngeal mucosa provide better conditions for fungal growth. These environmental characteristics are probably

(Continued)

the cerebellum and in 25% of cases it was the brainstem, while in 4% it was in the spinal cord.[63,68]

6.3.3 Treatment

Both meningitis and the parenchymatous form of neuroparacoccidioidomycosis are treated with antifungal drugs. The trimethoprim-sulfamethoxazole association can be used as the first therapeutic option. Sulfas are considered the drugs of choice for the treatment of neuroparacoccidioidomycosis, with amphotericin B being used only in cases with resistance or intolerance to sulfonamides. The association of sulfamethoxazole-trimethoprin is used because of the facility of its administration and its low toxicity.[63,65] The duration of treatment depends on the clinical response. Ketoconazole, itraconazole, and fluconazole have been used in the treatment of systemic paracoccidioidomycosis. The first two, particularly ketoconazole, penetrate the BBB poorly and are therefore not indicated for the treatment of neuroparacoccidioidomycosis. There are some reports of neuroparacoccidioidomycosis cases that have been successfully treated with itraconazole. Fluconazole could be an alternative for treatment because it has excellent penetration into the CNS.[63,70] Voriconazole has potential for use in neuroparacoccidioidomycosis because of its good BBB penetration. Despite clinical improvement, some patients present residual lesions in the brain CT scan, independent of the drug used for treatment. Neurosurgical procedures are indicated under specific circumstances: patients with granulomas that do not respond to clinical treatment, intracranial hypertension related to hydrocephalus, or a granuloma inducing signs and s-ymptoms of compression.[71,72] Due to the small number of cases reported, it is not clear whether intramedullary paracoccidioidomycosis granulomata should be surgically treated or if they are best treated with specific therapy. Some doctors consider surgical resection

Fig. 3—cont'd more similar to those supporting saprobic growth of *Paracoccidioides brasiliensis*. Mucosal lesions also can be linked to transitory fungemia during systemic infections, when yeast cells can attach to previously damaged tissues in the oropharyngeal mucosa. Bacterial infections (canies, gin givitis, gingivostomatites), local trauma (dental prosthesis, plants fragments for tooth cleaning), and smoking may cause local damage to the oral mucosa, shown is an ulcerated lesion with fine granulation resembling a mulberry (mulberry-like stomatitis). These lesions are commonly found in patients with the chronic form of PCM, evolve slowly, and are painful. (D) Cranial CT scan from a 45-year-old man with a progressive left hemiplegia due to a pseudotumoral form of neuroparacoccidioidomycosis. Involvement of the CNS by paracoccidioidomycosis (PCM) may be manifested by meningitis or, more frequently, parenchymal lesions. Clinical evidence of PCM in other organs or systems may be absent. CNS lesions have been reported in up to 20% of patients in necropsy series. However, less than 5% of patients in clinical series have CNS involvement. Paracoccidioidomycotic meningitis has a chronic course. Neuroparacoccidioidomycosis most commonly occurs in the pseudotumoral form with single or multiple lesions with mass effect causing sensory or motor deficits, seizures, mental status changes, and intracranial hypertension. Involvement of the brain stern, cerebellum, and spinal cord are less frequently reported than hemispheric lesions.[65]

of the lesion when there is progressive spinal cord compression.[73,74] Surgical excision of the mass lesions together with therapy with amphotericin-B (the drug of choice) offers the best chance of paracoccidioidomycosis treatment.[70] Therapy may also consist of long-term administration of itraconazole. Early diagnosis and adequate therapy may prevent extensive tissue destruction. Long-term follow-up is mandatory. Clinical improvement may be accompanied by diminishing *P. brasiliensis* antigen and antibody titers in cases of neuroparacoccidioidomycosis.[71,72]

6.4 Histoplasmosis

Histoplasmosis is considered one of the most common fungal respiratory infections in the world. It is endemic in the central and southeastern states of the United States, Latin America, Africa, and now parts of Asia, whereas in Europe its incidence is low.[75] The causative organism (2–5 μm yeast) is *H. capsulatum*. The fungus is a dimorphic fungus taking both mycelia and yeast forms. It feeds on nitrogen-rich soils, especially those replete with chicken, bird, and bat droppings. When inhabited soils like these are disturbed, spores, which are the reproducing body of the mycelia, can become airborne and inhaled. Once they are inhaled, infection begins. Due to the elevated temperature inside the lungs compared with room air, the spores grow as unicellular haploid yeast cells and are phagocytosed by macrophages. The fungus continues to live within the phagolysosomes of the macrophages where it increases the pH and potentially disturbs intracellular processes. It is known to invade the reticuloendothelial system. Lesions are found in the spleen, liver, and lymph nodes as well as the lungs. CNS involvement is rare. The primary focus is formed in the lungs, which become calcified, but may occur in the mouth, gastrointestinal tract, or skin. The risk of developing CNS histoplasmosis is increased in individuals with impaired cellular immunity, but not all patients with this condition are immunocompromised. In immunocompetent individuals, CNS histoplasmosis is extremely rare, but many patients with CNS involvement are immunocompromised by burns, antibiotics, steroids, and AIDS. Even though about 25% of the population in the United States has positive histoplasmin skin tests, CNS involvement is rare. Hematogenous dissemination may spread to the CNS. Neurologic involvement occurs in 10%–20% of patients with disseminated histoplasmosis.[75,76]

6.4.1 Neuropathology

Involvement of the CNS in histoplasmosis occurs in less than 1% of all patients with active histoplasmosis and includes diffuse leptomeningitis, discrete granuloma, periventricular granulomata, parenchymal granulomatosis, choroid plexus granulomata, and granulomatous arteritis. In diffuse basilar leptomeningitis, there is a thickening of the leptomeninges, thick yellow exudate with miliary granulomas along the blood vessels. In chronic cases, meningeal fibrosis with hydrocephalus develops. Meningitis is proportionally less prominent in histoplasmosis than in cryptococcosis and coccidioidomycosis. Histoplasmosis granulomas mimic sarcoidosis, other fungal or tubercular mass lesions consisting of central noncaseating granulomas, or small caseous areas surrounded by macrophages, lymphocytes, plasma cells, and Langhans giant cells. It is well shown with PAS. In sections stained with H&E, shrinkage of the organisms produces a halo and gives an impression of a capsule. Methenamine silver stain demonstrates macrophages packed with organisms, reactive gliosis, and fibrosis.[66]

6.4.2 Clinical Presentation

Fig. 4 shows pathological, radiographic, and CT scan findings of histoplasmosis. The disease has two peaks of incidence: one in early childhood and the other in middle age. Four forms are described as leptomeningitis, cerebritis, military granulomas, and spinal cord lesions. Leptomeningitis is common whereas other CNS lesions in histoplasmosis are rare. The condition usually presents as chronic meningitis with or without hydrocephalus. Mass lesions are rare, and occasionally chorioretinitis is seen. The skull and vertebrae may be involved by osteomyelitis with secondary spinal cord compression, especially with *H. duboisii.*[78]

6.4.3 Diagnosis

The clinician should maintain a high index of suspicion with patients who are from any area endemic for histoplasmosis. Definite diagnosis is made by demonstration of the organism by culture of sputum, CSF, and serum or histology. C fixation, agar gel, diffusion testing, and radioimmunoassay (urine or serum) are helpful. Biopsy supplies a more definitive answer. The CSF is usually under pressure, shows a moderate pleocytosis up to 300 cells/mm^3, and is more often mononuclear than polymorphonuclear in type. Proteins are elevated, glucose is somewhat reduced, and the organism is cultured from the CSF samples in about 50% of cases. Peripheral blood smear and bone marrow examination may yield early diagnosis. MRI shows enhancing mass lesions in the brain. On T1-weighted images, *Histoplasma* lesions appear as hypointense rims, and they show surrounding edema on T2-weighted images. CT scans show ring-enhancing lesions.[75,78]

6.4.4 Treatment

Untreated CNS histoplasmosis is nearly always fatal. In CNS histoplasmosis, amphotericin-B is given at dosages of 0.1–0.25 mg/kg/day to a total dosage of 30–35 mg/kg. It is then followed by fluconazole (800 mg/day) for 9–12 months. The therapy is monitored with periodic lumbar punctures and CSF estimations; a normal CSF picture should conclude treatment and keep relapse in check. Lipid formulations of amphotericin-B may be used long term. Neurosurgical interventions are undertaken wherever needed. In patients with infected shunts, initiation of antifungal therapy and removal of the shunt usually result in improvement. Still, in CNS histoplasmosis, the cure rate is about 50% in immunocompetent patients and far less in immunocompromised individuals.[75,78]

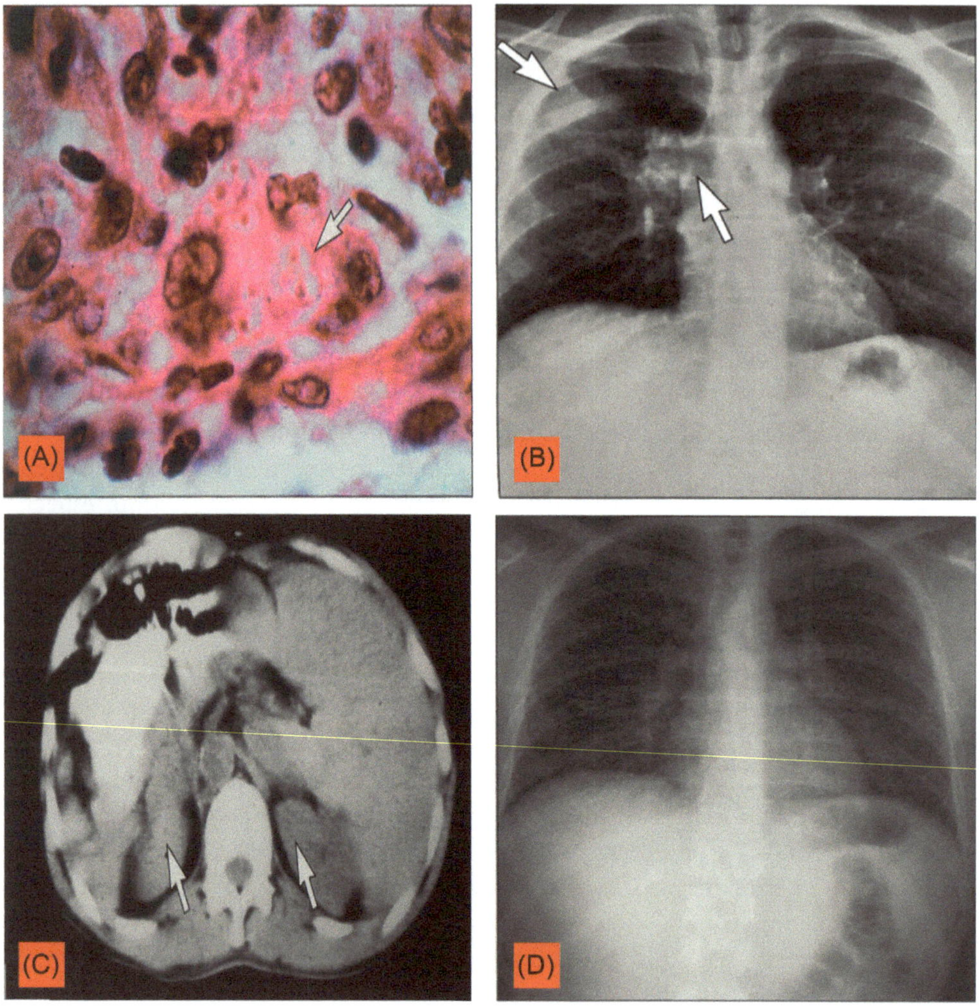

Fig. 4 (A) Hematoxylin-Eosin stain of liver biopsy specimen from a patient with progressive disseminated histoplasmosis. The stain shows heavy parasitization with the yeast phase of the organism. The organisms demonstrate an eosinophilic staining central nucleus surrounded by a clear halo *(arrow)*, which represents cytoplasmic shrinkage rather than a capsule. (B) Chest radiograph of low-intensity acute pulmonary histoplasmosis. Radiographic findings in acute pulmonary histoplasmosis include mediastinal lymphadenopathy and patchy pulmonary infiltrates. This radiograph shows enlargement of the right hilar lymph nodes *(bottom arrow)* associated with a faint right upper lobe infiltrate *(tap arrow)*. The patient complained of substernal chest pain and cough. He recovered over a few weeks without treatment. (C) Abdominal CT-scan showing bilateral adrenal masses *(arrows)*. A fine-needle aspiration biopsy revealed organisms consistent with *Histoplasma capsulatum* yeast on Gomori methenamine silver stain. A diagnosis of disseminated histoplasmosis with adrenal involvement was established. Adrenal involvement is common in patients with progressive disseminated histoplasmosis, although adrenal insufficiency is seen only in those with extensive disease. (D) Chest radiograph showing miliary infiltrate in a patient with disseminated histoplasmosis. The patient presented with fever and weight loss. Anemia, leukopenia, and thrombocytopenia prompted a bone marrow examination, which showed granulomatous changes and contained organisms resembling *Histoplasma capsulatum*. Although diffuse infiltrates may be seen in patients with acute pulmonary histoplasmosis following a heavy inoculum challenge, they suggest disseminated disease in the absence of extensive exposure.[77]

6.5 Sporotrichosis

Sporotrichosis is caused by *Sporothrix schenckii,* which is found in soil and on plants.[79] Domestic cats may be an important carrier of agents of sporotrichosis to humans.[80] The infection usually starts when the fungus is inoculated peripherally at the site of cutaneous injury or is inhaled. Lymphangitis or granulomatous pneumonitis is produced. The fungus may then disseminate hematogenously to other organs, including the CNS.[79,80] The organism is a Gram-positive, cigar-shaped body with occasional budding. In culture, branching septate hyphae bear conidia laterally or in groups. The conidia are ovoid to spherical and are 2–6 μm in size. Sporotrichosis is a chronic benign infection involving skin, mucous membranes, eyes, muscles, bones, and viscera following direct inoculation. Endophthalmitis can cause loss of vision. CNS infections occasionally result from direct traumatic implantation of the organisms and rarely from hematogenous spread causing meningitis and cerebritis. Meningitis caused by *S. schenckii* is usually indolent.[81]

6.5.1 Neuropathology

Cerebral sporotrichosis shows features of chronic meningitis, microabscesses, and granulomas. Microscopically, widespread cortical granulomatous microabscesses in the brain, spinal cord, or spinal nerve rootlets are seen.[82]

6.5.2 Clinical Presentation

Fig. 5 shows fungal morphology and clinicopathological presentation of sporotrichosis. Meningitis presents as chronic occipital headaches and dizziness. Cerebral involvement results in transient episodes of aphasia, limb weakness, and visual disturbances, progressing to lethargy and coma. Spinal involvement causes quadriplegia, paraplegia, and radicular pains.[79]

6.5.3 Diagnosis

With sporotrichosis, the CSF shows pleocytosis, mostly lymphocytes (300 cells/mm^3), a moderately increased protein level, and a reduced sugar level. Besides tissue cultures, repeated cultures of large volumes of CSF or serologic studies are required to make the diagnosis.[79,84]

6.5.4 Treatment

Patients with sporotrichosis are treated with itraconazole for 3–6 months or more prolonged courses, and this drug has become the primary therapy because of its consistent efficacy. Amphotericin-B is not consistently effective, and flucytosine may be needed with it for some response if other medications are not effective. Neurosurgical interventions are undertaken where indicated. Cutaneous and lymphocutaneous sporotrichosis may respond to ketoconazole and potassium iodide therapies.[79,80]

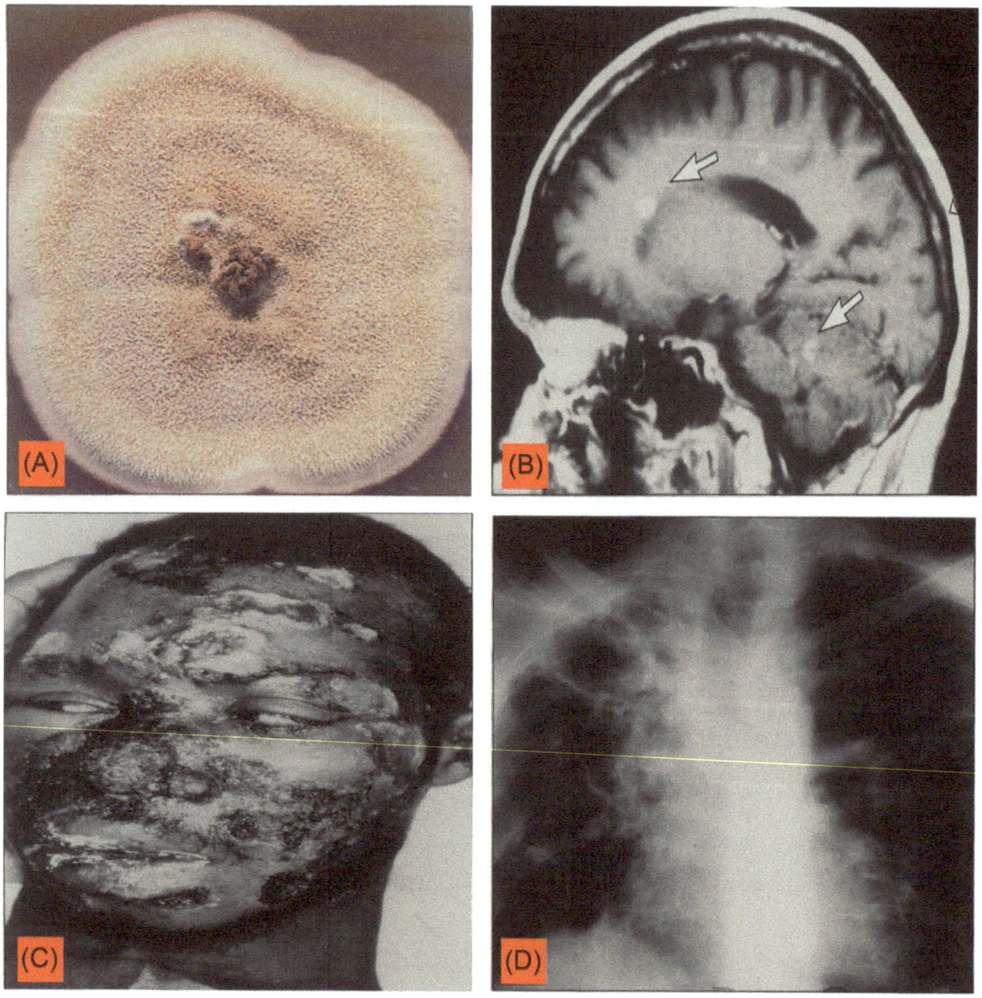

Fig. 5 (A) Typical appearance of *Sporothrix schenckii* growing in vitro at 25°C on malt extract agar. (B) MRI scan showing sporotrichosis. MRI scan of the brain of a 40-year-old patient with AIDS (CD4 count, 80/mm^3) who developed vomiting and dysequilibrium. Cerebrospinal fluid culture yielded *Sporothrix schenckii.* The scan shows multiple enhancing intracraniallesions *(arrows)* and enhancement of the meninges *(arrowhead).* The patient died of the infection. (C) Sporotrichosis in a patient with HIV infection. This 22-year-old man with advanced HIV infection (CD4 count, 17/mm^3) developed painful ulcerated lesions on his face, trunk, and extremities; infiltrative lesions of the palate and tongue; and multiple CNS lesions seen on an MRI. Despite antifungal therapy, the patient died. Necropsy showed widespread infection with *Sporothrix schenckii.* (D) Chest radiograph from a patient with pulmonary and osteoarticular sporotrichosis. The patient also had chronic obstructive pulmonary disease. Extensive bilateral infiltrates with upper lobe cavitary lesions were due to sporotrichosis.[83]

7. OPPORTUNISTIC FUNGI CAUSING INFECTIONS OF THE CNS

A. fumigatus, *C. albicans*, *C. neoformans*, *R. arrhizus*, etc. cause CNS infections.[85,86] These organisms usually cause infections in immune-compromised hosts (opportunistic fungi).

7.1 Aspergillosis

In recent years, the incidence of invasive aspergillosis has risen sharply, with practical difficulties in its diagnosis as well as in its appropriate treatment, especially in immunocompromised patients. About 10%–20% of all cases of invasive aspergillosis develop CNS manifestations. Patients with persistent immunosuppression and disseminated malignant diseases as well as patients undergoing allogeneic stem cell transplants have very high incidence rates of CNS aspergillosis.[87,88] However, routinely the majority of cases of aspergillosis are reported from countries with a temperate climate where constant exposure to the high spore content of pathogenic *Aspergillus* species is present in the moldy work environment. Several species of *Aspergillus* can cause CNS infections, but most cases are due to *A. fumigatus* (primary pulmonary infection), *A. niger* (primary otitic infections), *A. flavus* (primary paranasal sinus infection), and *A. terreus*. These are saprophytic, opportunistic, ubiquitous fungi found in soil, plants, and decaying matter; they consist of branching septate hyphae varying from 4 to 12 μm in width. *Aspergillus* species have a worldwide distribution. The primary portal of entry for *Aspergillus* is the respiratory tract. Infection either reaches the brain directly from the paranasal sinuses via vascular channels or is bloodborne from the lungs and gastrointestinal tract. The infection may also be airborne, contaminating the operative field during a neurosurgical procedure.[89]

7.1.1 Neuropathology

Single or multiple abscess formation with blood vessel invasion leading to thrombosis is a characteristic feature of Aspergillosis on neuropathologic examination. *Aspergillus* has a marked tendency to invade arteries and veins (angiotropic) producing a necrotizing angitis, secondary thrombosis, and hemorrhage. Onset of cerebral aspergillosis is heralded by manifestations of focal neurologic deficits in the anterior and middle cerebral arterial distributions. The evolving hemorrhagic infarcts convert into septic infarcts with associated abscesses and cerebritis. The fungal hyphae are found in large, intermediate, and small blood vessels with invasion through vascular walls into adjacent tissue; invasion in the reverse direction can also occur. Purulent lesions may be chronic and have a tendency toward fibrosis and granuloma formation. Microscopically the most striking feature is the intensity of the vascular invasion with thrombosis. In purulent lesions, pus is seen in the center of the abscesses with abundant polymorphs at the periphery. Granulomas consist of lymphocytes, plasma cells, and fungal hyphae.[2] Intracranial spread of *Aspergillus* infection occurs more frequently by hematogenous routes and less frequently through direct or contiguous spread. Direct extension from the sinuses or orbit has been reported.

Dissemination of the *Aspergillus* fungi to the brain from a pulmonary source occurs in 10% of patients. Direct infiltration into the basal bones leads to the more commonly encountered skull base osteomyelitis. Intracranial infection can affect the parenchyma or the meninges. According to the site and nature of infection, the patient may present with features of meningitis, focal neurological signs, or symptoms of raised intracranial pressure. The cerebral vasculature can be involved by mycotic aneurysms or intra-arterial thrombosis. *Aspergillus* hyphae can invade directly into the vessel wall, which becomes weakened due to necrosis and polymorphonuclear infiltration, resulting in mycotic aneurysm formation. These patients may present with typical subarachnoid hemorrhage syndrome. Intraluminal extension of the hyphae can also initiate thrombus formation. Rarely, major arterial stenosis may occur following leptomeningeal infection. Steroids can inhibit the macrophage response to intracellular fungus and may permit enhanced germination.[90]

7.1.2 Clinical Presentation

Fig. 6 shows clinical and paraclinical presentations of aspergillosis. Aspergillosis should be considered in cases manifesting with acute onset of focal neurologic deficits resulting from suspected vascular or space-occupying lesions, especially in immunocompromised hosts. In patients with paranasal sinus disease, orbital extension with proptosis, ocular palsies, visual deterioration, and chemosis may occur. The symptoms frequently encountered are headache, vomiting, convulsions, hemiparesis, fever, cranial nerve deficits, paralysis, and sensory impairment to varying degrees. Features typical of meningitis and subarachnoid hemorrhage resulting from mycotic aneurysms may manifest.[2,91] Patients are often afebrile or have only a low-grade fever. Their symptoms are usually those of a cerebral mass lesion, although the propensity of the fungus to invade blood vessels may lead to extensive necrosis and sometimes to intracranial bleeding.[92] The disease is usually slowly progressive and symptoms may persist for months.

7.2 Candidiasis

Once considered rare, the true incidence of CNS candidiasis has significantly increased recently. Candidiasis accounts for about 80% of systemic fungal infections, which are most commonly acquired endogenously in the hospitals. Clinical risks associated with invasive candidiasis are well recognized and include recurrent/persistent fever, colonization of tubes and catheters in intensive care units (ICUs), hemodynamic instability, shock, renal impairment, and disseminated intravascular coagulation. As a complication of disseminated candidiasis, the brain parenchyma and meninges get involved, with a fatal outcome in many patients. In autopsy studies, cerebral mycosis was noted to be most commonly caused by candidiasis.[2,85] Candidiasis commonly presents as thrush in the oral cavity or vagina; less commonly, skin and visceral organs are involved. *Candida* organisms are found worldwide. Their pseudohyphae are associated with 2–3 μm spherical or oval

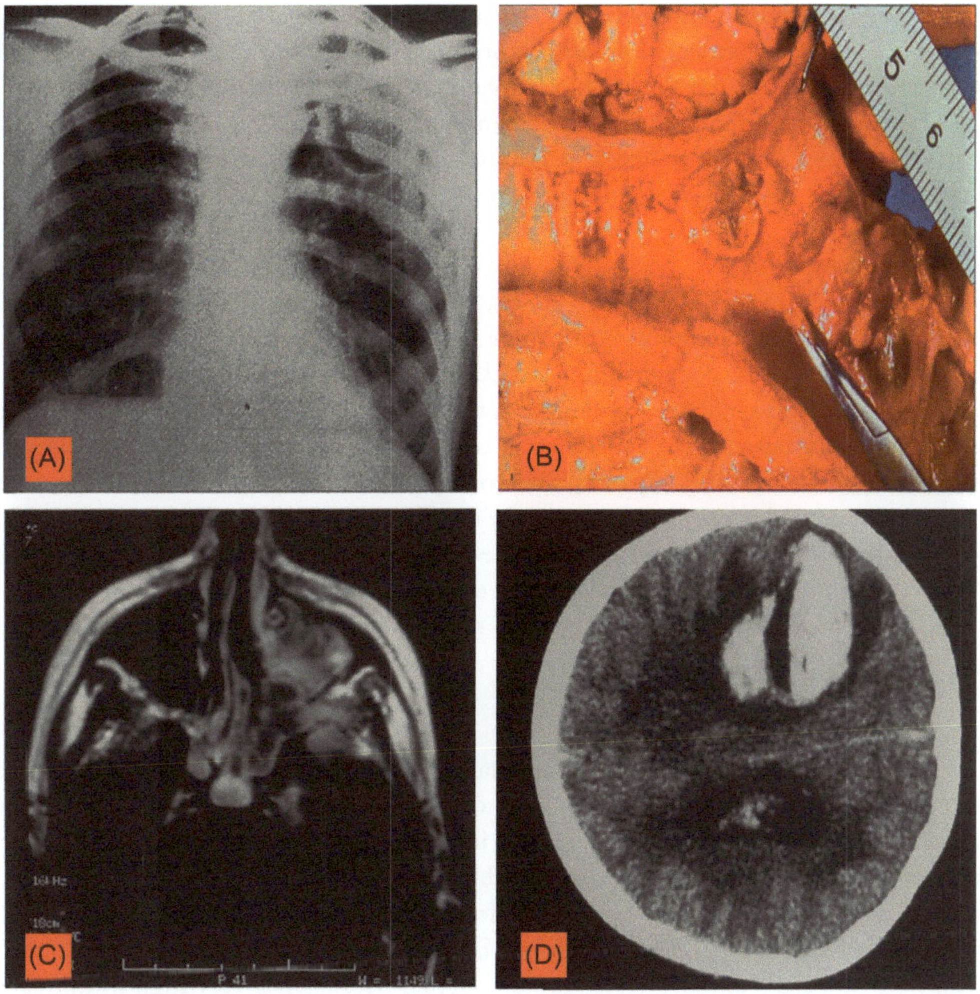

Fig. 6 (A) Aspergilloma of the left upper lung lobe. The posterioranterior chest radiograph shows two cavities in which a central opacity is partially surrounded by a crescent patch of air (Monod's sign). An aspergilloma (mycetoma, "fungus ball") typically develops in a preformed and poorly drained, mostly tuberculous lung cavity, and consists of a bati of saprophytically growing fungus. In most instances, the mycetomal lesion is caused by *Aspergillus* species but mycetomas owing to other filamentous fungi also have been described. Aspergilloma must not be confused with cavitating lesions in patients with invasive pulmonary aspergilosis who are recovering from neutropenia. Although these lesions look quite similar, they are pathophysiologically different and represent sequestration of necrotic lung tissue after hemorrhagic infarction during neutropenia. (B) Necrotizing tracheitis caused by *Aspergillus* species. Note the two circumscript acerations in the distal part of the trachea. (C) *Aspergillus* sinusitis in a patient with advanced HIV infection. The gadolinium-enhanced MRI scan shows pansinusitis sparing the right maxillary sinus. (D) Aspergillosis of the CNS presenting as intracerebral hemorrhage in a patient with acute leukemia and invasive pulmonary aspergillosis recovering from neutropenia.

blastospores. Candidiasis arises when the balance between *Candida* species and the host is altered in favor of the yeast. The *C. albicans* originally infects the gastrointestinal tract (oral cavity and esophagus) after the antibiotic treatment, then invades the submucosal blood vessels, and finally disseminates hematogenously to the CNS. Organisms also reach the CNS via colonization of ventricular drains, shunt tubing, and central venous lines. Recently, AIDS has been recognized as a predisposing factor for *Candida* meningitis.[93]

7.2.1 Neuropathology

In CNS, *Candida* invades small blood vessels, causing thrombosis and infarction. Disseminated granulomatous lesions may be scattered throughout the meninges and brain, causing meningitis or focal encephalitis. *Candida* meningitis can occur spontaneously, as a complication of disseminated candidiasis, or as a complication of an infected wound or ventriculostomies via direct inoculation of the organism into the CNS. At autopsy, gross lesions may not be apparent. Microscopically, multiple microabscesses, small macroabscesses, and microgranulomas in the distribution of anterior and middle cerebral vessels are found. The abscesses are composed of neutrophils, lymphocytes, and macrophages that evolve to a granuloma after a week. On histology, they are faintly basophilic when stained with H&E but are intensely stained with PAS and methenamine silver reaction.[94]

7.2.2 Clinical Presentation

Fig. 7 shows clinical and paraclinical presentations of candidiasis. The clinical symptomatology of CNS candidiasis is that of low-grade meningitis. A marked basal infiltrate may cause multiple cranial nerve palsies and deterioration in the level of consciousness, which is frequent in later stages of the disease, together with hydrocephalus. In acute meningitis, the signs of meningeal irritation are present. In newborns, particularly a low-birth-weight neonate, the diagnosis is difficult and delayed, leading to permanent neurologic sequelae. Symptoms and signs in these cases are nonspecific. Mortality is very high in untreated cases, whereas with antifungal therapy, it is reduced substantially nowadays.[2,96] Spinal candidiasis is rare. It can involve the vertebral body and/or the disc space by hematogenous spread in patients with *Candida* sepsis. It may occur by local invasion and as a postoperative complication of spinal surgery. Persistent low back pain is a common complaint; however, 50% of cases may have significant neurologic deficits. Neuroimaging studies show nonspecific spondylitis and discitis.[2]

7.2.3 Diagnosis

Candidiasis should be suspected in patients with external ventricular drains or blocked shunts. CSF examination, including cultures, should be routine. The CSF may show a pleocytosis with either neutrophils or lymphocytes predominating, and glucose is often low. Serial serum examinations (double diffusion, counterimmunoelectrophoresis,

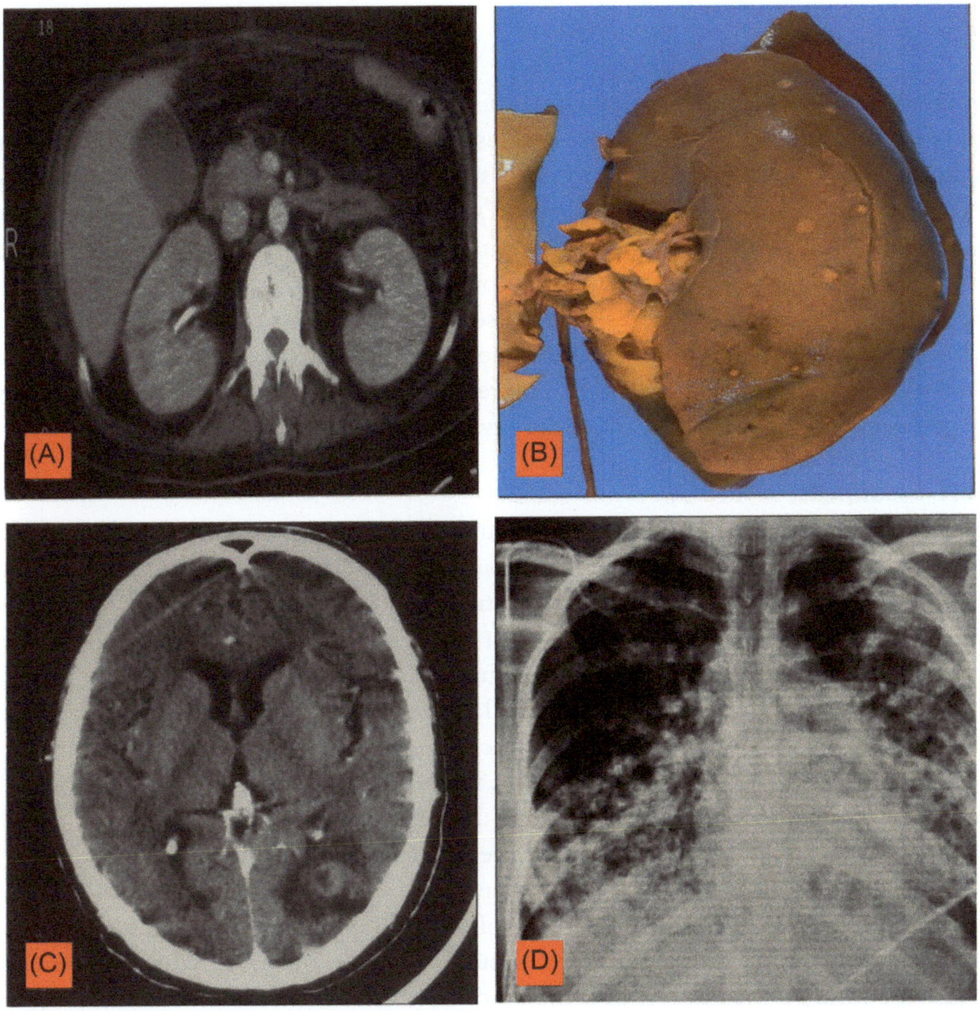

Fig. 7 (A) Computed tomogram from a patient with hematogenously disseminated candidiasis with renal involvement. Note the hypodense lesions, seen especially in the right kidney. (B) Postmortem findings from the same patient. Hematogenously acquired renal candidiasis typically leads to multiple micro- or macroabscesses scattered diffusely throughout the kidneys. This is the most common form of upper urinary tract candidiasis, When *Candida* species does disseminate, the kidney is one of the most common organs affected. Patients frequently present with candiduria, fever, chills, rigors, and shock, which are unresponsive to broad-spectrum antibiotics. (C) Computed tomograms (CTs) of patient with *Candida* brain abscess. CT from a 39-year-old patient with hematogenously disseminated candidiasis and acute lymphocytic leukemia who was undergoing chemotherapy. It is estimated that almost half of patients with disseminated candidiasis have CNS involvement. The lesions reported include meningitis, microabscesses, macroabscesses, vasculitis with thrombosis and infarction, mycotic aneurysm, subdural spinal fungal granuloma, demyelination and transverse myelitis, fungus ball, and hemorrhage or hemorrhagic necrosis. The most commonly reported CT pattern is that of multiple abscesses with ring enhancement and edema after intravenous contrast, as seen in this patient. (D) Chest radiograph from a patient with acute leukemia. neutropenia, and hematogenously disseminated candidiasis. Although *Candida albicans* is commonly isolated from sputum, it is rarely found invading the lung. Biopsy evidence of tissue invasion is required for definitive diagnosis. A diagnosis of *Candida* pneumonia cannot be made on the basis of sputum culture and radiographic findings alone. *Candida* pneumonia is usually seen in patients with neutropenia. Chest radiograph shows a finely nodular, diffuse interstitial infiltrate. Pathologic sect ion revealed numerous foci of hematogenously seeded *Candida* species.[95]

immunofluorescence, and latex agglutination tests) are helpful. Positive blood and CSF cultures provide definitive diagnosis. Funduscopic examination is invaluable, detecting endophthalmitis before permanent loss of vision occurs. Bone or disc biopsy, histology, and culture principally make the diagnosis of spinal candidiasis.[2] An MRI of fungal spinal osteomyelitis usually shows hypointensity of the vertebral bodies on T1-weighted images with enhancement on contrast administration. There is a lack of hyperintensity within the discs on T2-weighted images with preservation of the intranuclear cleft, which is in contrast with pyogenic osteomyelitis where the intervertebral disc may be hyperintense with loss of intranuclear cleft on T2-weighted images. The diagnosis may be made by percutaneous needle aspiration of the involved area.[97]

7.2.4 Treatment

Removal of infected artifacts, correction of underlying predisposing/risk factors, neurosurgical intervention for cerebral abscesses, and amphotericin-B with or without flucytosine form the mainstays of candidiasis treatment. For meningitis and for diffuse parenchymal infection, combined amphotericin-B and flucytosine therapy is the most rational treatment. In fulminant cases, antifungal drugs are given intrathecally. In neonates and infants with meningitis, liposomal amphotericin-B has been used with good outcomes. In some cases, newer azoles (voriconazole and posaconazole) and echinocandins are used.[2] Amphotericin-B, flucytosine, fluconazole, miconazole, and ketoconazole have been used successfully in spinal candidiasis. Decompressive surgery is usually required in patients with significant compressive spinal lesions. If there is evidence of pus collection on a spinal MRI, then preoperative image-guided aspirations to prove the fungal nature of the lesion should be followed by preoperative antifungal medications; surgical débridement and fixation should be carried out with a full postoperative course of antifungal medications to achieve good results in many cases. Fluconazole (with amphotericin-B in the initial phases) for 3–6 months postoperatively may lead to healing of fungal osteomyelitis with osseous consolidation. The overall prognosis is better than in earlier times due to combined medical and surgical therapies.[2,94]

7.3 Cryptococcosis (European Blastomycosis)

Cryptococcosis is one of the most common CNS fungal infections in immunocompromised patients.[98] It is ubiquitous but more common in Europe than elsewhere. It is a generalized systemic visceral mycosis affecting previously healthy people; however, in 50% of cases, it has been reported in immunocompromised subjects, children, and middle-aged and older males. Pigeon breeders are at special risk. The causative agent, *C. neoformans,* is a spherical budding capsulated yeast (5–20 μm). It is found in soil and wood contaminated with bird excreta.[2] It has been isolated from fruit juices and milk. *C. neoformans* var. *neoformans* causes disease in immunocompromised hosts, and *C. neoformans* var. *gattii* is the cause in immunocompetent hosts; the portal of entry is the respiratory system. The primary focus lies in the lungs, from which secondary systemic dissemination occurs via hematogenous spread. There is a strong neurotropic tendency to involve meninges and the brain. CNS cryptococcal infection commonly presents with meningitis (subacute or chronic) or meningoencephalitis. Acute fatal meningitis occurs rarely in cryptococcosis.[2,98] Usually, chronic meningitis presents with the periods of remission and relapses. Cryptococcosis is one of the common CNS fungal infections in immunocompromised patients: nearly one-tenth of patients with HIV develop cryptococcosis, and many HIV patients have cryptococcal meningitis as their initial clinical presentation.[99]

7.3.1 Neuropathology

Fig. 8 shows clinical and paraclinical findings of cryptococcosis. With cryptococcosis, the leptomeninges become infiltrated, thickened, and opaque. The Virchow-Robin spaces around penetrating vessels are distended with organisms. Granulomatous lesions can be found in the cerebral or spinal parenchyma. Spinal arachnoiditis may also be present. Chronic fibrosing leptomeningitis may cause hydrocephalus. Less commonly, intraparenchymatous cysts are seen (basal ganglionic regions), related to exuberant mucinous capsular material produced by the proliferating cryptococci. Rarely, fungi aggregate in an inflammatory lesion and produce small or large granulomas (cryptococcomas or torulomas) in the meninges, parenchyma, ependymal surfaces, or choroid plexuses.[2] Microscopic examination shows three types of tissue reactions: (1) disseminated leptomeningitis, (2) granulomas, and (3) intraparenchymal cysts. In meningitis, there is minimal inflammatory response. The capsule of the fungus seems to impede inflammation by masking surface antigen. Inflammatory responses consist of lymphocytes (mainly), plasma cells, eosinophils, fibroblasts, and multinucleated giant cells (studded with cryptococci). Glial reaction and associated cerebral edema are minimal. Granulomas are rarer late-tissue reactions mimicking tubercles. They are composed of fibroblasts, giant cells (with fungal organisms), and necrotic areas. Multiple intraparenchymal cysts related to exuberant capsular material produced by the proliferating cryptococci create honeycomb-like cystic cerebral changes, especially in the basal ganglia. No membrane or capsule surrounds these cysts, which are well delineated from the surrounding tissue. Inflammatory response (macrophages with fungi and giant cells) around these cystic lesions is minimal.[101]

7.3.2 Clinical Presentation

CNS cryptococcosis commonly presents with nonspecific manifestations. At onset, patients usually present with headaches, nausea, vomiting, visual impairment, and papilledema; at a later stage, neck stiffness develops, followed by fever, personality changes, seizures, deterioration in sensorium, cranial nerve palsies, and hydrocephalus. In many patients, there are no physical signs. Periods of remission and relapse are noted.

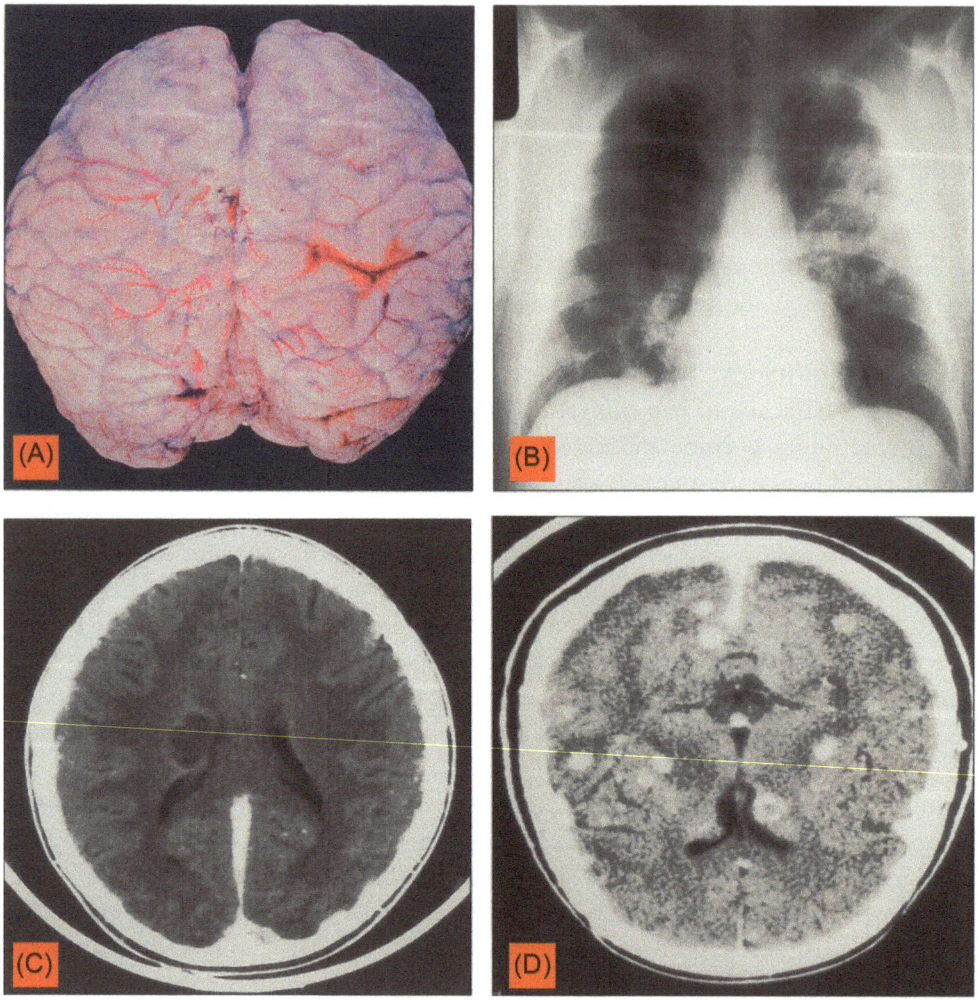

Fig. 8 (A) Leptomeningitis in cryptococcal meningitis. Leptomeningitis is the primary feature of CNS cryptococcosis. The leptomeningitis is usually devoid of a thick, exudate-like bacterial meningitis. Slight clouding of the leptomeninges is shown in this dorsal view of the cerebral hemisphere in a patient with cryptococcal meningitis. (B) Chest radiograph identifying a left upper lobar pneumonitis. This radiograph, combined with symptoms of a lower respiratory infection, may be clinically confused with a bacterial pneumonia. If *Cryptococcus neoformans* is considered a cause of pneumonia, sputum cultures must be observed for a longer period than routine bacterial cultures to detect the fungus. (C) CT showing cryptococcoma. Cryptococcoma may present as a single mass lesion. In this CT. a large cystic-appearing periventricular mass is shown with only minimal enhancement. When aspirated, the mass revealed *Cryptococcus neoformans.* However, it should be emphasized that in severely immunosuppressed patients such as those with AIDS, even with cryptococcal meningitis, solitary brain masses may represent a second etiologie infection such as toxoplasmosis or nocardiosis. (D) CT scan of the brain demonstrates multiple enhanced nodules throughout the brain parenchyma caused by cryptococcus. Brain lesions can be either multiple or single and appear to be more common in CNS infections with *Cryptococcus neoformans* var. *gattii* while *Cryptococcus neoformans* var. *neoformans* commonly produces a meningitis syndrome.[100]

Cryptococcomas may produce prolonged, unresolving, focal neurologic deficits. Acute fatal meningitis in cryptococcosis is extremely rare.

7.3.3 Diagnosis

The fungal capsule is transparent with cryptococcosis; therefore, the CSF appears clear, although it is mildly xanthochromic and under high pressure. The cell count may increase to 100 cells/mm^3 (mainly lymphocytes and polymorphs). The sugar and chloride levels are reduced, and total proteins may be raised. As the fungal capsule is transparent on routine microscopy, India ink preparation of the CSF can demonstrate a mucoid capsule. Organisms can be seen in tissues with PAS and methenamine silver stains. On mucicarmine and alcian blue stains, the fungal capsule is well recognized.[2] However, routinely CNS cryptococcosis is diagnosed by positive cryptococcal antigen titer and India ink staining of the CSF film preparation. Chest X-rays (pulmonary lesion) and CT/MRI brain scans (brain edema, hydrocephalus, basal meningitis, granuloma, and intraparenchymal cysts) are helpful. The most common patterns of CNS cryptococcosis are ventricular dilation in CT scans and Virchow-Robin space dilation in an MRI. MRI is more sensitive in detecting CNS cryptococcal infection such as Virchow-Robin space dilation and leptomeningeal enhancement. There was no significant pattern difference between immunocompromised and nonimmunocompromised patients with CNS cryptococcosis. The CSF culture should be done at 30°C for 5 days. This organism can be isolated in 95% of cryptococcal meningitis cases. Positive latex agglutination testing with rising serum titers of polysaccharide capsular antigen is of prognostic value.[101]

7.3.4 Treatment

Untreated cryptococcal meningitis is generally fatal. Early aggressive therapy with combined amphotericin-B and flucytosine offers the best chance of a cure. Alternatively, a 6-week course of amphotericin-B followed by maintenance oral fluconazole gives good results. Granulomas, cysts, and hydrocephalus are treated on their own merits. Predisposing factors should be carefully corrected. Failure of treatment raises the possibility of an underlying medical disorder, such as AIDS. Patients receiving treatment before CNS complications have a good prognosis.[2]

8. MELANIZED FUNGI CAUSING INFECTIONS OF THE CNS

Several varieties of infections of the CNS by melanized fungi can be eminent: (a) single-organ infection of the cerebrum, (b) extension into the cerebrum from adjacent cavities, (c) fungal presence in the cerebrospinal fluid, or (d) meningitis. The fungal order Chaetothyriales (containing *Exophiala*-like black yeasts and families) is particularly rich in fungi causing cerebral infections. Cases by the main agents such as *Exophiala dermatitidis*,

Cladophialophora bantiana, and *Ramichloridium mackenziei* published during recent years are studied. Most of these infections prove to be fatal. Resection of the lesions in combination with antimycotic therapy may reduce mortality.[102] Fig. 9 shows involvement of the CNS by melanized fungi.

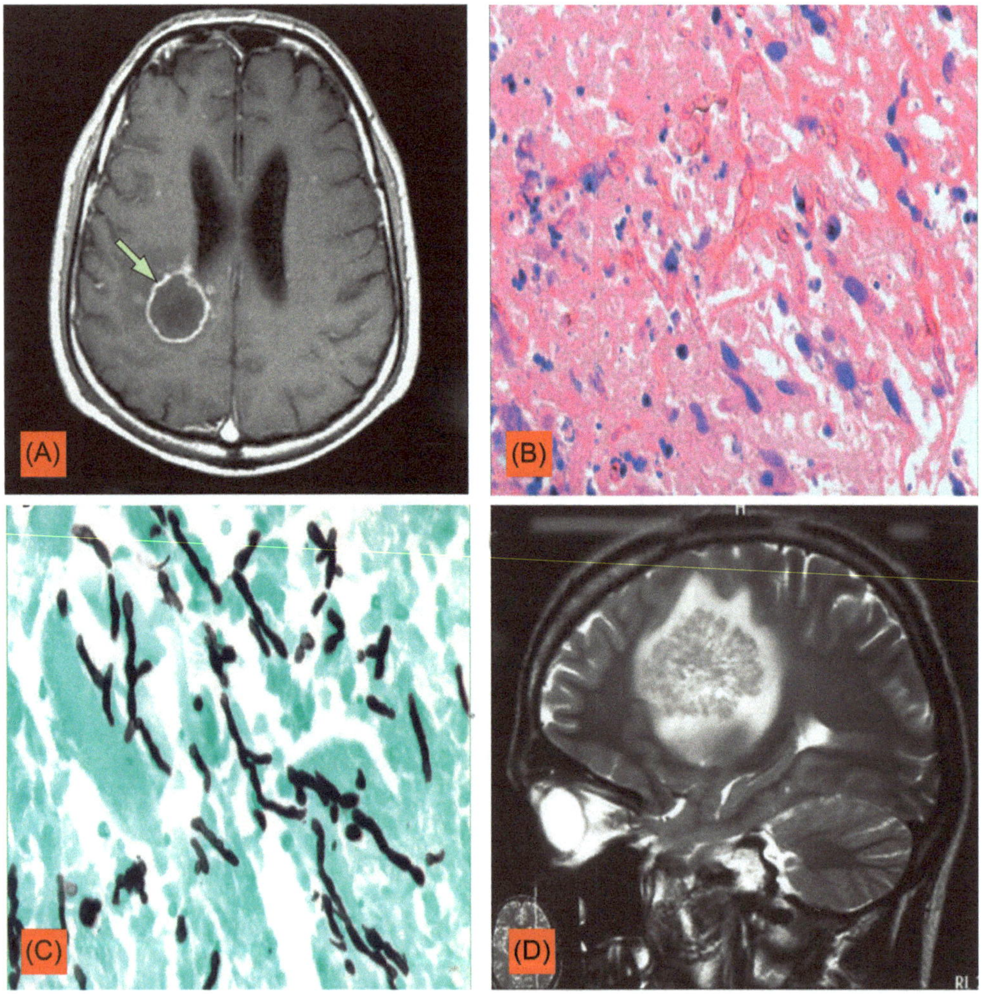

Fig. 9 (A) MRI showing ring-enhancing lesion in the right hemisphere of a brain infected with *Rhinocladiella mackenziei.* (B,C) showing histopathological features of *Cladophialophora bantiana* Pigmented fungal hyphae (B) elongated, and septate with unicellular lemon-shaped conidia (hematoxylin and eosin stain). The fungal elements (C) are intensely positive with Grocott's methenamine silver. Hyphae and conidia shown inside multinucleated giant cell (D) MRI (a 21-year-old male student) showing lesion in the left frontoparietal region of a brain infected with *Exophiala dermatitidis.*[103,104]

8.1 CNS Infection Due to *C. bantiana*

The species *Cladophialophora* comprises nearly exclusively human-pathogenic species, earlier classified in *Cladosporium* or *Xylohypha*.[105,106] *C. bantiana* is a recognizably neurotropic species that has seldom been isolated from sources other than living mammal tissue. The species has been found in cases of cerebritis worldwide.[102,106] A common preference for warmer climates with high average humidity is apparent, although the species is also known from Canada and the United Kingdom. Cases from hot, arid climatic zones such as the central United States, North Africa, the Middle East, and Australia are rare.[102,106] The fungus is probably introduced via inhalation and is fatal if untreated.[102,106] The infection is usually found in immunocompetent young males with a history of solid organ transplantation or drug abuse in about 40% of patients.[102,107,108] Nishimura and Miyaji demonstrated that suppression of innate cellular immunity by the use of cortisone promotes *C. bantiana* infection.[109] Infections nearly always provoke granulomatous reactions containing large numbers of giant cells. The fungal elements inside the phagocytes retain their viability.[110] The first line of defense probably includes complement system and PMNs, but also T cell-mediated immunity plays a role.[111] Of the previous 53 cases reviewed by Horré and de Hoog with *C. bantiana* involving the brain, the average was 35.8 years (6 months–76 years); 13 were female, 39 male, and the ratio of Caucasian to African was 9:16. Remarkably, in none of the cases where the race of the patient was mentioned, an Asian patient was concerned. Pharmacologic immunosuppression was not an important risk factor whereas cellular immune deficiencies proved to be significant.[102,106,112] Occupations with regular exposure to dust such as farming were abundantly represented among patients. The overall survival rate in patients with an infection caused by *C. bantiana* ranges between 28% and 35%. Acute, severe headaches, paralysis of the limbs, seizures, and somnolence were the most common clinical manifestations. Although most lesions were located in the cerebrum, a few cases involved the cerebellum and ventricles and their adjacent CSF-containing cavities, suggesting hematogenous spread. Fifteen patients had a solitary, 16 had multiple (ring) lesions. Fever was mostly absent or could be recognized only at the end of the disease. In about 25% of the cases, symptoms indicative of meningitis—such as neck stiffness—were noted, despite absence of meningeal involvement in histopathology. Death occurred between 3 weeks and 12 months after the first symptoms were noted.[102] Six additional cases caused by *C. bantiana* were reported, one of which was in an otherwise healthy individual.[104,113] Four were solid-organ transplant recipients combined with diabetic problems. One was a pediatric patient with a primary immunodeficiency of unknown origin.[114] Five patients were male and one was female. The age range was 8–72. One fatal case caused by *C. bantiana* in a lung transplant patient was mentioned in a retrospective global report on mycotic brain abscesses in solid organ and hematopoietic stem cell transplant recipients, without providing detailed patient data.[115] In most cases, a single cerebral lesion is present which progress to becoming multiple at the involving area after surgery or disease relapse. One patient had multiple abscesses upon first diagnosis. Among the most prevalent symptoms were headache, nausea, seizures, speech difficulties, and right hemiparesis. In one patient, motoric dysphasia was observed. In one patient antituberculosis treatment was started, based on an interpretation of a CT scan, but no clinical improvement was obtained after 2 weeks of therapy. Four cases were fatal despite antimycotic therapy with amphotericin B (AMB), liposomal amphotericin B (L-AMB), itraconazole (ITZ), or fluconazole (FLZ), either alone or as combination therapies. The pediatric patient with an inoperable cerebral abscess had a fatal outcome despite a combination treatment with voriconazole (VRZ), flucytosine (5-FC), and caspofungin. One patient responded favorably to FLZ therapy, but as a second patient under combined FLZ and AMB therapy did not survive, survival may have been because of successful surgery. Death occurred between 2 weeks and 4 months after the first symptoms were noted. The etiologic agent of one case was deposited in the culture collection of the Westerdijk Fungal Biodiversity Institute (Formerly; Centraalbureau voor Schimmelcultures) in the Netherlands, with accession number CBS 102586; its identity was confirmed by sequencing. The fungus from another case was confirmed by demonstrating the presence of a 558 bp intron considered to be specific for *C. bantiana* sensu stricto using an intron-specific primer and preserved in the culture collection of Duke University's Mycology Research Unit in the United States as DUMC 112.01.[106] Kantarcioglu et al. (2016) reported a 28-year-old coal-miner with *C. bantiana* infection. The first symptoms occurred about 8 weeks after known traumatic inoculation. Lesions were compatible with tuberculous granulomas, and the patient initially received antitubercular treatment. Melanized fungal cells were seen in a brain biopsy and abscess materials. Therapy was switched from empirical antitubercular treatment to amphotericin B (0.5 mg/kg/day), but was changed to voriconazole 200 mg/day through i.v. on the basis of antifungal susceptibility test results. The patient responded clinically, and gradually improved. The isolate was identified by sequencing of the Internal Transcribed Spacer domain of rDNA.[116]

8.2 CNS Infection Due to *R. mackenziei*

Primary cerebral phaeohyphomycosis due to *Rhinocladiella mackenziei* is an extremely rare infection carrying more than 80% mortality, with most cases reported from the Middle East.[112] This darkly pigmented black yeast is highly neurotropic, aggressive, and refractory to most antifungal agents. Cases diagnosed in the United States concerned visitors or immigrants from Saudi Arabia and Kuwait.[117,118] Infections were probably contracted in that region, although the fungus has neither been recorded in aereomycological studies nor has it been isolated from environmental sources. Restriction of human infections to the Middle East suggests a preference of the saprobic phase, if existent, for hot arid climates.[106,112] It is the only neurotropic fungus reported from this region; *E. dermatitidis* and *C. bantiana* are missing. In Saudi Arabia, it is the most common agent of infectious brain disease.[102,119] Eleven cases have been reported between 1983 and 1999. All cases were located in the cerebrum, often in otherwise healthy hosts. About half the patients were previously subjected to major surgery. The male-to-female ratio was 1:1. Mortality was 100% within 1 year after diagnosis for all reported cases, despite surgery and antifungal therapy.[120] None of the reported cases of brain abscess showed manifestations of pulmonary or other organ involvement. Tissue morphology of the fungus is similar to that of other dematiaceous filamentous fungi; therefore an unequivocal diagnosis is dependent upon evidence from the culture.[121] Seven additional cerebral infections were reported since 1998 in patients mostly with underlying diseases, all originating from or residing in the Middle East. One case occurred in a patient with renal failure, one in a patient with diabetes mellitus, one in a patient with chronic myelomonocytic leukemia, one with myelofibrosis and Hodgkin's lymphoma, one in a patient with chronic liver disease, one in a kidney transplant recipient, and one in an otherwise healthy host.[104,120] The age range was 42–71; four were male and three were female. Early symptoms were headaches, blurry vision or loss of vision, seizures, dizziness, convulsions, fever, occasional loss of consciousness, slurred speech and drowsiness. In one case, direct microscopic examination of the aspirate revealed many moniliform hyphal elements, with long, slender septate hyphae branching at acute angles and with conidia projecting from both sides. This was thought to be consistent with the *Aspergillus* species. *R. mackenziei* was subsequently cultured from biopsy material of the lesion and from a second brain aspirate. Six of the seven cases were fatal. Four patients expired despite AMB therapy; two cases had a fatal outcome despite aggressive antifungal therapy with AMB and/or ITZ with or without 5FC along with surgical intervention. One case showed radiographic and physical improvement by switching the therapy from AMB and ITZ to salvage therapy with posaconazole (PSZ). Most strains from human cases have been preserved in major culture collections and showed a high degree of sequence similarity (G.S. de Hoog, unpublished data). Additionally, an agent from an unpublished case from Pakistan is held in the CBS culture collection with accession number CBS 102592. Yusupov et al. (2017) reported an immunocompetent elderly male presenting with multiple brain abscesses with *R. mackenziei* confirmed by nuclear ribosomal repeat region sequencing, who was successfully treated by surgical debridement and intravenous voriconazole. This is the first case reported from the United Kingdom.[122]

8.3 CNS Infections Due to Other Melanized Fungi

Two secondary CNS infections as a result of members of the ascomycete order Pleosporales, *Bipolaris spicifera* and *Curvularia clavata*, have been reported. *B. spicifera* was observed as the etiological agent of meningitis in an 18-year-old male, probably secondary to preceding acoustic neuroma resection.[106] A sinusitis followed by cerebritis caused by *C. clavata* in a 46-year-old immunocompetent female was reported by Ebright et al.[123] They reported that the patient responded to complete surgical removal of the infected sinus tissue with subsequent AMB therapy followed by ITZ. A case by a member of the order Dothideales, *Cladosporium cladosporioides,* was reported in a 30-year-old Caucasian male.[124] Otherwise *C. cladosporioides* is an extremely common airborne saprobe. The nonmelanized fungus *Paecilomyces variotii* (order Sordariales) was encountered in a 60-year-old female cancer patient.[125] The case was fatal. Strain identities were confirmed by DNA sequencing and have been deposited in the collection of the Centraalbureau voor Schimmelcultures, Utrecht, The Netherlands, with accession numbers CBS 109501 and CBS 110036, respectively. A very remarkable infection by a *Nodulisporium* species was reported in an otherwise healthy 55-year-old male from India.[41] There was no sinusitis, no underlying disease prior to the infection, and no lesion outside the CNS. An earlier case by *Nodulisporium* sp. concerned just an allergic sinusitis without tissue invasion.[126] *Nodulisporium* belongs to an order of strictly wood-inhabiting fungi, the Xylariales, thus far without any medical representative known.[106,126,127]

8.4 In Vitro Antifungal Susceptibility Data of Melanized Fungi

Data on activities of antifungal agents against melanized fungi are still scant. An overview of published susceptibility data for *E. dermatitidis*, *C. bantiana*, *R. mackenziei*, *B. spicifera*, *C. clavata,* and *C. cladosporioides* was also performed. Testing methods differed slightly among published studies. A standard M38-A document by the US National Committee for Clinical Laboratory Standards is proposed as a reproducible procedure for the antifungal susceptibility testing of molds.[128] Modifications were proposed for determining MICs and minimum fungicidal concentrations (MFCs) for some species. Optimal testing condition was found to be the combination of RPMI 1640 broth with 48–72 h of incubation (72 h appeared the most reliable incubation time for *C. bantiana*) and 100% growth inhibition (MIC-0). Because black yeasts are rather slow-growing fungi, some investigators prefer to incubate somewhat longer. Particularly for *R. mackenziei*, the preference is to read the MICs at the first 24 h interval and 48 h when growth was observed in the drug-free growth control.[106] Several investigators reported separate results for each test condition.[106,128] Variations in medium, inoculum size, and incubation time or temperature have significant impact on the reproducibility of MIC data. The number of isolates tested to date is rather low. Thus, the susceptibility profiles of species of melanized fungi as yet have limited predictive value. In main traits, melanized Chaetothyriales such as *Exophiala* and *Cladophialophora* appear susceptible to AMB in vitro as well as to triazole antifungals, such as ITZ and VRZ.[106,128] Terbinafine was also found to be active in vitro against *E. dermatitidis* although it appeared inactive in an experimental mice model.[106] There is no clinical experience with this drug in treatment of patients with CNS infections. The susceptibility for *R. mackenziei* against AMB appeared to be strain-dependent;

MICs for FLZ were high while isolates yielded low-MIC values against ITZ and the newer triazoles VRZ and PSZ.[106,128] Some strain-dependent AMB susceptibility was also noted in other black yeasts by Vitale et al. (2003) Patients may expire despite therapy with AMB, a compound to which strains mostly showed in vitro susceptibility. This indicates a marked absence of correlation between in vitro results and in vivo response, possibly because of pharmacokinetics of the drug, general host factors, and site and progression of the infection at onset of therapy, combined with the virulence of the pathogen.[129]

9. CONCLUDING REMARKS

The CNS fungal infections were previously regarded as rare diseases. The frequency of such infections has increased dramatically in past decades mainly due to the increased awareness of clinicians about these conditions, widespread use of antibacterial agents, a rapid increase in the number of the immunocompromised population, the advances in imaging, and the availability of microbiological techniques to confirm the diagnosis from bodily fluids and specimens. Besides known yeasts and molds that have been introduced as documented pathogens involved in CNS fungal infections, the role of new immerging fungal pathogens in the etiology of such infections is established by the time. Early diagnosis is an important hallmark that determines the outcome of CNS fungal infections in debilitating patients. Overall, progress in the discovery of novel diagnostic tests and effective antifungal agents has improved the prognosis of the CNS fungal infections.

ACKNOWLEDGMENT

This work was supported financially by the Pasteur Institute of Iran.

REFERENCES

1. Murthy JM, Sundaram C. Fungal infections of the central nervous system. *Handb Clin Neurol* 2014;**121**:1383Y1401. https://doi.org/10.1016/B978-0-7020-4088-7.00095-X.
2. Sharma RR. Fungal infections of the central nervous system: current perspective and controversies in management. *Int J Surg* 2010;**8**:591–601.
3. Datta K, Bartlett KH, Baer R, Byrnes E, Galanis E, Heitman J, et al. Spread of *Cryptococcus gattii* into Pacific Northwest region of the United States. *Emerg Infect Dis* 2009;**15**:1185–91. https://doi.org/10.3201/ eid1508.081384.
4. Teeling JL, Perry VH. Systemic infection and inflammation in acute CNS injury and chronic neurodegeneration: underlying mechanisms. *Neuroscience* 2009;**158**:1062–73. https://doi.org/10.1016/j.neuroscience 2008.07.031.
5. Mathur M, Johnson CE, Sze G. Fungal infections of the central nervous system. *Neuroimaging Clin N Am* 2012;**22**:609–32. https://doi.org/10.1016/j.nic.2012.04.004.

6. Ramesha KN, Kate MP, Kesavadas C, Radhakrishnan VV, Nair S, Thomas SV. Fungal infections of the central nervous system in HIV-negative patients: experience from a tertiary referral center of South India. *Ann Indian Acad Neurol* 2010;**13**:112.
7. Sheikh SS, Amr SS. Fungal infections of the central nervous system. In: *Progress in Mycology*. Netherlands: Springer; 2010. p. 141–80. https://doi.org/10.1007/978-90-481-3713-8_5.
8. Jacobs CS, Etherton MR, Lyons JL. Fungal infections of the central nervous system. *Curr Infect Dis Rep* 2014;**16**. https://doi.org/10.1007/s11908-014-0449-2.
9. Pruitt AA. Nervous system infections in patients with cancer. *Neurol Clin* 2003;**21**:193–219.
10. Cunha BA. Central nervous system infections in the compromised host: a diagnostic approach. *Infect Dis Clin North Am* 2001;**15**:567–90. https://doi.org/10.1016/S0891-5520(05)70160-4.
11. Smith RM, Schaefer MK, Kainer MA, Wise M, Finks J, Duwve J, et al. Fungal infections associated with contaminated methylprednisolone injections. *N Engl J Med* 2013;**369**:1598–609. https://doi.org/10.1056/NEJMoa1213978.
12. Bauserman SC, Schochet Jr SS. Bacterial, fungal and parasitic diseases of the central nervous system. In: Nelson JS, Parisi JE, Schochet Jr SS, editors. *Principles and practice of neuropathology*. USA: Oxford University Press; 1993. p. 42–74.
13. Bazan C, Rinaldi MG, Rauch RR, Jinkins JR. Fungal infections of the brain. *Neuroimaging Clin N Am* 1991;**1**:57–8.
14. Bennett JE. Introduction to mycoses. In: *Principles and practice of infectious diseases*. 6th ed. Philadelphia, PA: Churchill Livingstone; 2005.
15. Chimelli L, Mahler-Araújo MB. Fungal infections. *Brain Pathol* 1997;**7**:613–27. https://doi.org/10.1111/j.1750-3639.1997.tb01078.x.
16. Baradkar V, Mathur M, De A, Kumar S, Rathi M. Prevalence and clinical presentation of cryptococcal meningitis among HIV seropositive patients. *Indian J Sex Transm Dis* 2009;**30**:19–22.
17. Welsh O, Vera-Cabrera L, Rendon A, Gonzalez G, Bonifaz A. Coccidioidomycosis. *Clin Dermatol* 2012;**30**:573–91. https://doi.org/10.1016/j.clindermatol.2012.01.003.
18. Gonyea EF. The spectrum of primary blastomycotic meningitis: a review of central nervous system blastomycosis. *Ann Neurol* 1978;**3**:26–39. https://doi.org/10.1002/ana.410030106.
19. Binford CH, Thompson RK, Gorham ME. Mycotic brain abscess due to *Cladosporium trichoides*, a new species. *Am J Clin Pathol* 1952;**22**:535–42. https://doi.org/10.1093/ajcp/22.6.535.
20. Wheat LJ, Batteiger BE, Sathapatayavongs B. *Histoplasma capsulatum* infections of the central nervous system: a clinical review. *Medicine* 1990;**69**:244.
21. Ameen M, Talhari C, Talhari S. Advances in paracoccidioidomycosis. *Clin Exp Dermatol* 2010;**35**:576–80. https://doi.org/10.1111/j.1365-2230.2009.03647.x.
22. Boiron P, Provost F, Chevrier G, Dupont B. Review of nocardial infections in France 1987 to 1990. *Eur J Clin Microbiol Infect Dis* 1992;**11**:709–14. https://doi.org/10.1007/BF01989975.
23. Wong B, Pfeiffer CD. *Diagnosis and treatment of fungal infections*. Springer International Publishing; 2015. https://doi.org/10.1007/9783319130903_5.
24. Hibbett DS, Binder M, Bischoff JF, Blackwell M, Cannon PF, Eriksson OE, et al. A higher-level phylogenetic classification of the Fungi. *Mycol Res* 2007;**111**:509–47. https://doi.org/10.1016/j.mycres.2007.03.004.
25. Neofytos D, Fishman JA, Horn D, Anaissie E, Chang CH, Olyaei A, et al. Epidemiology and outcome of invasive fungal infections in solid organ transplant recipients. *Transpl Infect Dis* 2010;**12**:220–9. https://doi.org/10.1111/j.1399-3062.2010.00492.x.
26. Biller J, Ferro JM, editors. *Neurologic aspects of systemic disease*. Elsevier; 2014.
27. Scheld MW, Marra CM, Whitley RJ. *Infections of the central nervous system*. Philadelphia, USA: Lippincott Williams & Wilkins; 2014.
28. McCarthy M, Rosengart A, Schuetz AN, Kontoyiannis DP, Walsh TJ. Mold infections of the central nervous system. *N Engl J Med* 2014;**371**:150–60. https://doi.org/10.1056/NEJMra1216008.
29. Sundaram C, Umabala P, Laxmi V, Purohit AK, Prasad VS, Panigrahi M, et al. Pathology of fungal infections of the central nervous system: 17 years' experience from Southern India. *Histopathology* 2006;**49**:396–405. https://doi.org/10.1111/j.1365-2559.2006.02515.x.
30. Honda H, Warren DK. Central nervous system infections: meningitis and brain abscess. *Infect Dis Clin North Am* 2009;**23**:609–23. https://doi.org/10.1016/j.idc.2009.04.009.
31. Lyons RW, Andriole VT. Fungal infections of the CNS. *Neurol Clin* 1986;**4**:159–70.

32. Chakrabarti A. Epidemiology of central nervous system mycoses. *Neurol India* 2007;**55**:191. https://doi.org/10.4103/0028-3886.35679.
33. Walsh TJ, Groll A, Hiemenz J, Fleming R, Roilides E, Anaissie E. Infections due to emerging and uncommon medically important fungal pathogens. *Clin Microbiol Infect* 2004;**10**:48–66. https://doi.org/10.1111/j.1470-9465.2004.00839.x.
34. Martino P, Girmenia C. Diagnosis and treatment of invasive fungal infections in cancer patients. *Support Care Cancer* 1993;**1**:240–4. https://doi.org/10.1007/BF00366042.
35. Jensen HE, Chandler FW. Histopathological diagnosis of mycotic diseases. In: *Topley and Wilson's microbiology and microbial infections*; 2005. https://doi.org/10.1002/9780470688618.taw0133.
36. Love GL, Boyd GD, Greer DL. Large *Cryptococcus neoformans* isolated from brain abscess. *J Clin Microbiol* 1985;**22**:1068–70.
37. Lipton SA, Hickey WF, Morris JH, Loscalzo J. Candidal infection in the central nervous system. *Am J Med* 1984;**76**:101–8. https://doi.org/10.1016/0002-9343(84)90757-5.
38. Pappas PG, Kauffman CA, Andes D, Benjamin DK, Calandra TF, Edwards JE, et al. Clinical practice guidelines for the management candidiasis. 2009 update by the Infectious Diseases Society of America. *Clin Infect Dis* 2009;**48**:503–35. https://doi.org/10.1086/596757.
39. Enzmann DR, Brant-Zawadzki M, Britt RH. CT of central nervous system infections in immunocompromised patients. *Am J Roentgenol* 1980;**135**:263–7. https://doi.org/10.2214/ajr.135.2.263.
40. Lowe J, Bradley J. Cerebral and orbital *Aspergillus* infection due to invasive aspergillosis of ethmoid sinus. *J Clin Pathol* 1986;**39**:774–8. https://doi.org/10.1136/jcp.39.7.774.
41. Ribes JA, Vanover-Sams CL, Baker DJ. Zygomycetes in human disease. *Clin Microbiol Rev* 2000;**13**:236–301. https://doi.org/10.1128/CMR.13.2.236-301.2000.
42. Bariola JR, Perry P, Pappas PG, Proia L, Shealey W, Wright PW, et al. Blastomycosis of the central nervous system: a multicenter review of diagnosis and treatment in the modern era. *Clin Infect Dis* 2010;**50**:797–804. https://doi.org/10.1086/650579.
43. Pappas PG. Blastomycosis. In: *Atlas of infectious diseases*. Current Medicine Group; 2000. p. 42–8. https://doi.org/10.1007/978-1-4757-9313-0_3.
44. Salaki JS, Louria DB, Chmel H. Fungal and yeast infections of the central nervous system: a clinical review. *Medicine* 1984;**63**:108.
45. Gupta J, Patel G, Epelbaum O. Reversal of fortune: central nervous system blastomycosis. *Am J Med* 2016;**129**:109–12. https://doi.org/10.1016/j.amjmed.2015.08.037.
46. Castillo CG, Kauffman CA, Miceli MH. Blastomycosis. *Infect Dis Clin North Am* 2016;**30**:247–64. https://doi.org/10.1016/j.idc.2015.10.002.
47. Wheat LJ, Knox KS, Hage CA. Approach to the diagnosis of histoplasmosis, blastomycosis and coccidioidomycosis. *Curr Treat Option Infect Dis* 2014;**6**:337–51. https://doi.org/10.1007/s40506-014-0020-6.
48. Quinones-Hinojosa A. *Schmidek and Sweet: operative neurosurgical techniques: indications, methods and results*. Philadelphia, USA: Elsevier Health Sciences; 2012.
49. Chapman SW, Dismukes WE, Proia LA, Bradsher RW, Pappas PG, Threlkeld MG, et al. Clinical practice guidelines for the management of blastomycosis: 2008 update by the Infectious Diseases Society of America. *Clin Infect Dis* 2008;**46**(12):1801. https://doi.org/10.1086/588300.
50. Panackal AA, Williamson PR. Fungal infections of the central nervous system. *Continuum: Lifelong Learn Neurol* 2015;**21**:1662–78.
51. Goldstein EJ, Johnson RH, Einstein HE. Coccidioidal meningitis. *Clin Infect Dis* 2006;**42**:103–7. https://doi.org/10.1086/497596.
52. Fierer J. *Coccidioidomycosis*. Springer Berlin Heidelberg; 2009. https://doi.org/10.1007/978-3-540-29676-8_3261.
53. Garcia Garcia SC, Salas Alanis JC, Flores MG, Gonzalez Gonzalez SE, Vera Cabrera L, Ocampo Candiani J. Coccidioidomycosis and the skin: a comprehensive review. *An Bras Dermatol* 2015;**90**:610–9. https://doi.org/10.1590/abd1806-4841.20153805.
54. Yau AA. Risk factors and epidemiology of coccidioidomycosis demonstrated by a case of spontaneous pulmonary rupture of cavitary coccidioidomycosis. *Case Rep Infect Dis* 2016;**2016**: https. //doi.org/10.1155/2016/8165414.
55. Ampel NM, Einstein HE, Galgiani JN, Wieden MA. Coccidioidomycosis. In: *Atlas of infectious diseases*. Current Medicine Group; 2000. p. 24–36. https://doi.org/10.1007/978-1-4757-9313-0_2.

56. Saubolle MA, McKellar PP, Sussland D. Epidemiologic, clinical, and diagnostic aspects of coccidioidomycosis. *J Clin Microbiol* 2007;**45**:26–30. https://doi.org/10.1128/JCM.02230-06.
57. Blair JE. Coccidioidal meningitis: update on epidemiology, clinical features, diagnosis, and management. *Curr Infect Dis Rep* 2009;**11**:289–95. https://doi.org/10.1007/s11908-009-0043-1.
58. Hartmann K, Pennisi MG, Ferrer L, Addie D, Belak S, Boucraut-Baralon C, et al. Rare systemic mycoses in cats: blastomycosis, histoplasmosis and coccidioidomycosis: ABCD guidelines on prevention and management. *J Feline Med Surg* 2013;**15**(7):624.
59. Thompson III G, Wang S, Bercovitch R, Bolaris M, Van Den Akker D, Taylor S, et al. Routine CSF analysis in coccidioidomycosis is not required. *PLoS One* 2013;**8**: https. //doi.org/10.1371/journal.pone.0064249.
60. Rodriguez-Cerdeira C, Arenas R, Moreno-Coutiño G, Vásquez E, Fernández R, Chang P. Systemic fungal infections in patients with human inmunodeficiency virus. *Acta Dermosifiliogr* 2014;**105**:5–17. https://doi.org/10.1016/j.adengl.2012.06.032.
61. D'Avino A, Di Giambenedetto S, Fabbiani M, Farina S. Coccidioidomycosis of cervical lymph nodes in an HIV-infected patient with immunologic reconstitution on potent HAART: a rare observation in a nonendemic area. *Diagn Microbiol Infect Dis* 2012;**72**:185–7. https://doi.org/10.1016/j.diagmicrobio.2011.10.002.
62. Stevens DA, editor. *Coccidioidomycosis: a text*. New York, USA: Springer Science & Business Media; 2013.
63. de Almeida SM. Central nervous system paracoccidioidomycosis: an overview. *Braz J Infect Dis* 2005;**9**:126–33. https://doi.org/10.1590/S1413-86702005000200002.
64. de Almeida SM, Roza TH. HIV immune recovery inflammatory syndrome and central nervous system paracoccidioidomycosis. *Mycopathologia* 2017;**182**:393–6. https://doi.org/10.1007/s11046-016-0076-6.
65. Colombo AL, Queiroz-Telles F, Graybill JR. Paracoccidioidomycosis. In: *Atlas of infectious diseases*. Current Medicine Group; 2000. p. 53–63. https://doi.org/10.1007/978-1-4757-9313-0_4.
66. Bonifaz A, Vázquez-González D, Perusquía-Ortiz AM. Endemic systemic mycoses: coccidioidomycosis, histoplasmosis, paracoccidioidomycosis and blastomycosis. *J Dtsch Dermatol Ges* 2011;**9**:705–15. https://doi.org/10.1111/j.1610-0387.2011.07731.x.
67. Wanke B, Aidê MA. Paracoccidioidomycosis. *J Bras Pneumol* 2009;**35**:1245–9. https://doi.org/10.1590/S1806-37132009001200013.
68. Ribeiro CM, Caixeta CA, de Carli ML, Sperandio FF, de Sá Magalhães EM, Pereira AA, et al. Photodynamic inactivation of oral paracoccidioidomycosis affecting woman with systemic lupus erythematosus: an unusual case report. *Photodiagnosis Photodyn Ther* 2017;**17**:160–3. https://doi.org/10.1016/j.pdpdt.2016.12.006.
69. Huntington RW. Pathology of coccidioidomycosis. In: Stevens DA, editor. *Coccidioidomycosis: A text*. New York, USA: Springer Science + Business Media; 1980.
70. Peçanha PM, de Souza S, Falqueto A, Grão-Veloso TR, Lírio LV, Ferreira CU, et al. Amphotericin B lipid complex in the treatment of severe paracoccidioidomycosis: a case series. *Int J Antimicrob Agents* 2016;**48**:428–30. https://doi.org/10.1016/j.ijantimicag.2016.06.011.
71. Kappe R. Antifungal activity of the new azole UK-109, 496 (voriconazole). *Mycoses* 1998;**42**:83–6.
72. Kappe R. Antifungal activity of the new azole UK-109 496 (voriconazole): addition and clarification. *Mycoses* 2001;**44**:432. https://doi.org/10.1046/j.1439-0507.2001.00680.x.
73. Montenegro MR, Franco M, Lacaz CS, Restrepo-Moreno A, Del Negro G. Paracoccidioidomycosis. *Mycoses* 1995;**38**:245. https://doi.org/10.1111/j.1439-0507.1995.tb00060.x.
74. Colli BO, Assirati Jr JA, Machado HR, Figueiredo JF, Chimelli L, Salvarani CP, et al. Intramedullary spinal cord paracoccidioidomycosis: report of two cases. *Arq Neuropsiquiatr* 1996;**54**:466–73. https://doi.org/10.1590/S0004-282X1996000300017.
75. Hariri OR, Minasian T, Quadri SA, Dyurgerova A, Farr S, Miulli DE, et al. Histoplasmosis with deep CNS involvement: case presentation with discussion and literature review. *J Neurol Surg Rep* 2015;**76**:167–72. https://doi.org/10.1055/s-0035-1554932.
76. Nyalakonda H, Albuerne M, Hernandez LP, Sarria JC. Central nervous system histoplasmosis in acquired immunodeficiency syndrome. *Am J Med Sci* 2016;**351**:177–86. https://doi.org/10.1016/j.amjms.2015.11.016.
77. Colombo AL, Queiroz-Telles F, Graybill JR. Histoplasmosis. In: *Atlas of infectious diseases*. Current Medicine Group; 2000. p. 1–21. https://doi.org/10.1007/978-1-4757-9313-0_1.

78. Wheat LJ, Freifeld AG, Kleiman MB, Baddley JW, McKinsey DS, Loyd JE, et al. Clinical practice guidelines for the management of patients with histoplasmosis: 2007 update by the Infectious Diseases Society of America. *Clin Infect Dis* 2007;**45**:807–25. https://doi.org/10.1086/521259.
79. de Lima Barros MB, de Almeida Paes R, Schubach AO. *Sporothrix schenckii* and sporotrichosis. *Clin Microbiol Rev* 2011;**24**:633–54. https://doi.org/10.1128/CMR.00007-11.
80. Lopes-Bezerra LM, Schubach A, Costa RO. *Sporothrix schenckii* and sporotrichosis. *An Acad Bras Ciênc* 2006;**78**:293–308. https://doi.org/10.1590/S0001-37652006000200009.
81. Rodrigues AM, De Hoog S, De Camargo ZP. Emergence of pathogenicity in the *Sporothrix schenckii* complex. *Med Mycol* 2013;**51**:405–12. https://doi.org/10.3109/13693786.2012.719648.
82. Ellison D, Love S, Chimelli LM, Harding B, Lowe JS, Vinters HV, et al. *Neuropathology: a reference text of CNS pathology*. New York, USA: Elsevier Health Sciences; 2012.
83. Kauffman CA. Sporotrichosis. In: *Atlas of infectious diseases*. Current Medicine Group; 2000. p. 66–76. https://doi.org/10.1007/978-1-4757-9313-0_5.
84. Freitas DF, Lima MA, de Almeida-Paes R, Lamas CC, do Valle AC, Oliveira MM, et al. Sporotrichosis in the central nervous system caused by *Sporothrix brasiliensis*. *Clin Infect Dis* 2015;**61**(4):663. https://doi.org/10.1093/cid/civ361.
85. May RC, Stone NR, Wiesner DL, Bicanic T, Nielsen K. *Cryptococcus*: from environmental saprophyte to global pathogen. *Nat Rev Microbiol* 2016;**14**:106–17. https://doi.org/10.1038/nrmicro.2015.6.
86. Pfaller MA, Diekema DJ. Rare and emerging opportunistic fungal pathogens: concern for resistance beyond *Candida albicans* and *Aspergillus fumigatus*. *J Clin Microbiol* 2004;**42**:4419–31. https://doi.org/10.1128/JCM.42.10.4419-4431.2004.
87. Beal MF, O'carroll CP, Kleinman GM, Grossman RI. Aspergillosis of the nervous system. *Neurology* 1982;**32**:473. https://doi.org/10.1212/WNL.32.5.473.
88. Koutsouras GW, Ramos RL, Martinez LR. Role of microglia in fungal infections of the central nervous system. *Virulence* 2017;**18**:1–14. https://doi.org/10.1080/21505594.2016.1261789.
89. Kerr SC, Fischer GJ, Sinha M, McCabe O, Palmer JM, Choera T, et al. Flea expression in *Aspergillus fumigatus* is recognized by fucosylated structures on mucins and macrophages to prevent lung infection. *PLoS Pathog* 2016;**12**: https. //doi.org/10.1371/journal.ppat.1005555.
90. Nadkarni T, Goel A. Aspergilloma of the brain: an overview. *J Postgrad Med* 2005;**51**:37–41.
91. Chakrabarti A, Chatterjee SS, Das A, Shivaprakash MR. Invasive aspergillosis in developing countries. *Med Mycol* 2011;**49**:35–47. https://doi.org/10.3109/13693786.2010.505206.
92. Metta H, Corti M, Redini L, Bruggesser F, Arechavala A, Negroni R, et al. Renal abscess due to *Aspergillus fumigatus* as the only sign of disseminated aspergillosis in a patient with AIDS. *Rev Iberoam Micol* 2010;**27**:136–9. https://doi.org/10.1016/j.riam.2009.09.004.
93. Casado JL, Quereda C, s Oliva J, Navas E, Moreno A, Pintado V, et al. Candidal meningitis in HIV-infected patients: analysis of 14 cases. *Clin Infect Dis* 1997;**25**:673–6. https://doi.org/10.1086/513746.
94. Faix RG. Systemic *Candida* infections in infants in intensive care nurseries: high incidence of central nervous system involvement. *J Pediatr* 1984;**105**:616–22. https://doi.org/10.1016/S0022-3476(84)80433-3.
95. Chu LA, Filler SG. Paracoccidioidomycosis. In: *Atlas of infectious diseases*. Current Medicine Group; 2000. p. 97–114. https://doi.org/10.1007/978-1-4757-9313-0_7.
96. Singh N, Victor LY, Gayowski T. Central nervous system lesions in adult liver transplant recipients: clinical review with implications for management. *Medicine* 1994;**73**:110.
97. Sánchez-Portocarrero J, Pérez-Cecilia E, Corral O, Rome-Rovivas J, Picazo JJ. The central nervous system and infection by Candida species. *Diagn Microbiol Infect Dis* 2000;**37**:169–79. https://doi.org/10.1016/S0732-8893(00)00140-1.
98. Aye C, Henderson A, Yu H, Norton R. Cryptococcosis—the impact of delay to diagnosis. *Clin Microbiol Infect* 2016;**22**:632–5. https://doi.org/10.1016/j.cmi.2016.04.022.
99. Jarvis JN, Harrison TS. HIV-associated cryptococcal meningitis. *AIDS* 2007;**21**:2119–29. https://doi.org/10.1097/QAD.0b013e3282a4a64d.
100. Perfect JR. Cryptococcosis. In: *Atlas of infectious diseases*. Current Medicine Group; 2000. p. 81–90. https://doi.org/10.1007/978-1-4757-9313-0_6.
101. Kourbeti IS, Mylonakis E. Fungal central nervous system infections: prevalence and diagnosis. *Expert Rev Anti-Infect Ther* 2014;**12**:265–73. https://doi.org/10.1586/14787210.2014.874282.

102. Horré R, De Hoog GS. Primary cerebral infections by melanized fungi: a review. *Stud Mycol* 1999;**43**:176–93.
103. Sood S, Vaid VK, Sharma M, Bhartiya H. Cerebral phaeohyphomycosis by *Exophiala dermatitidis*. *Indian J Med Microbiol* 2014;**32**:188–90. https://doi.org/10.4103/0255-0857.129830.
104. Li DM, de Hoog GS. Cerebral phaeohyphomycosis—a cure at what lengths? *Lancet Infect Dis* 2009;**9**:376–83. https://doi.org/10.1016/S1473-3099(09)70131-8.
105. de Hoog GS, Guarro J, Gené J, Figueras MJ. Atlas of clinical fungi, 2nd ed. Centraalbureau voor Schimmelcultures, Utrecht, The Netherlands.
106. Kantarcıoğlu AS, Hoog GD. Infections of the central nervous system by melanized fungi: a review of cases presented between 1999 and 2004. *Mycoses* 2004;**47**:4–13. https://doi.org/10.1046/j.1439-0507.2003.00956.x.
107. Dixon DM, Walsh TJ, Merz WG, McGinnis MR. Infections due to *Xylohypha bantiana* (*Cladosporium trichoides*). *Rev Infect Dis* 1989;**11**:515–25. https://doi.org/10.1093/clinids/11.4.515.
108. Walz R, Bianchin M, Chaves ML, Cerski MR, Severo LC, Londero AT. Cerebral phaeohyphomycosis caused by *Cladophialophora bantiana* in a Brazilian drug abuser. *J Med Vet Mycol* 1997;**35**:427–31. https://doi.org/10.1080/02681219780001541.
109. Nishimura K, Miyaji M. Tissue responses against *Cladosporium trichoides* and its parasitic forms in congenitally athymic nude mice and their heterozygous littermates. *Mycopathologia* 1985;**90**:21–8. https://doi.org/10.1007/BF00437269.
110. Aravysky RA, Aronson VB. Comparative histopathology of chromomycosis and cladosporiosis in the experiment. *Mycopathologia* 1968;**36**:322–40. https://doi.org/10.1007/BF02050379.
111. Lobetti RG. Leukogram and serum globulin values in two dogs with systemic *Xylohypha bantiana* infection. *J S Afr Vet Assoc* 1996;**67**:91–2.
112. Keyser A, Schmid FX, Linde HJ, Merk J, Birnbaum DE. Disseminated *Cladophialophora bantiana* infection in a heart transplant recipient. *J Heart Lung Transplant* 2002;**21**:503–5. https://doi.org/10.1016/S1053-2498(01)00357-6.
113. Silveira ER, Resende MA, Mariano VS. Brain abscess in a renal transplant patient due to *Cladophialophora* (*Xylohypha*) *bantiana*. *Med Online Rev Virt Med* 2001.
114. Trinh JV, Steinbach WJ, Schell WA, Kurtzberg J, Giles SS, Perfect JR. Cerebral phaeohyphomycosis in an immunodeficient child treated medically with combination antifungal therapy. *Med Mycol* 2003;**41**:339–45.
115. Baddley JW, Salzman D, Pappas PG. Fungal brain abscess in transplant recipients: epidemiologic, microbiologic, and clinical features. *Clin Transplant* 2002;**16**:419–24. https://doi.org/10.1034/j.1399-0012.2002.02033.x.
116. Kantarcioglu AS, Guarro J, de Hoog GS, Apaydin H, Kiraz N, Balkan II, et al. A case of central nervous system infection due to *Cladophialophora bantiana*. *Rev Iberoam Micol* 2016;**33**:237–41. https://doi.org/10.1016/j.riam.2016.01.004.
117. Sutton DA, Slifkin M, Yakulis R, Rinaldi MG. US case report of cerebral phaeohyphomycosis caused by *Ramichloridium obovoideum* (*R. mackenziei*): criteria for identification, therapy, and review of other known dematiaceous neurotropic taxa. *J Clin Microbiol* 1998;**36**:708–15.
118. Podnos YD, Anastasio P, De La Maza L, Kim RB. Cerebral phaeohyphomycosis caused by *Ramichloridium obovoideum* (*Ramichloridium mackenziei*). *Neurosurgery* 1999;**45**:372–5 [Case report].
119. Jamjoom AB, Al-Hedaithy SA, Jamjoom ZA, Al-Hedaithy M, El-Watidy SF, Rahman N, et al. Intracranial mycotic infections in neurosurgical practice. *Acta Neurochir* 1995;**137**:78–84. https://doi.org/10.1007/BF02188786.
120. Al-Abdely HM, Alkhunaizi AM, Al-Tawfiq JA, Hassounah M, Rinaldi MG, Sutton DA. Successful therapy of cerebral phaeohyphomycosis due to *Ramichloridium mackenz*iei with the new triazole posaconazole. *Med Mycol* 2005;**43**:91–5. https://doi.org/10.1080/13693780400011104.
121. Khan ZU, Lamdhade SJ, Johny M, Al-Khalidi J, Thussu A, Yossef HN, et al. Additional case of *Ramichloridium mackenziei* cerebral phaeohyphomycosis from the Middle East. *Med Mycol* 2002;**40**:429–33. https://doi.org/10.1080/mmy.40.4.429.433.
122. Yusupov N, Merve A, Warrell CE, Johnson E, Curtis C, Samandouras G. Multiple brain abscesses caused by *Rhinocladiella mackenziei* in an immunocompetent patient: a case report and literature review. *Acta Neurochir* 2017. https://doi.org/10.1007/s00701-017-3141-0.

123. Ebright JR, Chandrasekar PH, Marks S, Fairfax MR, Aneziokoro A, McGinnis MR. Invasive sinusitis and cerebritis due to *Curvularia clavata* in an immunocompetent adult. *Clin Infect Dis* 1999;**1**:687–9.
124. Kantarcioğlu AS, Yücel A, Hoog GD. Isolation of *Cladosporium cladosporioides* from cerebrospinal fluid. *Mycoses* 2002;**45**:500–3. https://doi.org/10.1046/j.1439-0507.2002.00811.x [Case report].
125. Kantarcıoğlu AS, Hatemi G, Yücel A, De Hoog GS, Mandel NM. *Paecilomyces variotii* central nervous system infection in a patient with cancer. *Mycoses* 2003;**46**:45–50. https://doi.org/10.1046/j.1439-0507.2003.00837.x.
126. Ju YM, Rogers JD. A revision of the genus Hypoxylon. Mycologia memoir no. 20. St. Paul, MN.
127. Umabala P, Lakshmi V, Murthy AR, Prasad VS, Sundaram C, Beguin H. Isolation of a *Nodulisporium* species from a case of cerebral phaeohyphomycosis. *J Clin Microbiol* 2001;**39**:4213–8. https://doi.org/10.1128/JCM.39.11.4213-4218.2001.
128. National Committee for Laboratory Standards. *Reference method for broth dilution antifungal susceptibility testing of conidium-forming filamentous fungi; proposed standard* [Document M38-P]. Wayne: National Committee for Laboratory Standards; 1998.
129. Vitale RG, De Hoog GS, Verweij PE. *In vitro* activity of amphotericin B, itraconazole, terbinafine and 5-fluocytosine against *Exophiala spinifera* and evaluation of post-antifungal effects. *Med Mycol* 2003;**41**:301–7. https://doi.org/10.1080/13693780310001600822.

FURTHER READING

1. Saccente M, Woods GL. Clinical and laboratory update on blastomycosis. *Clin Microbiol Rev* 2010;**23**:367–81. https://doi.org/10.1128/CMR.00056-09.
2. Starkey J, Moritani T, Kirby P. MRI of CNS fungal infections: review of aspergillosis to histoplasmosis and everything in between. *Clin Neuroradiol* 2014;**24**:217–30. https://doi.org/10.1007/s00062-014-0305-7.
3. Groll AH, Walsh TJ. Aspergillosis. In: *Atlas of infectious diseases*. Current Medicine Group; 2000. p. 137–56. https://doi.org/10.1007/978-1-4757-9313-0_9.
4. Salama AD, Rogers T, Lord GM, Lechler RI, Mason PD. Multiple Cladosporium brain abscesses in a renal transplant patient: aggressive management improves outcome. *Transplantation* 1997;**63**:160–2.

CHAPTER 13

Cryptococcal Meningitis and Other Opportunistic Fungal Infections of the Central Nervous System: Epidemiology, Pathogenesis, Diagnosis, and Treatment

Andrés F. Henao-Martínez, Daniel Vela-Duarte
University of Colorado Denver, Denver, CO, United States

1. INTRODUCTION

Fungal organisms are a significant source of disability and death. They act as opportunistic pathogens in the immunocompromised host. As ubiquitous organisms, fungi are environmentally in proximity to us; therefore we are continually exposed to them. Their capacity to invade and cause severe disease largely depends on the host's immune system status. We often see cellular immune debilitating conditions such as HIV-AIDS, solid-organ transplantation (SOT), immunosuppressive medications, and malignancy among others as primary risk factors for the disease. Central nervous system (CNS) invasion usually begins with lung exposure followed by dissemination. CNS pathology can be broad. Fungal organisms such as *Cryptococcus*, *Histoplasma*, *Blastomyces*, and *Coccidioides* cause meningitis more frequently than molds such as *Aspergillus* and *Mucor*, which cause more brain abscesses. Management of fungal CNS infections is a challenge to the clinician since it carries a high mortality while responses to antifungal therapy are often suboptimal. Herein, we present an overview of the epidemiology and clinical management of the most common fungal pathogens causing human disease.

2. CRYPTOCOCCAL MENINGITIS

2.1 Epidemiology

Cryptococcus spp. is a saprophytic yeast with global distribution known to be a constituent of the human microbiota. Two species—*Cryptococcus neoformans* and *Cryptococcus gatti*—are the main ones responsible for human disease. Although these pathogens can infect

The Microbiology of Central Nervous System Infections
https://doi.org/10.1016/B978-0-12-813806-9.00013-5

immunocompetent hosts, their surge as an opportunistic infection over recent decades has been thanks to the emergence of the HIV infection. HIV-related Cryptococcal meningitis affects approximately 1 million people each year and causes nearly 625,000 deaths worldwide.[1] High mortality in cryptococcosis—up to 20%—is seen in immunosuppressive conditions.[2] SOT, systemic lupus erythematosus (SLE), malignancy, sarcoidosis, and cirrhosis are conditions of functional immunosuppression known to increase the risk for *Cryptococcus* dissemination and neuroinvasion.[3–8] Other host factors associated with extrapulmonary cryptococcosis are the male gender, fungemia, smoking, diabetes, and Hispanic ethnicity.[8–13] Finally, primary immunodeficiencies such as idiopathic CD4+ lymphopenia and hyper IgE syndrome also increase the risk for Cryptococcal meningitis.[14] In the United States, Cryptococcal infection is responsible for approximately 3400 hospitalizations every year and carries an overall annual age-adjusted mortality of 0.07 per 100,000 population.[15,16] Cryptococcosis accounts for up to 7% of the invasive fungal infections in solid-organ transplant recipients[17] and follows only Aspergillus and Candida in frequency for invasive fungal infections.

2.2 Pathogenesis

Exposure initiates after inhalation of the yeast forms of *Cryptococcus*. Desiccated yeast cells (~3 μm) or basidiospores (1–2 μm) are considered infectious particles due to their small size and ability to reach the alveoli. Some findings suggest that yeast cells may be more resistant to the innate immunity response than spores.[18] In the lungs, yeast forms encounter alveolar macrophages as one of the first lines of defense. At this point, three possible scenarios can take place: (A) total clearance of the pathogen without any host repercussion, (B) yeast produces a small lung-lymph node complex and remains dormant in tissues, or (C) lung parenchyma and tissue invasion following initial infection or reactivation. Pathogenesis following reactivation is similar to that seen with tuberculosis and may be the most important source of infection.[19] It is known that *C. neoformans* and *C. gatti* yeast cells proliferate and survive in alveolar macrophages.[20,21] Among the most prominent pathogen virulent factors present are the capsule that possesses an antiphagocytic property, melanin that confers protection against phagocytosis and oxidative killing, and urease that enhances CNS invasion.

C. neoformans can cross the blood-brain barrier (BBB) by transcytosis through the endothelial cells or by the passage of yeast inside phagocytic cells.[22] An acapsular or poorly encapsulated *C. neoformans* may be the predominant variant that adheres to the vascular endothelial cells to initiate its escape and CNS invasion.[23] Initial central cryptococcal invasion following fungemia consists of leptomeningeal and brain parenchyma inflammation (Fig. 1). This inflammation includes CD4+/CD8+ T cells and macrophages as well as microglial activation in an animal model. Likewise, associated infiltrates in acute human cryptococcal disease were mainly CD8+ T cells and macrophages.

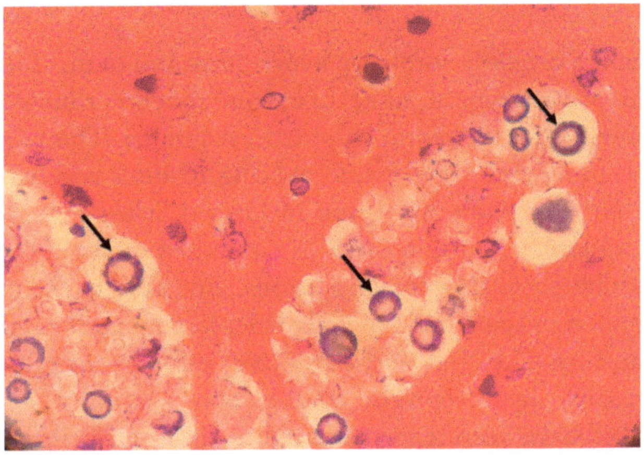

Fig. 1 Cryptococcus yeast forms in brain tissue from an autopsy specimen. Cryptococcus yeast in brain tissue (*arrows*). 400 ×; H&E stain.

In a model of chronic human cryptococcal infection, microglial activation was increased and related to reactive astrocytosis.[24]

2.2.1 Clinical Manifestations

Cryptococcosis has a wide range of possible clinical presentations. Because its main port of entry is through the lungs, patients initially can present with respiratory complaints. Patients complain of fever, weight loss, cough, shortness of breath, hemoptysis, and chest pain.[25] However, imaging studies often find asymptomatic incidental cases.[26] Cryptococcal meningitis can be manifested as part of a disseminated disease affecting other organs, including the lungs, or as the only manifestation without evidence of pulmonary involvement. Cryptococcal meningitis has a subacute presentation and symptoms may not be typical; therefore, a high index of suspicion is always necessary. Among patients with CNS involvement, headaches, constitutional symptoms such as fatigue or tiredness, altered mental status, and fever are more common than in patients with pulmonary involvement only.[8] Other manifestations include cranial nerve palsies and memory loss. A high burden of yeasts—especially among HIV infected patients—can lead to an increase in intracranial pressure and cerebral edema causing lethargy and coma. Immunosuppressed patients can develop a paradoxical immune-mediated reaction when their immune system recovers—enhanced immune response to previously recognized infections (paradoxical) or to an otherwise latent unrecognized infection (unmasking)—called immune reconstitution syndrome (IRS). IRS has been described mainly in HIV-infected patients after antiretroviral therapy initiation and in SOT recipients (when clinicians reduce the immunosuppressive regimens). IRS manifests by worsening of the disease while on appropriate treatment and fungal clearance or by new revealing symptoms such as fever and headaches after immune recovery.

2.3 Diagnosis

The first step to make a diagnosis is to keep a high suspicion of the disease. Symptoms of subacute meningitis are not typical, more so in immunocompromised hosts. The preferable initial screening test is the serum cryptococcal antigen. This test has a >90% sensitivity and fast lab processing times while proving to be a very effective method of screening in endemic areas with AIDS and in United States centers with a relatively low prevalence of disease.[8,27] Imaging such as computed tomography (CT) or MRI is essential to rule out intracranial occupying lesions, especially in the presence of altered mental status. Common imaging findings include leptomeningeal enhancement, cryptococcomas (nodular, mass-like, and ring-enhancing lesions), enlarged Virchow-Robin spaces, diffuse atrophy, hydrocephalus, and cerebral edema[28,29] (Fig. 2). Strokes, although uncommon, have been reported as a complication of cryptococcal meningitis associated-vasculopathy[30] (Fig. 2). Normal findings do not exclude disease. Imaging also can help to diagnose concomitant or alternative conditions such as toxoplasma encephalitis, CNS lymphoma, and brain abscesses. A lumbar puncture (LP) has two roles: (1) measure and relieve the opening pressure by drainage if necessary, and (2) obtain cerebrospinal fluid (CSF) samples for analysis—cell counts and fungal cultures (the gold standard for diagnosis). CSF cryptococcal antigen and blood cultures follow the serum cryptococcal antigen in sensitivity. CSF findings in patients with cryptococcal meningitis

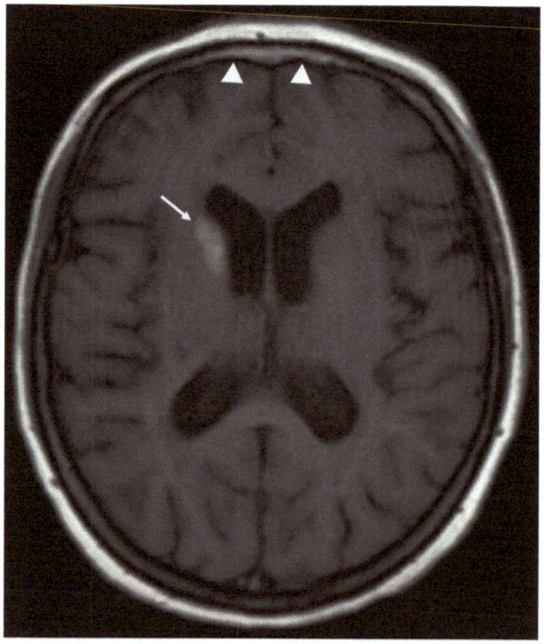

Fig. 2 MRI changes in Cryptococcal meningitis. MRI of a brain in a patient with Cryptococcal meningitis. Parenchymal infarct involving the left frontal periventricular white matter, anterior body of the corpus callosum (*arrow*), and leptomeningeal enhancement (*arrowheads*).

include mild lymphocytic predominant pleocytosis, hypoglycorrhachia, and elevated CSF proteins. IRS criteria consist of: (1) new or worsening clinical findings, (2) presence of symptoms through an optimal antifungal therapy that cannot be explained by an alternative process or infection, and (3) evidence of negative microbiology results.[31]

2.4 Treatment

One of the cornerstones of treatment is to release intracranial pressure when CSF pressure $\geq$25 cm H_20 and symptoms are present. The goal of CSF drainage is to obtain a 50% reduction in pressure or a pressure of $\leq$20 cm H_20. Therapeutic LPs have shown to decreased mortality among HIV-infected patients.[32] Antifungal therapy should be initiated as soon as the diagnosis is suspected. The preferred regimen for cryptococcal meningitis consists of intravenous amphotericin B plus flucytosine.[33] The treatment course consists of induction, consolidation, and maintenance phases of therapy. Table 1 lists

Table 1 Antifungal regimens for cryptococcal meningitis

	Therapy of choice per phase	Duration
Induction	*Preferred*: Liposomal AmB (3–4 mg/kg per day IV) or ABLC (5 mg/kg per day IV)[a] *plus* Flucytosine (100 mg/kg per day in 4 divided doses)[b]	2 weeks (SOT and HIV)
	Alternative (in the absence of flucytosine): Lipid formulation of AmB	4–6 weeks (Non-HIV-infected, nontransplant hosts or in the absence of flucytosine)
Consolidation	Fluconazole PO 400–800 mg (6–12 mg/kg) daily (SOT) Fluconazole PO 200 mg daily (HIV)	8 weeks
	Salvage therapy in relapses: Fluconazole 800–1200 mg PO daily OR Voriconazole 200–400 mg PO daily OR Posaconazole 200 mg PO QID	10–12 weeks
Maintenance	Fluconazole PO 200–400 mg daily[c]	6–12 months

AmB, amphotericin B; *ABCL*, amphotericin B lipid complex; *SOT*, solid-organ transplant.

[a]6 mg/kg per day can be considered in relapses or with a high cryptococcal burden. Amphotericin B deoxycholate (0.7–1.0 mg/kg per day intravenously) can be used in HIV-infected patients.

[b]A dosage reduction is required with renal impairment: creatinine clearance (CrCl) >50 mL/min, 100–150 mg/kg/day; CrCl 26–50 mL/min, 75 mg/kg/day; CrCl 13–25 mL/min, 37 mg/kg/day; CrCl <13 mL/min, avoid flucytosine. Dose monitoring is also recommended in the first 72 h of therapy with a goal trough concentration of 20–40 mg/L; peak concentration should not exceed 100 mg/L.

[c]For HIV-infected patients, consider discontinuing suppressive therapy during ART in patients with a CD4 cell count >100 cells/μL and an undetectable or very low HIV RNA level sustained for $\geq$3 months (minimum of 12 months of antifungal therapy).

the dosing and preferred treatment regimens. Use of flucytosine along with a recommended regimen and dosing are associated with lower attributable mortality rates.[34] In cases of relapse, we recommend restarting the induction phase and considering a higher dose of amphotericin B. Additionally, we advise obtaining isolate susceptibility testing. Salvage consolidation therapy often includes higher doses of fluconazole or an alternative azole (voriconazole or posaconazole). Guidelines instruct performing a follow-up LP at 2 weeks after onset of induction therapy. If CSF cultures remain positive, the induction phase should be continued for 2 additional weeks. Management of IRS is not well defined. Treatment-naïve HIV-infected patients should delay the start of antiretroviral therapy for at least 2–10 weeks after antifungal therapy—especially in the presence of increased intracranial pressure and low CSF white blood cell counts.[35] We recommend corticosteroids with IRS and signs of elevated intracranial pressure.

3. ASPERGILLUS AND OTHER MEDICALLY IMPORTANT MOLDS

3.1 CNS Aspergillosis

3.1.1 Epidemiology

Aspergillus spp. is a ubiquitous fungus with worldwide distribution. The four more-common species causing invasive disease are *Aspergillus fumigatus*, *Aspergillus terreus*, *Aspergillus flavus*, and *Aspergillus niger*. Since the fungus lives in water, plants, food, air, and soil, humans are constantly exposed. The main port of entry is the lungs. Exposure to construction dust, pepper, and marijuana have been linked to disease.[36–38] The principal factor favoring invasion is immunosuppression. Prolonged neutropenia associated with hematologic malignancies, chemotherapy, and patients sustaining a bone marrow or stem cell transplant are the most common risk factors for invasive aspergillosis.[39] Other immunosuppressive states such as corticosteroids, use of biologic medications, SOT (especially lung transplants), AIDS, and primary immunodeficiencies are also predisposing factors. Cerebral aspergillosis has a very high mortality rate (>90% in some series).[40] Brain invasion occurs after spread from a sinus infection or hematogenous seeding of the disseminated disease. Concomitant pulmonary disease is often present but not always.

3.1.2 Clinical Manifestations

Aspergillosis of the CNS presents as a brain abscess and rarely as meningitis. As any other space-occupying intracranial lesion, an *Aspergillus* brain abscess can manifest with headaches, seizures, focal neurologic deficits, and altered mental status.[41] A CT or MRI of the brain usually shows solitary or multiple brain abscesses with ring enhancement. Often, the lesion can be hemorrhagic at the gray-white junction and can also affect the corpus callosum.[42] The differential includes pyogenic brain abscess, toxoplasma encephalitis, and lymphoma.[43]

3.1.3 Diagnosis and Treatment

Diagnosis of cerebral aspergillosis is challenging. The presence of lung disease with brain lesions should alert clinicians to the diagnosis. A biopsy of the brain lesion or abscess remains the gold standard. Histopathology shows a dichotomous branching septate fungi with surrounding granulomas with epithelioid histiocytes, and multinucleated giant cells.[44] Clinicians should perform a fungal culture on any CNS specimen (brain tissue or CSF). Additionally, a nucleic acid amplification test and galactomannan detection can be done in CSF.[45,46] The treating physician should institute antifungal therapy early without awaiting diagnostic workup results. The three bases for treatment consist of (1) surgical debridement as amenable, (2) antifungal therapy, and (3) reduction of immunosuppression. Primary antifungal treatment should be started with voriconazole—6 mg/kg IV every 12 h for 1 day, followed by 4 mg/kg IV every 12 h. Oral therapy can be used at 200–300 mg every 12 h or weight-based dosing on an mg/kg basis.[47,48] Lipid formulation of amphotericin B is used as an alternative for cases intolerant or refractory to voriconazole. However, amphotericin remains a poor option for cerebral aspergillosis.[49] Itraconazole and posaconazole have been used successfully in some cases. Finally, combination therapy with echinocandins has had limited beneficial evidence.

3.2 Cerebral Mucormycosis

Fungi in the order of Mucorales are the origin of Mucormycosis, formerly known as zygomycosis. These ubiquitous organisms cause life-threatening infections in the immunocompromised host. The *Rhizopus* species are the most common organisms within this group. Most recognized risk factors include poorly controlled diabetes mellitus, high-dose steroid therapy, burns, penetrating trauma, hematologic malignancy, SOT, and neutropenia.[50] Rhinocerebral infection is the most common presentation of the disease—mostly seen with diabetic ketoacidosis.[51] The usual reported symptoms are sinus pressure, purulent nasal discharge or ulceration, periorbital swelling, headaches, and fever. Sinus tissue necrosis and angioinvasion can produce marked swelling and black scars, which are characteristic of the infection. Symptoms of invasive cerebral disease including cranial nerves palsies, visual loss, and altered mental status; these carry a bad prognosis. Radiographic imaging—although not diagnostic—can reveal the extension of the disease, including sinus tissue destruction, bone erosion, and extra-orbital muscle thickening. Endoscopic examination is crucial to assess the extension of the infection, debride necrotic tissue, and obtain samples for culture. The gold standard for diagnosis remains confirmation on the tissue fungal culture. Therapy involves aggressive surgical debridement of the affected tissues and antifungal therapy. Lipid amphotericin B is the first line of treatment against cerebral mucormycosis. Consolidation or step-down therapy can be achieved with oral posaconazole or isavuconazole. This condition is almost universally fatal without treatment and mortality can reach up to 30% even with surgery and antifungal therapy.[52] The treatment duration depends on radiographic and clinical improvement and often takes several months.

3.3 CNS Fusarium Infection

Fusariosis is a relatively rare opportunistic infection. Most cases are seen in prolonged and profound neutropenia, hematologic malignancies, and hematopoietic stem cell transplant recipients. CNS infection—manifested mainly as a brain abscess or endophthalmitis—is infrequent, and it is usually followed by hematogenous dissemination or progression through the nasal sinus tissue.[53] *Fusarium* spp. grows easily on blood cultures and microbiologists can easily identify it by its banana shape (Fig. 3). Mortality passed the 80% mark. There is no optimal antifungal therapy. Treatment recommendations include surgical debridement, combination antifungal therapy with amphotericin B plus voriconazole, and measures aimed to correct the neutropenia.[54]

3.4 Scedosporium

Two species are recognized as causing human disease—the asexual form of *Pseudallescheria boydii*, *Scedosporium apiospermum*, and *Scedosporium prolificans*. As other molds, they are ubiquitous and have global distribution. Ports of entry are the lungs, sinuses, or the skin after penetrating trauma. Studies have identified drowning as a risk factor for *S. apiospermum* brain abscesses.[55] Other immunocompromised conditions such as solid-organ transplant, prolonged neutropenia, trauma, steroid therapy, and bone marrow transplant are

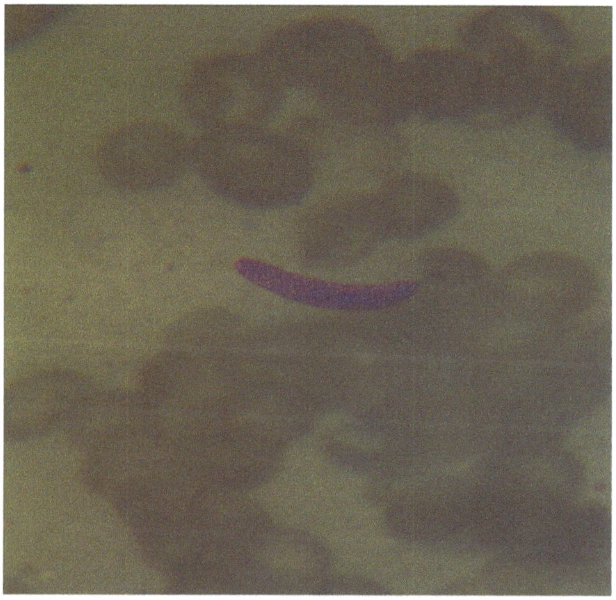

Fig. 3 Blood smear with a positive blood culture. Blood smear (magnification 1000×; Giemsa stain).

also predisposition conditions. CNS infections are more commonly seen with *S. apiospermum*. Manifestations include multiple brain abscesses, meningitis, and endophthalmitis. Diagnosis is established when the mold is recovered from an affected tissue. The first line of therapy for *S. apiospermum* is voriconazole along with surgical debridement and immunosuppression correction. A combination therapy of voriconazole with terbinafine may be beneficial in cases of partial or limited surgical debridement or unresponsiveness to voriconazole monotherapy.[56] *S. prolificans* is intrinsically resistant to most antifungals. Disseminated disease and CNS infections carry a very high mortality. We recommend surgical debridement and immune system optimization as possible. Voriconazole has had some partial response in some *S. prolificans* infection cases.

4. CANDIDA

4.1 Epidemiology and Pathogenesis

The genus *Candida* encompasses over 100 species with fewer than 10 species being clinically relevant. *Candida* is part of the normal human flora, and it becomes invasive when the host cell-mediated immune response is impaired. It is the most common fungal pathogen affecting immunocompromised patients. *Candida albicans* is the predominant cause of invasive disease in the United States, followed by *C. glabrata*.[57] Other species, including *C. parapsilosis*, *C. tropicalis*, *C. krusei*, and the rarest, *C. dubliniensis* and *C. lusitaniae*, have also been reported. The nonalbicans species tends to exhibit pharmacologic resistance to usual therapies such as triazoles and amphotericin B.[58] Systemic candida infections implicate several risk factors (Table 2).[59,60] Oral and intestinal colonization

Table 2 Risk factors for invasive candidiasis

- Extremes of age
- Indwelling catheters
- Injectable drug use
- Prolonged ICU stay
- Multiple intravenous catheters
- Central venous catheters
- Gastrointestinal surgery
- Prior systemic antifungal exposure
- Extensive burns
- Intravenous hyper alimentation
- Hematologic malignancy
- Neutropenia
- Immunosuppressive drugs
- Broad-spectrum antibiotics
- Acute renal failure requiring hemodialysis
- Mucosal disruption
- Glucocorticoid

and translocation to the blood stream is common in immunosuppressed patients. The fungus leads to the formation of biofilm on indwelling catheters resulting in a persistent source of fungemia, which can cause infective endocarditis and septic emboli. Endophthalmitis is a clinically silent complication of candidemia in critically ill patients—often unable to relay visual symptoms—resulting from hematogenous seeding of the eye.[61] The compromise of the CNS by candida—although uncommon—is due to hematogenous spreading from catheter-related infections, endocarditis or direct (exogenous) contamination of surgical wounds (i.e., craniotomy, ventriculostomy, LP), or hardware material such as extraventricular drains and ventriculoperitoneal shunts tubing.[62]

4.2 Clinical Manifestations

The disease tends to show a chronic and indolent course in adults while presenting as a severe form of meningitis in neonates. Affected newborns are usually premature with associated risk factors, including severe respiratory infections, central catheters, and prolonged antibiotic therapy. Candida meningitis and meningoencephalitis can present with subtle alterations of mental status. Focal motor deficits and papilledema are rarely observed whereas meningeal signs are common. The presence of microabscesses is a pathologic hallmark of postmortem specimens and nonautopsy cases.[63] Chronic cases can present with a dense exudate at the base of the brain and clinical manifestations of basal meningitis similarly to cryptococcosis or tuberculosis. Clinicians have also reported microabscesses, angiitis, myelitis, arachnoiditis, and cranial nerve palsies.[64] Infections associated with neurosurgical procedures and external devices present with fever and hydrocephalus.

4.3 Diagnosis and Treatment

We recommend LP and CSF cultures in patients with known Candidemia and any signs of CNS involvement. CT images lack the resolution to identify microabscesses. Individuals with *Candida* microabscesses present moderate CSF pleocytosis and increased CSF protein levels while in *Candida* meningeal forms, the CSF pleocytosis can be greater than 500 white blood cells/mm^3, is lymphocytic predominant and hypoglycorrhachia is present. A profile of chronic meningitis with persistent neutrophilic predominance and hypoglycorrhachia has also been described.

MRI studies exhibit meningeal enhancement that can take a nodular appearance in the subarachnoid space. The presence of microabscesses can manifest as T1-weighted hypointensities and nodular or ring-enhancing lesions on postcontrast images.[65,66] In patients with confirmed CNS candidiasis, we recommend liposomal amphotericin B, 5 mg/kg daily monotherapy or in addition to oral flucytosine, 25 mg/kg four times daily. If the patient has responded well to initial therapy, fluconazole, 400–800 mg (6–12 mg/kg) daily for susceptible isolates is continued until symptoms and radiographic and

CSF abnormalities have been resolved. Echinocandins lack adequate CNS penetration and therefore are not used to treat this type of infection. We strongly recommend the removal of any infected neurosurgical device, as *Candida* biofilms are very difficult to eradicate.

5. ENDEMIC FUNGI INFECTING THE CNS

5.1 Histoplasmosis

Histoplasma capsulatum is endemic in some parts of North America such as the Ohio and Mississippi River basins, the Great Lakes, and the St. Lawrence River valley. It is also prevalent in Mexico, Central America, and Taiwan.[67] Progressive disseminated histoplasmosis occurs in immunosuppressed patients (e.g., HIV infection), typically with CD4 lower than 150–200 cells/mm^3. It manifests with infection and granulomatous lesions in multiple organ systems. Bone marrow infection leads to anemia, leukopenia, and thrombocytopenia. CNS histoplasmosis can present with meningitis, brain or spinal cord lesions (single or multiple known as histoplasmomas), and cerebritis. Meningitis is the most common presentation. Symptoms may have been present for weeks and include altered mental status, headaches, cranial nerve palsies, seizures, meningismus, ataxia, focal neurologic deficits, and, rarely, strokes.[68] CSF analysis reveals persistent pleocytosis, lymphocytic predominant, low glucose, and elevated proteins.[69] The most sensitive test to diagnose disseminated disease remains the antigen test in urine, blood, or CSF. Additionally, fungal cultures can recover the organisms from CSF or blood (Fig. 4). Isolation of the fungi in any other sterile sites along with CNS abnormalities should prompt the diagnosis as well. The role of serologies in immunocompromised patients and polymerase chain

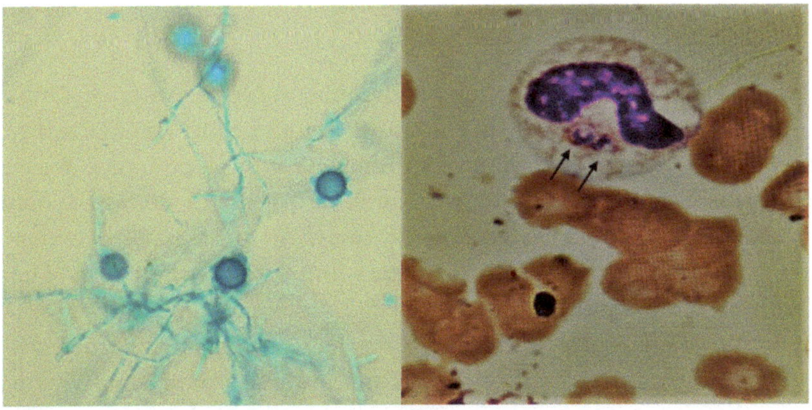

(A) (B)

Fig. 4 Histoplasmosis. (A) Hyphae in a specimen culture; original magnification 400×. (B) White blood cell Histoplasma inclusions (*arrows*) in a peripheral smear.

reaction is uncertain. Also, in low resource settings—although sensitivity is low—detection of organisms inside white blood cells in a peripheral smear makes the diagnosis (Fig. 4). Intravenous liposomal amphotericin B, 5 mg/kg daily, for 4–6 weeks is the treatment of choice, followed by itraconazole, 400–600 mg orally daily for at least one year or more and until CSF abnormalities normalize or lesions are no longer present.

5.2 Coccidioidomycosis

Coccidioides immitis is the most common cause of chronic meningitis in endemic areas — the southwest United States, Northern Mexico, and Central and South America. Inhalation of arthroconidia, the asexual spore, is the initial step in developing coccidioidomycosis. The primary infection has an indolent course accompanied by cough, shortness of breath, malaise, chest pain, fevers, and night sweats. Dissemination occurs in up to 8% of new infections, of which 17% involve the CNS. Neurological deficits can present weeks to months after pulmonary infection and typically include persisten and protracted headaches, fever, blurry vision, nausea, vomiting, and changes in mental status. Meningitis compromises the leptomeninges of the basal cisterns, developing communicating hydrocephalus and vasculitis of small to medium vessels that leads to vascular occlusion.[70,71] Complement-fixation antibody detection of antibodies in CSF is the preferred confirmatory test. The sensitivity can reach 95%. However, it varies in early stages of the disease. Neuroimaging studies can be suggestive of the infection, given a predilection for the basilar leptomeninges. Enhancement of the basal cisterns, the Sylvian fissure, the craniocervical junction, and the pericallosal region can be observed.[72] Fluconazole, 400 mg per day, is the recommended treatment. We recommend higher doses of 800–1200 mg of fluconazole for more severe cases and intrathecal amphotericin in patients with inadequate response to azoles.

5.3 Blastomycosis

Blastomyces dermatitidis is a dimorphic fungus endemic to North America. Outside North America, it is endemic in South Africa, Zimbabwe, and India. Individuals exposed to wet earth enriched with droppings of sick animals, rotting wood, and decaying vegetable matter can become infected by inhalation of infectious conidia. Constitutional symptoms are common, and up to half of cases can present with extrapulmonary involvement. The most common affected extrapulmonary site is the skin. The compromise of the CNS is often present in disseminated forms but, unlike other opportunistic agents, *B. dermatitidis* does not seem more prevalent in patients with defective cellular immunity. Blastomycosis of the CNS is uncommon but typically presents with meningitis, brain abscesses, and mass lesions, predominantly in the posterior fossa.[67] The CSF shows lymphocytic pleocytosis, elevated proteins, and usually, normal glucose. The clinical picture resembles tuberculosis and cultures of CSF may not yield the microorganism, in most cases. Spinal infection affects the thoracic and lumbar vertebrae. Antibody testing tends to be of low sensitivity and specificity. Antigen detection reaches a sensitivity of 93% in analyzed urine samples but has poor specificity due to crossed-reactions with antigens of *Histoplasma* and *Paracoccidioides*.[73,74] Treatment for CNS disease consists of lipid amphotericin B, 5 mg/kg per day for 4–6 weeks, followed by an oral azole, itraconazole, 200 mg twice or three times per day, fluconazole 800 mg per day, or voriconazole 200–400 mg twice per day for at least 1 year until CSF abnormalities normalize and lesions subside. We recommend consideration for surgical drainage of the brain and epidural or spine abscesses with neurological deficits.[75]

5.4 Paracoccidioidomycosis

Paracoccidioidomycosis is a systemic mycosis endemic to Latin America—particularly Brazil—caused by *Paracoccidioides brasiliensis*. The disease is acquired by inhalation of conidia, which undergoes subsequent transformation to a yeast form in pulmonary tissues. Paracoccidioidomycosis affects the CNS between 10% and 25% of the time. Tissue granulomas, also known as pseudotumors, can be simple or multiple and affect the brainstem, cerebellum, and spinal cord. A review of 20 patients collected over 17 years suggested that the pseudotumoral form—seen in more than 50% of cases—was more frequent than a ring-enhancement appearance on neuroimaging studies.[76] Diagnosis is made through culture isolation or direct visualization of the organism in biopsy samples. Serologic testing can be useful. Medical treatment of CNS infections consists of trimethoprim-sulfamethoxazole or amphotericin B. A combination of oral fluconazole and sulfamethoxazole/trimethoprim for 24–48 months was shown to be effective both for systemic and CNS disease.[77]

6. OTHER CNS MEDICALLY IMPORTANT FUNGI

6.1 Talaromyces

Talaromyces marneffei, formerly known as *Penicillium marneffei* and endemic to areas in Southeast Asia, causes penicilliosis. The mechanism of infection is unclear, although inhalation has been implied. The disease develops mostly in immunocompromised hosts, especially HIV/AIDS patients. Patients present with fever, altered mental status, headaches, facial nerve palsy, seizures, vision loss, and constitutional symptoms. Brain bilateral subcortical abscesses can occur. In a large series of cases of penicilliosis,[78],approximately 3% of individuals presented with CNS involvement, confirmed by CSF culture. Patients were young, had HIV, and had a low CD4 count (11 cells/μL (interquartile range: 10–12 cells/μL). Observed mortality was very high, and many died within 48–72 h of admission. The survivors received 14 days of amphotericin B followed by itraconazole or fluconazole.

6.2 Cladosporium

Cladosporium spp. is being recognized as a potential pathogen in immunocompromised patients and belongs to a rare and emerging cluster of fungi, denominated Phaeohyphomycosis. Several species of the genus *Cladosporium* can cause subcutaneous abscesses and CNS infections. One of the most frequently reported species is *C. bantiana*, followed by *C. sphaerospermum*.[79] The fungus is neurotropic; however, the invasive mechanisms are not well understood. The disease courses invariably with cerebral abscesses and meningitis, affecting immunocompetent and immunosuppressed individuals.[80,81] Authors observed the previous use of corticosteroids in 52% of cases on a systematic review.[82] They noticed extensive cerebritis with necrosis of the cerebrum or cerebellum in the majority of cases (97.5%) while seeing meningeal compromise in only 12.5%. Antifungal treatment of *Cladosporium* remains a matter of debate but surgical excision of the lesions has been shown to be of utmost importance in determining survival. Voriconazole and posaconazole have been used in vitro, and, in a few cases, in clinical settings. The addition of amphotericin B—a second-line option—is employed in the event of medication intolerance or resistance.[83]

6.3 Sporothrix

Sporothrix schenckii is a rare cause of meningitis as the infection is usually restricted to the skin and lymph nodes. Few cases of CNS involvement have been reported.[84,85] In a cohort of 1697 patients diagnosed with sporotrichosis and followed for 10 years, only 2 patients developed meningitis.[84] These individuals were HIV positive and acquired the infection through a cat scratch. Demonstration of the microorganism in cultures of CSF can take several months and several LPs, which often delays initiation of treatment. Coinfection with HIV carries complications and a higher mortality. These individuals tend to present with elevated intracranial pressure, persistent CSF abnormalities, and recurrent meningitis. The CSF profile commonly shows pleocytosis with lymphocytic predominance, hypoglycorrhachia, and elevated protein. IgM specific latex agglutination and serum and CSF antibodies (IgG ELISA) are used for diagnosis. IDSA guidelines recommend lipophilic amphotericin B, 5 mg/kg daily for 4–6 weeks, followed by itraconazole, 200 mg twice daily for at least 1 year. In patients with AIDS, lifelong maintenance therapy with itraconazole is advised.[86] The lack of knowledge and awareness about *S. schenckii* as a cause of chronic meningitis can delay the diagnosis and mislead antimicrobial therapy.[87]

REFERENCES

1. Park BJ, Wannemuehler KA, Marston BJ, Govender N, Pappas PG, Chiller TM. Estimation of the current global burden of cryptococcal meningitis among persons living with HIV/AIDS. *AIDS* 2009;**23**:525–30.

2. Brizendine KD, Baddley JW, Pappas PG. Predictors of mortality and differences in clinical features among patients with cryptococcosis according to immune status. *PLoS One* 2013;**8**.
3. Baddley JW, Perfect JR, Oster RA, Larsen RA, Pankey GA, Henderson H, et al. Pulmonary cryptococcosis in patients without HIV infection: factors associated with disseminated disease. *Eur J Clin Microbiol Infect Dis* 2008;**27**:937–43.
4. Wang LR, Barber CE, Johnson AS, Barnabe C. Invasive fungal disease in systemic lupus erythematosus: a systematic review of disease characteristics, risk factors, and prognosis. *Semin Arthritis Rheum* 2014;**44**:325–30.
5. Cancelli I, Merlino G, Serafini A, Valente M, Gigli GL. Sarcoidosis as risk factor for cryptococcal meningitis in an apparently immunocompetent patient. *Neurol Sci* 2008;**29**:33–5.
6. Lin YY, Shiau S, Fang CT. Risk factors for invasive *Cryptococcus neoformans* diseases: a case-control study. *PLoS One* 2015;**10**.
7. Henao-Martinez AF, Beckham JD. Cryptococcosis in solid organ transplant recipients. *Curr Opin Infect Dis* 2015;**28**:300–7.
8. Henao-Martinez AF, Gross L, McNair B, et al. Risk factors for *Cryptococcal meningitis*: a single United States center experience. *Mycopathologia* 2016;**181**:807–14.
9. Sorvillo F, Beall G, Turner PA, Beer VL, Kovacs AA, Kerndt PR. Incidence and factors associated with extrapulmonary cryptococcosis among persons with HIV infection in Los Angeles County. *AIDS* 1997;**11**:673–9.
10. Khan ZU. Smoking, melanization, and cryptococcosis: is there a connection? *J Clin Microbiol* 2006;**44**:1207.
11. Osawa R, Alexander BD, Lortholary O, et al. Identifying predictors of central nervous system disease in solid organ transplant recipients with cryptococcosis. *Transplantation* 2010;**89**:69–74.
12. Chuang YM, SC K, Liaw SJ, et al. Disseminated *Cryptococcus neoformans* var. grubii infections in intensive care units. *Epidemiol Infect* 2010;**138**:1036–43.
13. Lin KH, Chen CM, Chen TL, et al. Diabetes mellitus is associated with acquisition and increased mortality in HIV-uninfected patients with cryptococcosis: a population-based study. *J Infect* 2016;**72**(5): 608–14.
14. Williamson PR, Jarvis JN, Panackal AA, et al. Cryptococcal meningitis: epidemiology, immunology, diagnosis and therapy. *Nat Rev Neurol* 2017;**13**:13–24.
15. Barragan NC, Sorvillo F, Kuo T. Cryptococcosis-related deaths and associated medical conditions in the United States, 2000–2010. *Mycoses* 2014;**57**:741–6.
16. Pyrgos V, Seitz AE, Steiner CA, Prevots DR, Williamson PR. Epidemiology of cryptococcal meningitis in the US: 1997–2009. *PLoS One* 2013;**8**:.
17. Neofytos D, Fishman JA, Horn D, et al. Epidemiology and outcome of invasive fungal infections in solid organ transplant recipients. *Transpl Infect Dis* 2010;**12**:220–9.
18. Giles SS, Dagenais TR, Botts MR, Keller NP, Hull CM. Elucidating the pathogenesis of spores from the human fungal pathogen *Cryptococcus neoformans*. *Infect Immun* 2009;**77**:3491–500.
19. Saha DC, Goldman DL, Shao X, et al. Serologic evidence for reactivation of cryptococcosis in solid-organ transplant recipients. *Clin Vaccine Immunol* 2007;**14**:1550–4.
20. Feldmesser M, Kress Y, Novikoff P, Casadevall A. *Cryptococcus neoformans* is a facultative intracellular pathogen in murine pulmonary infection. *Infect Immun* 2000;**68**:4225–37.
21. Kronstad JW, Attarian R, Cadieux B, et al. Expanding fungal pathogenesis: cryptococcus breaks out of the opportunistic box. *Nat Rev Microbiol* 2011;**9**:193–203.
22. Bicanic T, Harrison TS. Cryptococcal meningitis. *Br Med Bull* 2004;**72**:99–118.
23. Ibrahim AS, Filler SG, Alcouloumre MS, Kozel TR, Edwards Jr. JE, Ghannoum MA. Adherence to and damage of endothelial cells by *Cryptococcus neoformans* in vitro: role of the capsule. *Infect Immun* 1995;**63**:4368–74.
24. Chretien F, Lortholary O, Kansau I, Neuville S, Gray F, Dromer F. Pathogenesis of cerebral *Cryptococcus neoformans* infection after fungemia. *J Infect Dis* 2002;**186**:522–30.
25. Warr W, Bates JH, Stone A. THe spectrum of pulmonary cryptococcosis. *Ann Intern Med* 1968;**69**:1109–16.
26. Mueller NJ, Fishman JA. Asymptomatic pulmonary cryptococcosis in solid organ transplantation: report of four cases and review of the literature. *Transpl Infect Dis* 2003;**5**(3):140.

27. Longley N, Jarvis JN, Meintjes G, et al. Cryptococcal antigen screening in patients initiating ART in South Africa: a prospective cohort study. *Clin Infect Dis* 2015;**62**(5):581–7.
28. Arnder L, Castillo M, Heinz ER, Scatliff JH, Enterline D. Unusual pattern of enhancement in cryptococcal meningitis: in vivo findings with postmortem correlation. *J Comput Assist Tomogr* 1996;**20**(6):1023.
29. Popovich MJ, Arthur RH, Helmer E. CT of intracranial cryptococcosis. *AJR Am J Roentgenol* 1990;**154**:603–6.
30. Rosario M, Song SX, McCullough LD. An unusual case of stroke. *Neurologist* 2012;**18**:229–32.
31. Singh N, Perfect JR. Immune reconstitution syndrome associated with opportunistic mycoses. *Lancet Infect Dis* 2007;**7**:395–401.
32. Rolfes MA, Hullsiek KH, Rhein J, et al. The effect of therapeutic lumbar punctures on acute mortality from cryptococcal meningitis. *Clin Infect Dis* 2014;**59**:1607–14.
33. Perfect JR, Dismukes WE, Dromer F, et al. Clinical practice guidelines for the management of cryptococcal disease: 2010 update by the infectious diseases Society of America. *Clin Infect Dis* 2010;**50**:291–322.
34. Bratton EW, El Husseini N, Chastain CA, et al. Approaches to antifungal therapies and their effectiveness among patients with cryptococcosis. *Antimicrob Agents Chemother* 2013;**57**:2485–95.
35. Masur H, Brooks JT, Benson CA, Holmes KK, Pau AK, Kaplan JE. Prevention and treatment of opportunistic infections in HIV-infected adults and adolescents: updated guidelines from the Centers for Disease Control and Prevention, National Institutes of Health, and HIV Medicine Association of the Infectious Diseases Society of America. *Clin Infect Dis* 2014;**58**:1308–11.
36. Nolard N. Links between risks of aspergillosis and environmental contamination. Review of the literature. *Pathol Biol* 1994;**42**:706–10.
37. Szyper-Kravitz M, Lang R, Manor Y, Lahav M. Early invasive pulmonary aspergillosis in a leukemia patient linked to aspergillus contaminated marijuana smoking. *Leuk Lymphoma* 2001;**42**:1433–7.
38. Haiduven D. Nosocomial aspergillosis and building construction. *Med Mycol* 2009;**47**(Suppl. 1):S210–6.
39. Baddley JW. Clinical risk factors for invasive aspergillosis. *Med Mycol* 2011;**49**(Suppl. 1):S7–s12.
40. Lin SJ, Schranz J, Teutsch SM. Aspergillosis case-fatality rate: systematic review of the literature. *Clin Infect Dis* 2001;**32**:358–66.
41. Mukoyama M, Gimple K, Poser CM. Aspergillosis of the central nervous system. Report of a brain abscess due to *A. fumigatus* and review of the literature. *Neurology* 1969;**19**:967–74.
42. Marzolf G, Sabou M, Lannes B, Cotton F, Meyronet D, Galanaud D, et al. Magnetic resonance imaging of cerebral aspergillosis: imaging and pathological correlations. *PLoS One* 2016;**11**:.
43. Sidani C, Freiser ME, Saigal G, Sklar E. Unusual case of cerebral aspergillosis with clinical and imaging findings mimicking lymphoma. *Neuroradiol J* 2013;**26**:290–6.
44. Phuttharak W, Hesselink JR, Wixom C. MR features of cerebral aspergillosis in an immunocompetent patient: correlation with histology and elemental analysis. *AJNR Am J Neuroradiol* 2005;**26**:835–8.
45. Hummel M, Spiess B, Kentouche K, et al. Detection of Aspergillus DNA in cerebrospinal fluid from patients with cerebral aspergillosis by a nested PCR assay. *J Clin Microbiol* 2006;**44**:3989–93.
46. Viscoli C, Machetti M, Gazzola P, et al. Aspergillus galactomannan antigen in the cerebrospinal fluid of bone marrow transplant recipients with probable cerebral aspergillosis. *J Clin Microbiol* 2002;**40**:1496–9.
47. Patterson TF, Thompson 3rd GR, Denning DW, Fishman JA, Hadley S, Herbrecht R, et al. Executive summary: practice guidelines for the diagnosis and management of aspergillosis: 2016 update by the Infectious Diseases Society of America. *Clin Infect Dis* 2016;**63**:433–42.
48. Schwartz S, Ruhnke M, Ribaud P, Corey L, Driscoll T, Cornely OA, et al. Improved outcome in central nervous system aspergillosis, using voriconazole treatment. *Blood* 2005;**106**:2641–5.
49. Schwartz S, Ruhnke M, Ribaud P, Reed E, Troke P, Thiel E. Poor efficacy of amphotericin B-based therapy in CNS aspergillosis. *Mycoses* 2007;**50**:196–200.
50. Ribes JA, Vanover-Sams CL, Baker DJ. Zygomycetes in human disease. *Clin Microbiol Rev* 2000;**13**:236–301.
51. McNulty JS. Rhinocerebral mucormycosis: predisposing factors. *Laryngoscope* 1982;**92**:1140–3.
52. Roden MM, Zaoutis TE, Buchanan WL, Knudsen TA, Sarkisova TA, Schaufele RL, et al. Epidemiology and outcome of zygomycosis: a review of 929 reported cases. *Clin Infect Dis* 2005;**41**:634–53.

53. Garcia RR, Min Z, Narasimhan S, Bhanot N. Fusarium brain abscess: case report and literature review. *Mycoses* 2015;**58**:22–6.
54. McCarthy M, Rosengart A, Schuetz AN, Kontoyiannis DP, Walsh TJ. Mold infections of the central nervous system. *N Engl J Med* 2014;**371**:150–60.
55. Gopinath M, Cherian A, Baheti NN, Das A, Antony M, Sarada C. An elusive diagnosis: *Scedosporium apiospermum* infection after near-drowning. *Ann Indian Acad Neurol* 2010;**13**:213–5.
56. Henao-Martinez AF, Castillo-Mancilla JR, Barron MA, Nichol AC. Combination antifungal therapy in the treatment of *Scedosporium apiospermum* central nervous system infections. *Case Rep Infect Dis* 2013;**2013**.
57. Pfaller MA, Diekema DJ. Epidemiology of invasive candidiasis: a persistent public health problem. *Clin Microbiol Rev* 2007;**20**:133–63.
58. Garnacho-Montero J, Díaz-Martín A, García-Cabrera E, Ruiz Pérez de Pipaón M, Hernández-Caballero C, Aznar-Martín J, et al. Risk factors for fluconazole-resistant candidemia. *Antimicrob Agents Chemother* 2010;**54**:3149–54.
59. Chow JK, Golan Y, Ruthazer R, et al. Risk factors for albicans and non-albicans candidemia in the intensive care unit. *Crit Care Med* 2008;**36**:1993–8.
60. Leroy O, Gangneux JP, Montravers P, et al. Epidemiology, management, and risk factors for death of invasive Candida infections in critical care: a multicenter, prospective, observational study in France (2005–2006). *Crit Care Med* 2009;**37**:1612–8.
61. Durand ML. Bacterial and fungal endophthalmitis. *Clin Microbiol Rev* 2017;**30**:597–613.
62. Raman Sharma R. Fungal infections of the nervous system: current perspective and controversies in management. *Int J Surg* 2010;**8**:591–601.
63. Fennelly AM, Slenker AK, Murphy LC, Moussouttas M, DeSimone JA. Candida cerebral abscesses: a case report and review of the literature. *Med Mycol* 2013;**51**:779–84.
64. Sanchez-Portocarrero J, Perez-Cecilia E, Corral O, Romero-Vivas J, Picazo JJ. The central nervous system and infection by Candida species. *Diagn Microbiol Infect Dis* 2000;**37**:169–79.
65. Khandelwal N, Gupta V, Singh P. Central nervous system fungal infections in tropics. *Neuroimaging Clin N Am* 2011;**21**:859–66. viii.
66. Shih RY, Koeller KK. Bacterial, fungal, and parasitic infections of the central nervous system: radiologic-pathologic correlation and historical perspectives. *Radiographics* 2015;**35**:1141–69.
67. Jacobs CS, Etherton MR, Lyons JL. Fungal infections of the central nervous system. *Curr Infect Dis Rep* 2014;**16**:449.
68. Nguyen FN, Kar JK, Zakaria A, Schiess MC. Isolated central nervous system histoplasmosis presenting with ischemic pontine stroke and meningitis in an immune-competent patient. *JAMA Neurol* 2013;**70**:638–41.
69. Ramireddy S, Wanger A, Ostrosky L. An instructive case of CNS histoplasmosis in an immunocompetent host. *Med Mycol Case Rep* 2012,**1**:69–71.
70. Williams PL, Johnson R, Pappagianis D, et al. Vasculitic and encephalitic complications associated with *Coccidioides immitis* infection of the central nervous system in humans: report of 10 cases and review. *Clin Infect Dis* 1992;**14**:673–82.
71. de Carvalho CA, Allen JN, Zafranis A, Yates AJ. Coccidioidal meningitis complicated by cerebral arteritis and infarction. *Hum Pathol* 1980;**11**:293–6.
72. Erly WK, Bellon RJ, Seeger JF, Carmody RFMR. imaging of acute coccidioidal meningitis. *AJNR Am J Neuroradiol* 1999;**20**:509–14.
73. Durkin M, Witt J, Lemonte A, Wheat B, Connolly P. Antigen assay with the potential to aid in diagnosis of blastomycosis. *J Clin Microbiol* 2004;**42**:4873–5.
74. Madigan T, Fatemi Y, Theel ES, Moodley A, Boyce TG. Central nervous system blastomycosis in children – a case report and review of the literature. *Pediatr Infect Dis J* 2016;**36**(7):679–84.
75. Hadjipavlou AG, Mader JT, Nauta HJ, Necessary JT, Chaljub G, Adesokan A. Blastomycosis of the lumbar spine: case report and review of the literature, with emphasis on diagnostic laboratory tools and management. *Eur Spine J* 1998;**7**:416–21.
76. Elias Jr. J, dos Santos AC, Carlotti Jr. CG, et al. Central nervous system paracoccidioidomycosis: diagnosis and treatment. *Surg Neurol* 2005;**63**(Suppl. 1):S13–21 [discussion S].

77. Francesconi F, da Silva MT, Costa RL, et al. Long-term outcome of neuroparacoccidioidomycosis treatment. *Rev Soc Bras Med Trop* 2011;**44**:22–5.
78. Le T, Huu Chi N, Kim Cuc NT, et al. AIDS-associated *Penicillium marneffei* infection of the central nervous system. *Clin Infect Dis* 2010;**51**:1458–62.
79. Chen CY, PL L, Lee KM, et al. Acute meningitis caused by *Cladosporium sphaerospermum*. *Am J Med Sci* 2013;**346**:523–5.
80. Roche M, Redmond RM, O'Neill S, Smyth E. A case of multiple cerebral abscesses due to infection with *Cladophialophora bantiana*. *J Infect* 2005;**51**:e285–8.
81. Chakrabarti A, Kaur H, Rudramurthy SM, et al. Brain abscess due to *Cladophialophora bantiana*: a review of 124 cases. *Med Mycol* 2016;**54**:111–9.
82. Kantarcioglu AS, Guarro J, De Hoog S, Apaydin H, Kiraz N. An updated comprehensive systematic review of *Cladophialophora bantiana* and analysis of epidemiology, clinical characteristics, and outcome of cerebral cases. *Med Mycol* 2016;**55**(6):579–604.
83. Chowdhary A, Meis JF, Guarro J, et al. ESCMID and ECMM joint clinical guidelines for the diagnosis and management of systemic phaeohyphomycosis: diseases caused by black fungi. *Clin Microbiol Infect* 2014;**20**(Suppl. 3):47–75.
84. Galhardo MC, Silva MT, Lima MA, et al. *Sporothrix schenckii* meningitis in AIDS during immune reconstitution syndrome. *J Neurol Neurosurg Psychiatry* 2010;**81**:696–9.
85. Hardman S, Stephenson I, Jenkins DR, Wiselka MJ, Johnson EM. Disseminated *Sporothix schenckii* in a patient with AIDS. *J Infect* 2005;**51**:e73–7.
86. Kauffman CA, Bustamante B, Chapman SW, Pappas PG, Infectious Diseases Society of A. Clinical practice guidelines for the management of sporotrichosis: 2007 update by the Infectious Diseases Society of America. *Clin Infect Dis* 2007;**45**:1255–65.
87. Hessler C, Kauffman CA, Chow FC. The upside of bias: a case of chronic meningitis due to *Sporothrix schenckii* in an immunocompetent host. *Neurohospitalist* 2017;**7**:30–4.

CHAPTER 14

The Challenge of Finding New Therapies for Sleeping Sickness

Sbaraglini Maria Laura, Bellera Carolina Leticia, Talevi Alan
University of La Plata, La Plata, Argentina

1. INTRODUCTION

The Sleeping Sickness disease, also called human African trypanosomiasis (HAT), is a vector-borne parasitic disease caused by the *Trypanosoma brucei* (*T. brucei*), a protozoan that belongs to the Trypanosomatidae family. It is included within a group of diseases called neglected tropical diseases (NTD), characterized by a lack of proportion between the investment in diagnostic tools and drug discovery and the disease's social and economic burden. Other parasites of this family that cause human diseases are *Trypanosoma cruzi,* that causes the Chagas disease or American Trypanosomiasis, transmitted by the bite of triatomine bugs; and *Leishmania* spp. causing the cutaneous or visceral Leishmaniasis transmitted by sandflies.

The transmission of HAT to humans occurs through the bite of the tsetse fly (*Glossina* spp.), which has acquired its infection from human beings or from animals harboring human pathogenic parasites.[1] Two different subspecies of the parasite could produce the HAT disease: *Trypanosoma brucei gambiense* (gHAT) and *Trypanosoma brucei rhodesiense* (rHAT). These human pathogens do not only differ in geographical distribution but also biologically, clinically, therapeutically, and epidemiologically, causing separate diseases.[2]

According to the World Health Organization (WHO), **gHAT** is found in about 24 countries in West and Central Africa. More than 98% of the reported cases of **gHAT** correspond to this type of the disease, which causes a chronic infection. The majority (97%) of the cases are spread in only five countries, mostly in the Democratic Republic of Congo (80%). The infection can be "silent" with minor and unspecific symptoms for months or even years. When more evident symptoms come out, the patient is frequently in an advanced phase of the disease (see the stages of the disease later in the text). Reported cases of **gHAT** have been falling with a 76% decrease in prevalence since 2000; 7000 cases have been reported between 2010 and 2013.[3] On the other hand, **rHAT** is found in 13 countries in Eastern and Southern Africa. This form represents fewer than 2% of the reported cases and causes an acute infection. In this case, the signs and symptoms are observed a few weeks or months after the infection, with a quick development of the infection that reaches the central nervous system (CNS) very fast.[1,3,4]

The Microbiology of Central Nervous System Infections
https://doi.org/10.1016/B978-0-12-813806-9.00014-7

Besides the difference in location and epidemiology, the clinical progression of both forms of HAT is divided into two phases. **Phase 1** is the early stage (also called hemolymphatic, because the parasite invades the blood and the lymphatic system).[1] At this point, the trypanosomes are in the systemic circulation, tissue spaces, and lymphatic system. The number of parasites increases fast and in a few days, there is a high circulating parasitemia. The localization of *T. brucei* is always extracellular. This situation alarms the immune system and leads to the activation of defenses against these parasites (primarily induced by trypanosome molecules following cell lyses) followed by a debilitation of the host immune system and multiple organ complications.[5] The duration of this phase and the severity of the symptoms are variable, depending on the subspecies of *T. brucei* involved. In **Phase 2**, also named late stage or the meningoencephalitic phase, parasites cross the blood-brain barrier (BBB), probably at the choroid plexus; mechanistic details of this process remain uncertain. **Phase 2** is accompanied by bipolar behavioral changes, sleeping problems, lethargy, coma and, ultimately, death. Disease progression depends on both the parasite subspecies and host factors, and disease pathology is so complex that the mechanism by which parasite-host interactions leads to organ damage and immune suppression remains unclear.[5] In this stage, many functions are affected and some patients suffer from sleeping issues, which is the reason for one of the names of this disease.[1]

Some other subspecies of this parasite also infect animals causing Nagana (*Trypanosoma brucei brucei*, *Trypanosoma congolense*, and *Trypanosoma vivax*), Surra (*Trypanosoma evansi*), or Dourine (*Trypanosoma equiperdum*), thereby restricting agricultural development and significantly contributing to poverty in afflicted areas. Sleeping Sickness has an overwhelming impact on human health and prosperity as Nagana has a significant effect on the nutrition of millions of people living in the endemic areas, and on the agriculture economics of their countries, resulting in an estimated annual economic cost of about US $4 billion.[2]

2. AN OVERVIEW ON THE DISCOVERY OF THE TSETSE FLY-TRYPANOSOME AND ITS DISTRIBUTION

The first in identifying that Nagana was caused by the bite of tsetse flies was the Scottish missionary David Livingston.[6] He described the occurrence of a disease in the valleys of the Limpopo and Zambezi rivers and at the banks of lakes Nyasa and Tanganyika, causing the death of livestock after tsetse flies bit them.

In 1895, David Bruce identified *T. brucei* as the cause of cattle trypanosomiasis. The first observation of trypanosomes in human blood was made in 1901 by the British physician Robert Michael Forde when examining a steamboat captain in Gambia. He believed that the organisms he found were worms but soon after the English physician

Joseph Everett Dutton identified them as trypanosomes, proposing the species name *Trypanosoma gambiense* (now *T. b. gambiense*). In the same year, the Italian physician Aldo Castellani found trypanosomes in the cerebrospinal fluid of Sleeping Sickness patients and suggested that this parasite produced neuronal degeneration that triggered a sleep disorder. One year later, decisive evidence that tsetse flies were vectors for the Sleeping Sickness was provided.[6]

Along history, HAT went through several cycles of epidemic, intensive control in answer to high numbers of victims, epidemic control, and subsequent neglect, restarting the cycle. Initial actions focused on vector control (mainly in English colonies), chemoprophylaxis (predominantly in French colonies), and mobile teams for active case finding and treatment. The latter still constitutes the basis of the efforts to control *T. b. gambiense*.[7] The WHO describes a small number of epidemics in Africa over the last century: one between 1896 and 1906, mostly in Uganda and the Congo Basin; another in 1920 in a few African countries; and the most recent, which started in 1970 and lasted until the end of the 20th century. The 1920 epidemic was controlled thanks to mobile teams that performed the screening of millions of people at risk. By the 1960s, HAT was under control with less than 5000 cases reported on the whole continent. After this achievement, vigilance was relaxed, and the disease resurfaced, reaching epidemic proportions by 1970. The WHO made several efforts to reverse the scenario during the 1990s and early 21st century, including national control programs, promotion of bilateral cooperation, and cooperation with nongovernmental organizations.[1] In 2014, the WHO reported 3796 recorded cases of Sleeping Sickness.[1]

The current epidemiological situation of HAT is heterogeneous. Fig. 1 illustrates the areas where *T. b. gambiense and T. b. rhodesiense* are prevalent. For gHAT those are:

(a) West Africa, with little prevalence.

(b) Central Africa, where the transmission is active with low or very low prevalence. Foci are well identified and limited with a very low (Cameroon, Equatorial Guinea, Gabon) or low prevalence (Angola, Central African Republic, Chad, Congo, South Sudan, North West Uganda), showing, in general, a diminishing tendency. Control and surveillance activities still continue.

(c) Democratic Republic of the Congo, which includes the most important burden of the disease (>80% of total HAT cases).

For *T. b. rhodesiense,* the distribution is a little different and the prevalence region is East Africa, where the number of cases is low (less than 200 cases notified/year, about 3% of total HAT cases reported) but the risk of becoming an epidemic is not. It is possible to differentiate two areas, one with disease mostly related to cattle (Uganda, Kenya) and another with disease related to wild animals (Tanzania, Malawi, Zimbabwe, Zambia). Cases have not been reported in Botswana, Burundi, Ethiopia, Mozambique, Namibia, Rwanda, and Swaziland for more than 15 years.

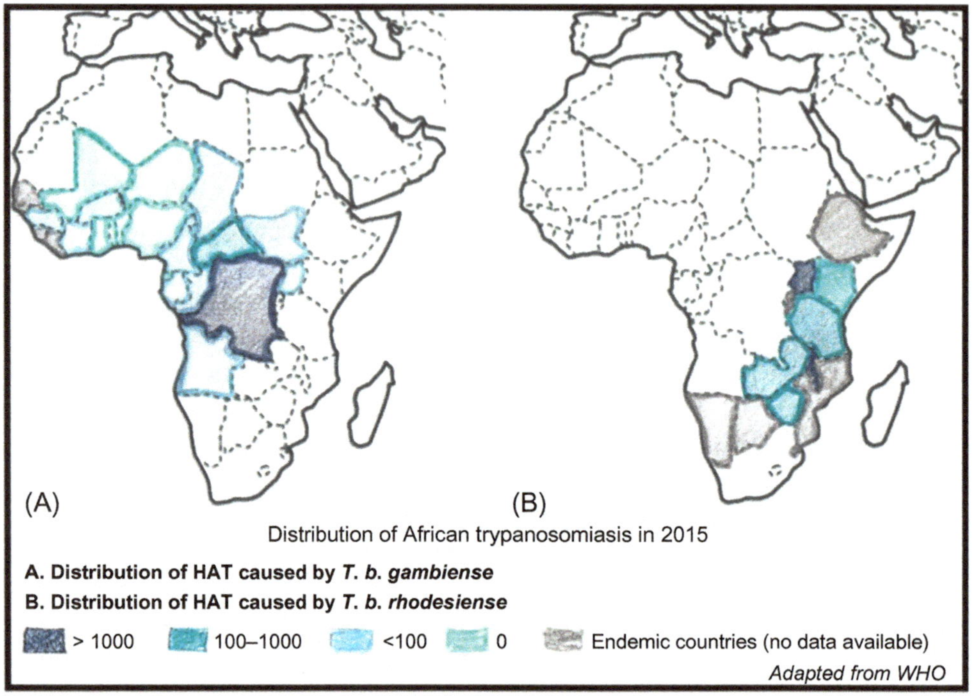

Fig. 1 Distribution of HAT caused by *T. b gambiense* (A) and *T. b. rhodesiense* (B).

3. CELL ARCHITECTURE OF *T. BRUCEI*

The family of Trypanosomatidae belongs to the order Kinetoplastida, which includes free-living and parasitic species.[5] The trypanosome cell is elongated and has a spindle shape which results from a spiraled array of interlinked microtubules that adapts to growth and cell division. The ability to remodel the cytoskeleton gives raise to the distinct morphologies of the various developmental forms. The single-copy organelles in the trypanosome cell, such as the flagellar pocket (FP), flagellum, kinetoplast, mitochondrion, and nucleus, are exactly positioned in a cytoskeletal corset and are concentrated between the posterior end and the center of the cell (Fig. 2).[8]

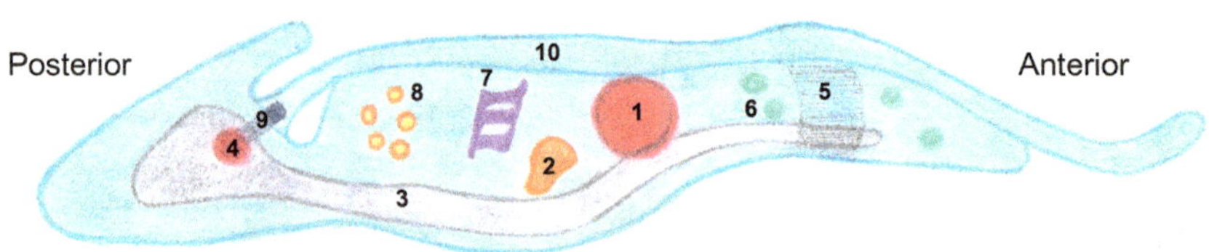

Fig. 2 The *T. brucei* architecture. A simplified scheme of the principal organelles in the trypanosome cell. (1) Nucleus, (2) lysosome, (3) mitochondria, (4) kinetoplastids, (5) microtubules cytoskeleton, (6) glycosome, (7) golgi, (8) endosome, (9) basal body, (10) flagellum. *(Adapted from Matthews KR. The developmental cell biology of* Trypanosoma brucei. J Cell Sci *2005;**118**:283–90.)*

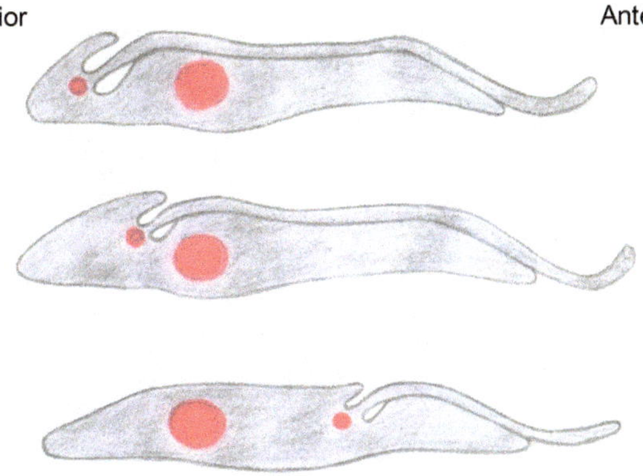

Fig. 3 Distinct positions of the kinetoplast during the trypanosome life cycle. The relative position of the kinetoplastids respect to the nucleus in bloodstream, procyclic, and epimastigotes of *T. brucei*. *(Adapted from Matthews KR. The developmental cell biology of* Trypanosoma brucei. J Cell Sci *2005;**118**:283–90.)*

T. brucei, as other kinetoplastids, presents one flagellum that emerges from the cytoplasm through the FP at the posterior end of the cell (Fig. 2).[9] This region (the FP) is a marked invagination of the plasma membrane that indicates the limit between the flagellar membrane and the rest of the plasma membrane. The FP is close to the basal body, which is attached to the kinetoplast (the mitochondrial genome). The trypanosome FP is a key host-parasite entrance because it is the only site of endocytosis, where the uptake of growth factors takes place, and it is a necessary transit point for proteins to or from the cell surface.[9] These parasites experience morphological changes depending on the environment. The most obvious morphological difference between the different life cycle stages of *T. brucei* is the location of the kinetoplast relative to the posterior end of the cell. In **bloodstream** forms, the kinetoplast locates close to the posterior end of the cell, and each offspring kinetoplast remains in this region during the cell cycle until cytokinesis. In contrast, the kinetoplast locates midway between the cell nucleus and posterior in **procyclic** forms. In **epimastigote** forms (which appear in the tsetse salivary gland), the kinetoplast is anterior to the central nucleus, as shown in Fig. 3.[8]

Remarkably, the basic cellular architecture is highly conserved among other Kinetoplastids, as *Leishmania* spp. and *T. cruzi*.[5]

4. LIFE CYCLE

The *T. brucei* life cycle includes several developmental forms, among which we find proliferative stages in both mammalian and insect hosts. Each transition is preceded by a preadaptation that includes growth arrest.[5] Both transitions (insect to mammal and vice versa) are accompanied by remodeling of the cell architecture, intracellular transport,

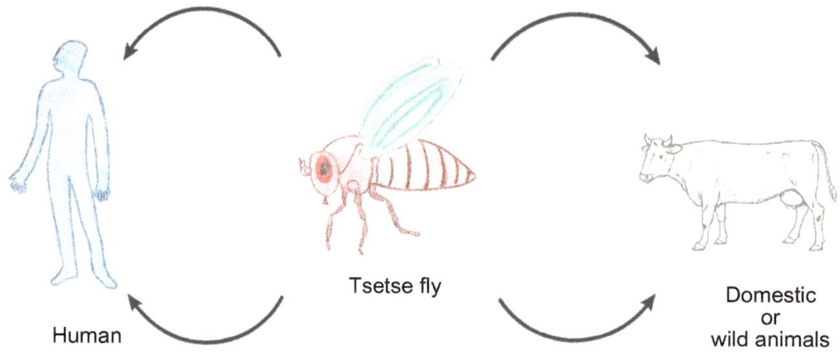

Fig. 4 Mammalian hosts and the vector.

primary metabolism, and major changes in gene expression. Mammalian forms have high endocytic and recycling activity while in most forms of the insect the levels of these processes are lower.

T. brucei parasites cycle between the tsetse fly vector (*Glossina* spp.) and the mammalian host (Fig. 4). In each host, parasites undergo many life cycle changes (i.e., in the tsetse fly as procyclic/epimastigote/metacyclic forms and in the mammalian host as bloodstream forms) with concomitant morphological and metabolic changes, which are precisely programmed to adapt to different growth conditions and nutrient availability imposed by the different microenvironments in which they live.[2] *T. b. gambiense* and *T. b. rhodesiense* are initially established in the midgut of the fly after a bite (blood meal). Next, parasites migrate to the salivary glands in preparation for transmission to a new mammalian host. In mammals, parasites survive free in the bloodstream, being able to evade antibody responses through antigenic variation (which will be discussed later). In the mammalian blood, parasites multiply as **slender forms**; as parasite numbers increase, nonproliferative **stumpy forms** replace them (Fig. 5). One reason for this could be that the accumulation of division-arrested forms limits the increase in parasite numbers, extending host survival and thus increasing the probability of disease transmission. Additionally, the uniform arrest of stumpy forms in the G1 phase of the cell cycle ensures that the morphological changes that occur upon transmission to the tsetse fly can be synchronized with re-entry into the cell cycle.[8] This is significant in the cycle of this microorganism because, as we mentioned earlier, the position of some organelles changes with the different stages taken during the cycle, being crucial for successful achievement of the cell cycle of tsetse midgut procyclic forms.

After a blood meal, bloodstream trypanosomes differentiate to early procyclic forms and replace the variant surface glycoproteins (VSG) coat (that helps survival in the mammal blood) with a less dense surface coat composed of procyclins. Energy generation is also modified from being exclusively based on glycolysis in the mammal host to a mitochondrion-based respiratory system.[8] Many flies can clear the infection at this point.

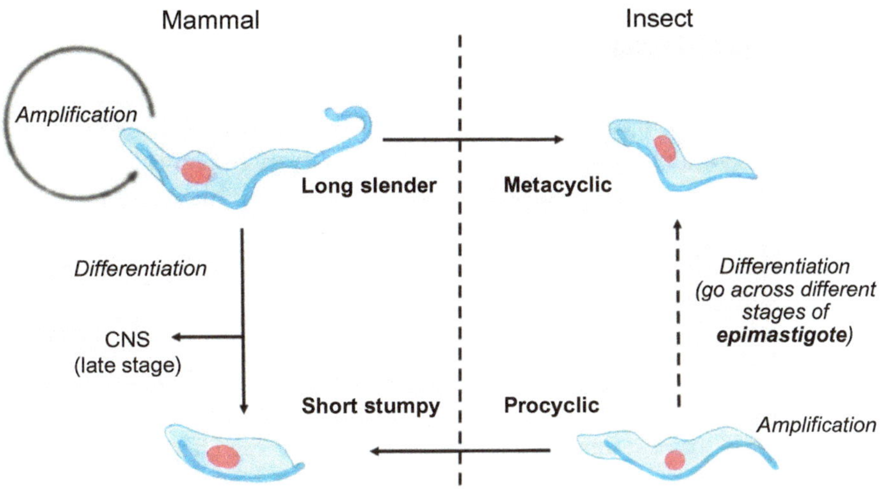

Fig. 5 Different morphological stages in mammals and flies.

To survive and colonize the midgut of the insect, the trypanosomes must migrate to the other side of the peritrophic matrix (PM), a chitinous structure that functions as a physical barrier between the insect blood and the insect gut epithelium. Six days post-infection, most trypanosomes have crossed the barrier and located at the ectoperitrophic space of infected flies. By then, most of these parasites are in late procyclic forms (Fig. 5). Surprisingly, even when colonization of the midgut is successful, many infections fail to progress beyond this stage. To complete the life cycle, parasites must occupy the salivary glands. Only a few cells colonize the glands, indicating that migration represents a bottleneck.[10] New morphological changes occur and the epimastigote forms are generated. After further multiplication, the parasite undergoes division arrest, reacquires a VSG coat, and is released into the salivary gland lumen in preparation for inoculation into a new mammalian host.[8] The transmission cycle is finished in the salivary glands with the formation of infectious metacyclic forms that are transferred to a new mammalian host.[10]

5. WHAT MECHANISMS DOES *T. BRUCEI* USE TO EVADE THE IMMUNOLOGICAL SYSTEM?

The protozoan blood parasite that causes human Sleeping Sickness across sub-Saharan Africa is a model system for **antigenic variation**, persisting in mammals by periodically replacing a monolayer of VSGs that covers its cell surface.[11,12] This process is essential to evade the host immune response by continual switching of surface proteins.

The *T. brucei* genome includes many hundreds of VSGs that encode a transcriptional silent reservoir of variant antigens, and each cell expresses just a single gene from a specialized telomeric expression site at a given time point.[12] Several VSGs are jointly expressed in the parasite population; when the host becomes immune to the prevailing VSG, clones expressing alternative copies multiply in a frequency-dependent manner. In consequence, the infection results in characteristic "waves of parasitemia." To survive in the long term, *T. brucei* requires novel VSG sequences through recombination. The system of antigenic variation in *T. brucei* results in an enormous range of antigens (approximately 2000 VSGs per gene).[11] This process is fundamental to the host-trypanosome relationship and closely linked to the course and severity of disease, parasite transmission, and host range (and therefore to disease epidemiology).[12]

There are other strategies of immune-evasion as the development of an efficient mechanism to remove immune factors from the surface, including immunoglobulins, through capping and internalization.[5] Other strategies include restriction of invariant receptors in the FP, making them inaccessible to host immune effectors, and the expression of receptors that can kidnap molecules secreted by innate immune cells.

6. THE TRADITIONAL TREATMENT

Many institutions and organizations have made considerable efforts to find a suitable treatment for Sleeping Sickness. However, at the moment there is no completely appropriate drug. There are several facts that make the search for a new drug difficult and slow: pharmaceutical companies are less inclined to invest in drug discovery/development against diseases that affect the poorest people, because it is not economically profitable; the political instability of the affected regions; the fact that wild animals function as reservoirs of the parasite and, therefore, hamper disease control; and the inappropriate use of the available drugs resulting in the emergence of drug resistance.[2]

Until now, only five drugs have been approved for HAT treatment: **suramin**, **pentamidine**, **melarsoprol**, **eflornithine**, **and nifurtimox**. It is important to note that treatment choice critically depends on the disease stage. The drugs used at the first stage are safer and easier to administer than those used at the second.

The first one, **suramin**, was discovered in 1920, and since then it has been used for the treatment of the first stage of *T. b. rhodesiense* infection because it does not cross the BBB. It is an effective therapy but is associated with urticarial rash in about 90% of patients. The rash usually resolves without discontinuation of the drug. Other common side effects include pyrexia, nausea, and reversible nephrotoxicity.[13] **Pentamidine** is an aromatic diamidine that has been used to treat the first stage of gHAT since the 1940s. It presents several adverse effects such as hypotension, glucose perturbations, and leukopenia. Another disadvantage is the administration route (intramuscular) and, again, that it has low BBB permeability so it is ineffective against second-stage HAT.[14,15] In conclusion, pentamidine and suramin are used in the treatment of hemolymphatic stage *T. b. gambiense* and *T. b. rhodesiense* infections, respectively.

In 1949 **melarsoprol** was used for the first time against the disease; it is currently the most widely used drug for second stage HAT induced by *T. b. rhodesiense.*[16] It is also effective against *T. b. gambiense* but it presents severe safety issues. Melarsoprol is an organoarsenic derivative, so it is highly toxic to patients and adverse reactions are very common. In 5%–10% of cases, melarsoprol induces severe encephalopathic syndrome that causes death in 50% of these patients.[17] This encephalopathic syndrome itself requires careful, labor-intensive management and drug treatment.

Eflornithine, a repositioned drug that had previously been used as an anticancer agent, has been introduced as an alternative to melarsoprol.[18] It is effective at the second phase of the disease but against *T. b. gambiense* exclusively. It has less serious side effects than melarsoprol but due to its low bioavailability, it should be administered intravenously in treatments of 14 days. Such a treatment regimen is usually difficult to follow in patients with gHAT, due to inadequate health facilities.

The last drug approved for the treatment of HAT is **nifurtimox**. This molecule was originally approved for the treatment of American Trypanosomiasis; however, the drug has been shown to be effective against both the early and late stages of the *T. b. gambiense* infection. Efficacy was first demonstrated in rat models and then in a limited clinical trial of four European patients. Subsequent trials with various nifurtimox regimens revealed that cure rates were variable (30%–80%), with toxicity accompanying higher doses and prolonged treatment.[19] Consequently, nifurtimox was not approved as a monotherapy for second-stage HAT and is used primarily to treat patients displaying drug resistance to the other therapies. Despite the problems associated with its use, nifurtimox has a place in HAT chemotherapy as nifurtimox-eflornithine combination therapy (NECT).[20] Eflornithine acts as an inhibitor of the enzyme ornithine decarboxylase, blocking polyamine biosynthesis, while nifurtimox is converted to a toxic metabolite following activation by a type I nitroreductase.[21] After safety and efficacy data were verified in clinical trials, NECT has been included in the "WHO List of Essential Medicines" and it is currently recommended as a first-line treatment for **gHAT**.[1] A combination treatment presents safety advantages, is easier to administer (i.e., infusion every 12h for 7 days vs. every 6h for 14 days for eflornithine alone), and is also potentially protective against the emergence of resistant parasites.[22]

Various small molecules of a diverse chemical nature have been identified as potential new treatments against HAT. Some of these are presently undergoing clinical trials, including **fexinidazole** (phase I completed) and **SCYX-7158**, an oxaborole that is currently being finalized in Phase I clinical trials for treatment of HAT.[15,23]

Fexinidazole had been in clinical development since the 1980s as a broad-spectrum antimicrobial drug and was rediscovered as a potential treatment against HAT by joint efforts of the Drugs for Neglected Diseases initiative (DNDi) and Sanofi. This drug presents trypanocidal properties and has demonstrated the potential to become a safe, short-course oral treatment for both HAT stages. It has successfully completed preclinical development and phase I clinical trials, and is currently undergoing phase II/III studies.[15,24] **SCYX-7158** has been shown to be safe and efficacious to treat stage 2 of the disease in animal models.[25] Preclinical development progressed successfully to clinical trials (Phase I) in 2015. [23] Table 1 presents a summary of the available treatments for HAT in the early and late stages as well as the new compounds that are currently undergoing clinical trials.

Table 1 Drug use for the treatment of HAT in different stages of the disease

Drugs	Administration via and dosage	Adverse effect	References
First stage of human African trypanosomiasis			
Pentamidine	Intramuscular; 4 mg/kg/day for 7–10 days	Hyperglycaemia or hypoglycaemia, prolongation of the QT interval on electrocardiogram, hypotension, and gastrointestinal features	16
Suramin	Intravenous; 20 mg/kg once per week over 5 weeks (maximum dose 1 g per injection)	Renal failure, skin lesions, anaphylactic shock, bone marrow toxicity, and neurological complications such as peripheral neuropathy	16,26
Second stage of human African trypanosomiasis			
Melarsoprol	Intravenous; *T. b. gambiense*: 2.2 mg/kg/day for 10 days *T. b. rhodesiense*: three series of 3.6 mg/kg for 3 days, spaced by intervals of 7 days	Reactive arsenical encephalopathy (RAE) has been attributed to the toxic effect of melarsoprol, peripheral neuropathy, cutaneous reactions, renal or hepatic dysfunction, allergic or hypersensitivity reactions	16,27
Eflornithine	Only useful against *T. b. gambiense* Administration difficult—4 slow intravenous infusions per day for 14 days; increasingly used as 1st line for *T. b. gambiense* stage 2 HAT	Generally, are reversible after the end of treatment Convulsions, gastrointestinal symptoms like nausea, vomiting and diarrhea, bone marrow toxicity leading to anaemia, leukopenia and thrombocytopenia	28

Table 1 Drug use for the treatment of HAT in different stages of the disease—cont'd

Drugs	Administration via and dosage	Adverse effect	References
Nifurtimox + eflornithine	Eflornithine: 200 mg/kg intravenous infusion two times per day for 7 days Nifurtimox: 5 mg/kg three times per day oral for 10 days	Convulsions, gastrointestinal symptoms like nausea, vomiting and diarrhoea. Genotoxicity, neurotoxicity	28–32
Experimental drugs at clinical stage			
SCYX-7158	Oral administration First stage Second stage	Not informed	23–33
Fexinidazole	Oral administration First stage Second stage	Headache and vomiting	24,34–36

7. NEW ADVANCES IN THE DEVELOPMENT OF TREATMENTS AGAINST HAT

As previously discussed, despite the high prevalence of HAT, the treatment is based on very limited options. The antigenic variation remains an obstacle for vaccine development, and drug resistance against almost all available drug treatments has been observed.[37] At present, strategies for the development of new drug treatments are divided into two major groups: the optimization of actual formulations (mostly through pharmaceutical nanocarriers); and the search for new drugs—including drug repositioning, high-throughput screening (HTS), computer-aided screening, and design of hybrid compounds. We will present a brief overview of some relevant efforts in this area.

7.1 Advanced Drug Delivery Systems and HAT

The design of nanoscale devices for drug delivery has become a central goal in modern medical and pharmaceutical technology. Different nanoparticulate systems have been developed, including inorganic, lipidic, and polymeric nanoparticles (NPs), among others.[38,39] The potential advantages of these systems include the protection of drugs against oxidoreduction and other enzymatic and nonenzymatic reactions; the increase of bioavailability; and targeted drug distribution and release with subsequent reduction of the effective doses and off-target adverse effects. Nanoparticles show great promise in the treatment of infectious protozoan diseases such as HAT and malaria.[40–42] The highly active endocytosis process that occurs in protozoan parasites constitutes an advantage for the delivery of nanoencapsulated drugs.

For instance, Arias et al. designed a drug nanocarrier based on polyethylene-glycol (PEG), covalently attached to poly (D,L-lactide-co-glycolide acid) (PLGA) to generate PEG-PLGA nanoparticles.[38] This nanocarrier was coupled to a heavy chain antibody fragment that specifically recognized the surface of *T. brucei*. Nanoparticles were loaded with pentamidine, finding a marked decrease in the half-effective concentration (EC_{50}) (relative to free drug) in an in vitro assay. Using an in vivo model of HAT, they showed that the nanoformulation cured infected mice at a 10-fold lower dose than the minimal curative dose of free pentamidine. In this example, the NPs were designed with components already approved for their use in humans and loaded with a drug that is currently in use to treat the disease, possibly expediting the advance of this development to the clinical stage. Other authors developed a drug delivery system of diminazene (DMZ),[43] a trypanocidal drug registered for veterinary use.[19] DMZ is characterized by poor chemical stability, poor distribution to the brain, and off-target toxic events. Authors reported colloidal formulations based on porous cationic NPs with an oily core (DGNP). Such nanocarriers resulted in enhanced drug stability. DMZ loaded into DGNP was protected against oxidation and was stable for at least six months at 4°C.[43] The in vitro tests against *T. brucei* showed an increased efficacy of the encapsulated DMZ.

As shown in the previous examples, nanoparticulate systems seem a feasible strategy to overcome some of the limitations of currently available drugs against HAT; interestingly, the use of existing drugs in innovative pharmaceutical products could provide novel therapeutic solutions against the disease in the short- and mid-term.

7.2 Systematic Drug Repurposing

Besides the research on novel pharmaceutical carriers for already-known HAT drugs, a relatively modern strategy, *systematic drug repurposing* (or *drug repositioning*), is gaining increasing popularity for the development of novel therapeutics for neglected diseases.[44–48] Drug repurposing implies expanding the use of existing drugs (approved, discontinued, shelved, or even experimental drugs) to other therapeutic indications, making this process rapid and low-cost in comparison with de novo drug discovery. This is, in part, because toxicological and pharmacokinetic profiles of repositioned drugs have already been characterized; therefore, if the doses required for the new indication are compatible with those originally used, some preclinical and clinical studies can be bypassed. Additionally, adequate pharmaceutical formulations have typically already been developed for the original therapeutic indication of the repurposed drug. Building a new indication on already available data, thus shortening development times and increasing the probability of success at clinical trials, is particularly attractive when searching for novel solutions for neglected diseases for which funding is known to be very limited.

While the private pharmaceutical sector could frequently be extremely cautious regarding disclosing pharmaceutical formulation data, they may be more inclined to liberate or share information on old or abandoned drugs repurposed for the treatment of neglected, economically unprofitable diseases. Systematic drug repurposing could be implemented through in silico or in vitro HTS methods as well as other knowledge-based approaches (e.g., application of bioinformatic tools, network-based approximations, and exploitation of on-target and off-target effects for new medical uses).

There are several stories of successful repurposing in other therapeutic fields. For example, valproic acid (an antiepileptic drug) is now also used for the treatment of bipolar disorder and migraines; since 2009, bromocriptine (antiparkinsonian) has been approved for the treatment of type 2 diabetes. Sildenafil is used for treatment of erectile dysfunction but has also been approved for the treatment of pulmonary hypertension.

Recently, exploring systematic drug repositioning against trypanosomatid-caused diseases, Alberca et al. developed computational models capable of recognizing trypanocidal polyamine analogs.[49] The models were applied in the screening of DrugBank 3.0 (this compiles FDA-approved drugs and drugs in the clinical evaluation stage)[50] and Sweetlead (which contains drugs approved by other international regulatory agencies plus natural products used in traditional medicine) databases.[51] Five of the hits emerging from the in silico screening were evaluated against different pathogenic trypanosomatids (*T. cruzi*, *T. brucei*, and *Leishmania infantum*). Inhibition of putrescine and amino-acid uptake by some of the hits was also demonstrated. Moreover, two of them—triclabendazole and paroxetine—displayed inhibitory effects on the proliferation of infective *T. brucei*. Using a similar approach, Rashmi and Swati chose the conserved glycosyl phosphatidyl inositol (GPI) domain of the *N*-acetylglucosamine-phosphatidylinositol de-*N*-acetylase (GlcNAc-PI de-*N*-acetylase) (a zinc metalloenzyme responsible for the second step of GPI biosynthesis) as a molecular target. Interestingly, the predicted active site residues of the GPI domain seem to be ultra-conserved within the Trypanosomatidae family. The authors screened Drugbank and, according to the docking-predicted binding energy, concluded that ethambutol and metaraminol may prove useful in the treatment of HAT.[52] It is important to consider that many glycoconjugate compounds are anchored to the plasma membrane with the help of GPI, not only in protozoan but also in eukaryotes. Despite these anchors being very common in protozoans and less common in eukaryotes, the compounds that emerge as potential candidates for the treatment of sleep disease will have to go through strict selectivity filters.

Kayser et al. have recently reported the systematic phenotypic screening of the Malaria Box as potential treatments against major kinetoplastid diseases.[53] The Malaria Box is a screening tool assembled by the Medicines for Malaria venture. It consists of a structurally diverse set of 200 drug-like and 200 probe-like compounds selected from more than 20,000 antimalarial hits collected from corporate and academic libraries. The use of the Malaria Box has already resulted in the identification of novel active scaffolds against other parasitic diseases such as cryptosporidiosis and schistosomiasis.[54,55] From their in vitro assays, the authors chose six active hits against *T. brucei* that fulfilled a number of medicinal chemistry criteria: absence of toxicophores (undesirable chemical moieties, permanently charged molecules, PAINS), drug likeness (based on the Lipinski rule), scaffold novelty, and CNS penetration. Moreover, from a clustering analysis of the compounds in the Malaria Box, the authors chose the hits from the most hit-populated clusters. Four of the candidates were tested in an acute mice model of HAT. Unfortunately, none of them fulfilled the expectations at the in vivo model. The authors proposed that the disappointing results could be due to pharmacokinetic reasons (e.g., failure to reach effective plasma concentrations, high plasma protein binding leading to inadequate free drug exposure) and pharmacodynamics reasons (reversibility of action). The previous example illustrates how moving from in vitro models to in vivo ones represents a key step in the drug discovery process of new trypanocidal agents: frequently, results from biochemical and cellular models fail to translate into a positive outcome at animal model levels.

Because an important aspect of drug candidates intended to treat phase 2 Sleeping Sickness is their ability to cross the BBB, Thomas S. and coworkers used the carbazoles, molecules that are being evaluated, in parallel, as treatments for neurodegenerative disorders.[56] They discovered a carbazole derivative that, by blocking mitosis, displayed antitrypanosome activity both in vivo and in vitro. They tested antitrypanosomal effects of carbazole-derived compounds called curaxins; their in vitro assays revealed 22 compounds with nanomolar potency against axenically cultured bloodstream trypanosomes. Oral administration of one of these compounds resulted in a cure. Interestingly, some other carbazole derivatives are active against other protozoan parasites, such as *L. donovani* and *Plasmodium falciparum*. Thus, carbazoles emerge as promising chemical scaffolds to discover new and better antitrypanosomal drugs. Similarly, Jones et al. identified FTY720 (fingolimod, gilenya) as a potent, selective antitrypanosomal compound active against *T. b. gambiense* and *T. b. rhodesiense*. The advantage of this potential compound is that FTY720 is an oral drug registered for the treatment of relapsing multiple sclerosis, indicating that this drug can efficiently cross the BBB, thus being potentially useful for phase 2 HAT.[57]

Finally, in 2014 Amata et al. reported that cilomilast, a known inhibitor of the human phosphodiesterase that had previously been investigated by GlaxoSmithKline as treatment for respiratory disorders (asthma and chronic obstructive pulmonary disease), displayed weak inhibitory effects against *T. brucei* phosphodiesterase B1 (*Tbr*PDEB).[58] This example constitutes an example of drug rescue because the drug had been abandoned by GlaxoSmithKline after failure to gain FDA-approval for the original pursued therapeutic indication. The authors emphasized the need to check the selectivity over the human enzyme; they also reported the design and synthesis of novel cilomilast analogs, including more potent inhibitors *Tbr*PDEB. It should be highlighted, though, that molecular optimization of repurposed candidates eliminates most (if not all) of the advantages of the repurposing approach. To be repurposed, the exact molecular entity that was already used in therapeutics must display the new indication.

The former results illustrate the potential of systematic drug repositioning to provide novel solutions for neglected diseases.

7.3 HTS and HAT

HTS is today one of the most frequent methods of exploring the vast universe of known chemicals in search of novel active scaffolds. It is the modern version of the traditional trial-and-error, "exhaustive" screening. The great advantage of this method is the integration of automation and miniaturization to the screening process, which results in efficient exploration of the chemical space.[59] The rational basis of the approach has been greatly improved by the design of target-focused libraries[60] and the recognition of privileged scaffolds[61] (molecular frameworks/building blocks that are present in many biologically active ligands against a diverse array of targets). However, the big disadvantage in HTS is that it requires very expensive technological platforms that are not frequently found in the academic sector or in low- and middle-income countries, thus limiting the opportunities of researchers from countries affected by endemic tropical diseases to apply this approximation.

Many researchers have applied HTS to the screening of new therapies against HAT, including systematic drug repurposing initiatives.

The proteins involved in the biogenesis of the glycosome constitute potentially valuable targets for drug development. Between them, Pex5p is an import receptor in the cytosol that recognizes newly synthesized glycosomal matrix proteins and transports them into glycosomes through an interaction with Pex14p (a membrane-anchored protein central component in the glycosomal protein import machinery).[62] The interaction of the Pex5p in *T. brucei* (TbPex5p) with the Pex14p in this organism (TbPex14p) is important for the survival of the parasite. The overall amino acid sequence identity between *T. brucei* targeted proteins (TbPex5p-TbPex14p) and their human counter parts (HsPex5p-HsPex14p) is quite low (20% and 27%, respectively), thus suggesting an opportunity to develop selective treatments. Watanabe et al. designed a highly sensitive HTS system with the ability to explore protein-protein interactions in a rapid manner. They screened chemical compounds that inhibit the TbPex5p-TbPex14p interaction. By using this assay system, they found, among 20,800 molecules, one compound (4-{[2-(methoxycarbonyl)-5-(2-thienyl)-3- thienyl]amino}-4-oxo-2-butenoic acid) that specifically inhibits TbPex5p-TbPex14p interaction without any HsPex5p-HsPex14p interaction.

Modern HTS can be also applied to explore the effects of small-molecule libraries on whole-cell assays, which is one of the most frequently used methods in drug discovery against infectious diseases. Among the many proteins in the *T. brucei* kinome, it has been established that the extracellular signal-regulated kinase 8 (TbERK8) is essential for parasite survival and, interestingly, the human ortholog of ERK8 is not essential for mammalian cell survival. In one recent study, authors implemented a phenotypic screen against *T. brucei* using a kinase inhibitor library.[63].This is an example of the use of focused libraries, where the collection of compounds has favorable properties for the target of interest. The authors identified 12 hits with adequate trypanocidal potency. The inhibitor activities ranged from ~80 nM to 1 μM. Dinaciclib, an investigational anticancer drug, was the most potent inhibitor with an EC_{50} of 80 nM, but other hits were almost equally potent. However, only AZ960, a novel ATP competitive JAK2 inhibitor, showed selective inhibition against the parasite kinase. Another similar HTS study on a kinase-targeted library consisted of the screening of 42,444 compounds from the GlaxoSmithKline collection, which were tested against parasite cell cultures and counter-screened against human cells.[64] A total of 797 submicromolar inhibitors of *T. brucei* proliferation were identified; all of them presented at least 100-fold selectivity over human cells. After a complete study of in silico and in vitro physicochemical and drug metabolism properties and human kinase selectivity data, three compounds were selected for in vivo studies; one of them (NEU-1053) demonstrated parasitological cure of a murine bloodstream infection of *T. b. rhodesiense*.

In line with the previous examples, Urbaniak et al. screened a focused library of zinc-binding fragments and identified salicylic hydroxamic acid as a GlcNAc-PI de-*N*-acetylase inhibitor with high ligand efficiency.[65] Remarkably, the zinc-metalloenzyme GlcNAc-PI de-*N*-acetylase is essential for the biosynthesis of mature GPI anchors and has been genetically validated in the bloodstream form of *T. brucei*. This is the first small molecule inhibitor reported for the trypanosome GPI pathway.

Another very interesting HTS campaign was performed by Benitez et al., by screening a library of 144 compounds from seven different chemical families against the trypanothione synthetases (TrySs) from the three major trypanosomatid pathogen species (*T. brucei*, *T. cruzi*, and *L. infantum*).[66] TryS is responsible for trypanothione biosynthesis, which is indispensable for parasite survival. A number of novel chemical scaffolds were found with low micromolar and selective inhibition. Remarkably, some compounds displayed multispecies TryS inhibition (*N,N′*-bis(3,4-substituted-benzyl) diamine derivatives). An N5-substituted paullone stopped the proliferation of infective *T. brucei* ($EC_{50}=4.3\pm0.7$ μM) and *L. infantum* promastigotes ($EC_{50}=12.6\pm1.6$ μM).

Importantly, in most HTS campaigns,[67,68] the emerging hits will frequently require hit-to-lead and then lead optimization medicinal chemistry programs to translate into new clinical candidates, an option that is not available if the screening is focused on drug repurposing. For example, the promising study executed by Tatipaka et al. led to the identification of substituted 2-(3-aminophenyl) oxazolopyridines as a starting point for hit-to-lead medicinal chemistry.[69] They prepared and tested more than 100 analogs in

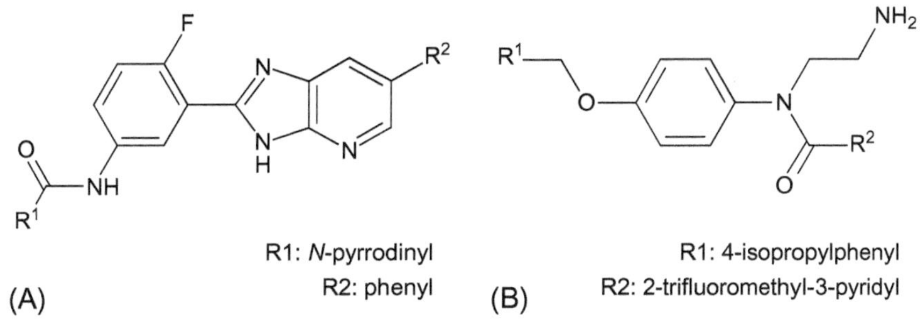

Fig. 6 Potentials leads for the development of drugs to treat HAT derived from 2-(3-aminophenyl) oxazolopyridines by Tatipaka et al. (A) and Buchynskyy et al. (B).

parasite cultures, finding the substituted 2-(3-aminophenyl) imidazopyridine as the best compound of the series (Fig. 6A) with an EC_{50} of 2nM, good oral bioavailablility, and adequate plasma and brain exposure. With these in vitro values, authors moved to in vivo assays, finding that this new molecule cured mice infected with *T. brucei* when administering daily doses of 2.5mg/kg. Later, other hits from the HTS screening were evaluated by Buchynskyy and coworkers.[70] The most powerful compound (Fig. 6B) of this series had an in vitro $EC_{50}=1$ nM, and when tested in an in vivo model, it cured two out of three mice infected with *T. brucei* (50mg/kg po once per day for 4 days). The potent antiparasitic effect and the relative simple synthesis make both compounds interesting leads for the development of drugs to treat HAT.

In 2016, Huang et al. found a compound active against methionyl-RNA synthetase after an HTS of the NIH Molecular Libraries Small Molecule Repository.[71] The hit was optimized using a structure-guided approach. The optimization led to the identification of two novel series of potent inhibitors, which were called by the authors the cyclic linker and linear linker series. Compounds of both series proved potent in in vivo assays and showed low toxicity to mammalian cells (human lymphoblast cell line—CRL8155—and a hepatocellular carcinoma cell line—Hep G2). The best compound of each series exhibited EC_{50}s of 39 and 22nM, respectively. They also displayed promising pharmacokinetics properties after oral dosing in mice. One of the compounds had moderate brain permeability, with a brain/plasma concentration ratio of 0.27 at 60min after intraperitoneal injection. This study provides new lead compounds for arriving at new treatments of HAT.

Other researchers presented an original approach to identify novel trypanothione reductase (TR) inhibitors as starting points for drug discovery, by combining target-focused in vitro screening and similarity-based in silico screening methods (inhibitory compounds from in vitro screening were used as queries for similarity-based in silico screening), leading to a considerably higher active enrichment ratio (10-fold enrichment) in comparison with pure in vitro screening.[72] The combined screening campaign resulted in the identification of 82 novel compounds more active than chlorhexidine, a well-known TR inhibitor. Four molecules showed to be active in the nM range (IC_{50} <1 μM). On parasites assays, three hits inhibited the proliferation of bloodstream *T. brucei* cell line 449 with EC_{50} values down to 2 μM.

It should also be once again underlined that, while in silico and wet high HTS are based on a common philosophy (both try to explore vast chemical libraries in an efficient manner), wet HTS is intrinsically more resource demanding. It not only requires physical samples of the whole library submitted to screening but also requires expensive screening platforms, which are in general not available in those countries where neglected diseases are endemic. In contrast, platforms to perform in silico are almost universally available at present.

7.4 Natural Products

As we mentioned repeatedly through the chapter, many of the modern strategies for drug discovery are based on the screening of synthetic chemical libraries. Natural products are a source of unique molecular diversity; for centuries, natural substances, particularly from plants, have been used to treat diseases and active compounds from natural origin have been crucial to developing most of the available pharmaceutical agents.[73]

By systematically screening 522 extracts from North American plant species, Surendra et al. found 125 dose-dependent active plant extracts against *T. brucei* (IC_{50} <10 μg/mL) with low toxicity against mammalian cells. Eight of their hits showed IC_{50} <1 ug/mL.[74] Another example described in the literature is the screening of 5976 microbial extracts of a natural product collection (including filamentous fungi, actinomycetes, and unicellular bacteria).[75] This subset of extracts was previously found to be nontoxic on HepG2. Using an HTS platform established for the identification of potentially novel natural product-based drug leads against 3 NTD (HAT, Chagas, and leishmaniasis), 228 extracts (3.82%) were initially identified as anti-*T. cruzi* hits, 134 (2.24%) as anti-*L. donovani* hits, and 91 (1.52%) as anti-*T. b. brucei* hits.

The number of studies focusing on natural products for the treatment of neglected conditions is growing. Hopefully, these studies will generate more interest in natural compounds as a source for novel chemotypes against NTD. It is worth mentioning that, depending on the availability of the active species, their yield of bioactive molecules, and their growing conditions, natural products could provide economic and sustainable solutions for neglected diseases, a nontrivial aspect if we take in mind the socioeconomic dimension of the affected populations.

7.5 Drug Design and Computer-Guided Drug Design

Wet and in silico screening, as described in previous sections, generally represent systematic ways of exploring the vast natural and/or synthetic chemical universe to find starting points to implement a drug design campaign. Drug design is intrinsically and unequivocally related to finding molecular novelty, that is, novel chemical entities. In its most common form, drug design encompasses modification of a known active scaffold (molecular optimization) or linking known active scaffolds, although de novo drug design (from scratch) is also viable (e.g., fragment growing approximations). Computer-assisted methods have gained a prominent role in making decisions on a rational basis regarding which chemical modifications are more opportune to introduce to active molecules. For that purpose, drug designers resort to available information on pursued molecular targets (direct approaches) or, alternatively, to the identification of structural patterns among a series of known active compounds (indirect approaches). It is worth emphasizing that chemical structural novelty is a crucial factor to drive investment from the pharmaceutical sector. Novelty is a fundamental requisite to obtain intellectual property rights on an invention (and thus exclusivity). Whereas investment return is not, in general, the main expectation when implementing drug discovery and development programs for neglected conditions (in fact, most of the drugs currently at clinical trials or already approved for the treatment of neglected diseases have emerged from repurposing campaigns),[15] the search for chemical novelty could foster private investment in treatments for neglected conditions, which is still rather marginal.

For instance, using a hybrid compound approach and a structure activity relationship (SAR) study, Havrylyuk et al. discovered new hits against *T. brucei.*[76] New 4-thiazolidinone-based compounds conjugated with a pyrazoline moiety were assayed for their trypanocidal activity. Two of these thiazolidinone-pyrazoline hybrids were found to be the most active derivatives. Later, an SAR study for the new anti-*T. brucei* agents was conducted, which included structural modifications such as modification of the substituents on the pyrazoline moiety, modifications of the thiazolidinone N3 position, and elongation of the linker between the two heterocycles. Authors noticed that when modifying thiazolidinone N3 position with methyl and aryl groups, the activity improved remarkably.

7.6 Avoiding Drug Resistance in HAT

Regarding the search for novel therapeutics to overcome drug-resistance issues, it is important to note that the resistance mechanisms developed by *T. brucei* are sometimes related, with a diminished uptake or increased efflux of the drug due to mutations or altered expression of membrane transporters.[37,77,78] One recent report involving melarsoprol includes the occlusion of the compound within a cyclodextrin (cyclic oligosaccharides, with a hydrophilic outer surface and a hydrophobic interior cavity).[37] The resulting complexes (drug-cyclodextrin) retained trypanocidal activities in vitro and cured the late stage of the disease in a murine model. The melarsoprol-cyclodextrin complexes enter into the parasite probably by endocytosis, unlike free melarsoprol, which uses the aqua glyceroporin-2-channel. This "Trojan horse" strategy (hiding/encapsulating the therapeutic agent from detoxicating mechanisms in the parasite or to provide alternative cell uptake routes) could prove useful in evading resistance issues of a pharmacokinetic nature (e.g., reduced cell uptake due to upregulation of efflux transporters or downregulation of influx transporters).

The hybrid agents are a modern approach for drug discovery of new drugs with less probability of target-related drug resistance. They are multitarget molecules obtained by linking two (or more) drugs or active scaffolds in a single molecule. This new perspective provides a novel structure molecule to attack two or more molecular targets simultaneously.[79] For instance, Gehrke et al. reported the synthesis of hybrid compounds by linking the 3-hydroxypyridin-4-one (HPO) to the 4-aminoquinoline ring system present in the antimalarial drug chloroquine (CQ).[77] They described inhibitory effects of these hybrids against four parasitic protozoa: *T. brucei*, *T. cruzi*, *L. infantum*, and *P. falciparum*. These new molecules can chelate iron, which has been proposed as a possible new strategy to combat parasitic infections.[80] Iron chelators can compromise the activity of Fe^{+3} containing enzymes in parasites, affecting the DNA synthesis. One of the new molecules shows a clearly improved activity against *T. brucei* (>25-fold compared with the reference compound, and about fourfold compared with CQ).[81] These results suggest that, at least against trypanosomes, the hybrid strategy leads to synergistic effects between the HPO and CQ fragments while retaining reasonable selectivity indices.

8. CONCLUSIONS

Throughout the chapter, we have detailed the major features of *T. brucei* as well as the disease it causes. It is known that this disease mainly affects low-income populations and explains why the pharmaceutical industry has historically been reluctant to invest time and material resources to find cures, due to the limited investment return. As a result, the major investors of HAT research have been public organizations, the academic sector, and nonprofit organizations. Regrettably, the barriers imposed by *T. brucei* for the development of vaccines and treatments that provide disease control in a sustainable manner have not yet been fully overcome. History suggests that any new attempt to eliminate the disease will only succeed if commitment is sustained through appropriate investments, efficacious tools and approaches, effective deployment, and program adaptation to local conditions.[7]

In order to make the most from the limited investment from the public and nonprofit sectors, research should mostly focus on translationally efficient approximations that enhance the chance of success at clinical trials, reducing the drug development timeframe, and building on already-known information on existing drugs. Among such methods, drug repurposing and the design of novel pharmaceutical carriers for already-approved HAT therapies seem particularly appealing. For instance, drug repurposing could result in the identification of new trypanocidal agents among off-patent, discontinued, or abandoned drugs. Novel pharmaceutical formulations, in turn, can provide improved pharmacokinetics for existing HAT drugs. Screening of natural products could also be useful, especially if they lead to identification of accessible active natural products (e.g., well-spread or easily growing natural drugs). On the other hand, implementing drug discovery technologies that lead to chemical novelty could result in interesting alternatives to attract private investment.

We would like to emphasize that, whereas many novel drug discovery technologies are appealing, some of them involve a greater technological gap than others. Because the involvement of endemic countries in the active search of therapeutic solutions is desirable and necessary, they should probably focus on accessible technologies and resources. In this regard, computer-guided and other knowledge-based approximations should be preferred over more technologically and resource-demanding ones (e.g., brute force approximations such as HTS).

Eradication of NTD requires research and actions from multiple fronts, including developing different approaches for vector control, efficient delivery of existing drugs, and overcoming or reducing drug-resistance issues.

REFERENCES

1. Word Health Organization. Available from: http://www.who.int/mediacentre/factsheets/fs259/en/; 2017.
2. Stijlemans B, Caljon G, Van Den Abbeele J, Van Ginderachter JA, Magez S, De Trez C. Immune evasion strategies of *Trypanosoma brucei* within the mammalian host: progression to pathogenicity. *Front Immunol* 2016;**7**:233. https://doi.org/10.3389/fimmu.2016.00233.
3. Welburn SC, Molyneux DH, Maudlin I. Beyond tsetse—implications for research and control of human African trypanosomiasis epidemics. *Trends Parasitol* 2016;**32**(3):230–41.
4. Perdomo D, Bonhivers M, Robinson D. The trypanosome flagellar pocket collar and its ring forming protein—TbBILBO1. *Cell* 2016;**5**(1):9.
5. Field MC, Carrington M. The trypanosome flagellar pocket. *Nat Rev Microbiol* 2009;**7**(11):775–86.
6. Steverding D. The history of African trypanosomiasis. *Parasit Vectors* 2008;**1**(1):3. https://doi.org/10.1186/1756-3305-1-3.
7. Steinmann P, Stone CM, Sutherland CS, Tanner M, Tediosi F. Contemporary and emerging strategies for eliminating human African trypanosomiasis due to *Trypanosoma brucei gambiense*. *Trop Med Int Health* 2015;**20**(6):707–18 [Review].
8. Matthews KR. The developmental cell biology of *Trypanosoma brucei*. *J Cell Sci* 2005;**118**:283–90.
9. Langousis G, Shimogawa MM, Saada EA, Vashisht AA, Spreafico R, Nager AR, et al. Loss of the BBSome perturbs endocytic trafficking and disrupts virulence of *Trypanosoma brucei*. *Proc Natl Acad Sci U S A* 2016;**113**(3):632–7.
10. Imhof S, Roditi I. The social life of African trypanosomes. *Trends Parasitol* 2015;**31**(10):490–8.
11. Morrison LJ, Vezza L, Rowan T, Hope JC. Animal African trypanosomiasis: time to increase focus on clinically relevant parasite and host species. *Trends Parasitol* 2016;**32**(8):599–607.
12. Jackson AP, Berry A, Aslett M, Allinson HC, Burton P, Vavrova-Anderson J, et al. Antigenic diversity is generated by distinct evolutionary mechanisms in African trypanosome species. *PNAS* 2012;**109**(9):3416–21. https://doi.org/10.1073/pnas.1117313109.
13. Nagle AS, Khare S, Kumar AB, Supek F, Buchynskyy A, Mathison CJN, et al. Recent developments in drug discovery for leishmaniasis and human African trypanosomiasis. *Chem Rev* 2014;**114**:11305–47.

14. Yang S, Wenzler T, Miller PN, Wu H, Boykin DW, Brun R, et al. Pharmacokinetic comparison to determine the mechanisms underlying the differential efficacies of cationic diamidines against first- and second-stage human African trypanosomiasis. *Antimicrob Agents Chemother* 2014;**58**(7):4064–74.
15. Sbaraglini M, Vanrell M, Bellera C, Benaim G, Carrillo C, Talevi A, et al. Neglected tropical protozoan diseases: drug repositioning as a rational option. *Curr Top Med Chem* 2016;**16**(19):2201–22.
16. Nok A. Arsenicals (melarsoprol), pentamidine and suramin in the treatment of human African trypanosomiasis. *Parasitol Res* 2003;**90**(1):71–9.
17. Kennedy P. Human African trypanosomiasis of the CNS: current issues and challenges. *J Clin Invest* 2004;**113**(4):496–504.
18. Nwaka S, Hudson A. Innovative lead discovery strategies for tropical diseases. *Nat Rev Drug Discov* 2006;**5**(11):941–55.
19. Bouteille B, Oukem O, Bisser S, Dumas M. Treatment perspectives for human African trypanosomiasis. *Fundam Clin Pharmacol* 2003;**17**(2):171–81.
20. Patterson S, Wyllie S. Nitro drugs for the treatment of trypanosomatid diseases: past, present, and future prospects. *Trends Parasitol* 2014;**30**(6):289–98.
21. O'Shea IP, Shahed M, Aguilera-Venegas B, Wilkinson SR. Evaluating 5-nitrothiazoles as trypanocidal agents. *Antimicrob Agents Chemother* 2016;**60**(2):1137–40.
22. Priotto G, Kasparian S, Mutombo W, Ngouama D, Ghorashian S, Arnold U, et al. Nifurtimox-eflornithine combination therapy for second-stage African *Trypanosoma brucei gambiense* trypanosomiasis: a multicentre, randomised, phase III, non-inferiority trial. *Lancet* 2017;**374**(9683):56–64. https://doi.org/10.1016/S0140-6736(09)61117-X.
23. Clinical Trials. *Human African trypanosomiasis: first-in-man clinical trial of a new medicinal product, SCYX-7158*. Available from:https://clinicaltrials.gov/ct2/show/NCT01533961; 2017.
24. Clinical Trails. *Pivotal study of fexinidazole for human African trypanosomiasis in stage 2*. Available from: https://www.clinicaltrials.gov/ct2/show/NCT01685827; 2017.
25. Jacobs R, Nare B, Wring S, Orr M, Chen D, Sligar J, et al. SCYX-7158, an orally-active benzoxaborole for the treatment of stage 2 human African trypanosomiasis. *PLoS Negl Trop Dis* 2011;**5**(6).
26. Sauter D, Goldfrank L, Charash B. Cardiopulmonary arrest following an infusion of calcium 2-amino ethanol phosphate. *J Emerg Med* 1990;**8**(6):717–20.
27. Eperon G, Balasegaram M, Potet J, Mowbray C, Valverde OCF. Treatment options for second-stage gambiense human African trypanosomiasis. *Expert Rev Anti-Infect Ther* 2014;**12**(11):1407–17. https://doi.org/10.1586/14787210.2014.959496.
28. Burri C. Chemotherapy against human African trypanosomiasis: is there a road to success? *Parasitology* 2010;**137**(14):1987–94.
29. Clinical Trials. *Eflornithine + nifurtimox late-stage human African trypanosomiasis (HAT) in West Nile, Uganda (NECS)*. Available from:https://clinicaltrials.gov/ct2/show/NCT00489658; 2017.
30. Clinical Trials. *Randomized clinical trial of three drug combinations for late-stage gambiense human African trypanosomiasis*. Available from:https://clinicaltrials.gov/ct2/show/NCT00330148; 2017.
31. Clinical Trials. *Clinical study to assess the tolerability, feasibility and effectiveness of nifurtimox and eflornithine (NECT) for the treatment of Trypanosoma brucei gambiense human African trypanosomiasis (HAT) in the meningo- encephalitic phase II*. Available from: https://clinicaltrials.gov/ct2/show/NCT00906880; 2017.
32. Yun O, Priotto G, Tong J, Flevaud L, Chappuis F. NECT is next: implementing the new drug combination therapy for *Trypanosoma brucei gambiense* sleeping sickness. *PLoS Negl Trop Dis* 2010;**4**(5).
33. Nare B, Wring S, Bacchi C, Beaudet B, Bowling T, Brun R, et al. Discovery of novel orally bioavailable oxaborole 6-carboxamides that demonstrate cure in a murine model of late-stage central nervous system african trypanosomiasis. *Antimicrob Agents Chemother* 2010;**54**(10):4379–88.
34. Tarral A, Blesson S, Mordt OV, Torreele E, Sassella D, Bray MA, et al. Determination of an optimal dosing regimen for fexinidazole, a novel oral drug for the treatment of human African trypanosomiasis: first-in-human studies. *Clin Pharmacokinet* 2014;**53**(6):565–80.
35. Tweats D, Bourdin Trunz B, Torreele E. Genotoxicity profile of fexinidazole a drug candidate in clinical development for human African trypanomiasis (sleeping sickness). *Mutagenesis* 2012;**27**(5):523–32. https://doi.org/10.1093/mutage/ges015.
36. Torreele E, Bourdin Trunz B, Tweats D, Kaiser M, Brun R, Mazué G, et al. Fexinidazole-a new oral nitroimidazole drug candidate entering clinical development for the treatment of sleeping sickness. *PLoS Negl Trop Dis* 2010;**4**(12).

37. Garcia-Salcedo JA, Unciti-Broceta JD, Valverde-Pozo J, Soriano M. New approaches to overcome transport related drug resistance in trypanosomatid parasites. *Front Pharmacol* 2016;**7**. https://doi.org/10.3389/fphar.2016.00351.
38. Arias JL, Unciti-Broceta JD, Maceira J, Del Castillo T, Hernández-Quero J, Magez S, et al. Nanobody conjugated PLGA nanoparticles for active targeting of African trypanosomiasis. *J Control Release* 2014;**197C**:190–8.
39. Soni S, Ruhela RK, Medhi B. Nanomedicine in central nervous system (CNS) disorders: a present and future prospective. *Adv Pharm Bull* 2016;**6**(3):319–35. https://doi.org/10.15171/apb.2016.044.
40. Olbrich G, Gessner A, Kayser O, Müller R. Lipid-drug-conjugate (LDC) nanoparticles as novel carrier system for the hydrophilic antitrypanosomal drug diminazenediaceturate. *J Drug Target* 2002;**10**(5):387–96. https://doi.org/10.1080/1061186021000001832.
41. Attasart P, Boonma S, Sunintaboon P, Tanwilai D, Pothikasikorn J, Noonpakdee WT. Inhibition of *Plasmodium falciparum* proliferation in vitro by double-stranded RNA nanoparticle against malaria topoisomerase II. *Exp Parasitol* 2016;**164**:84–90.
42. Murugan K, Panneerselvam C, Samidoss CM, Madhiyazhagan P, Suresh U, Roni M, et al. In vivo and in vitro effectiveness of *Azadirachta indica*-synthesized silver nanocrystals against *Plasmodium berghei* and *Plasmodium falciparum*, and their potential against malaria mosquitoes. *Res Vet Sci* 2016;**106**:14–22.
43. Kroubi M, Daulouede S, Karembe H, Jallouli Y, Howsam M, Mossalayi D, et al. Development of a nanoparticulate formulation of diminazene to treat African trypanosomiasis. *Nanotechnology* 2010;**21**(50). https://doi.org/10.1088/0957-4484/21/50/505102.
44. Sbaraglini ML, Bellera CL, Fraccaroli L, Larocca L, Carrillo C, Talevi A, et al. Novel cruzipain inhibitors for the chemotherapy of chronic Chagas disease. *Int J Antimicrob Agents* 2016;**48**(1):91–5. https://doi.org/10.1016/j.ijantimicag.2016.02.018.
45. Bellera CL, Balcazar DE, Alberca L, Labriola CA, Talevi A, Carrillo C. Application of computer-aided drug repurposing in the search of new Cruzipain inhibitors: discovery of amiodarone and bromocriptine inhibitory effects. *J Chem Inf Model* 2013;**53**(9):2402–8. https://doi.org/10.1021/ci400284v.
46. Bellera CL, Balcazar DE, Vanrell MC, Casassa AF, Palestro PH, Gavernet L, et al. Computer-guided drug repurposing: identification of trypanocidal activity of clofazimine, benidipine and saquinavir. *Eur J Med Chem* 2015;**93**:338–48.
47. Jamal R, Shimogawara R, Yamamoto K, Ohta N. Anti-trypanosome effects of nutritional supplements and vitamin D3: in vitro and in vivo efficacy against *Trypanosoma brucei brucei*. *Trop Med Health* 2016;**44**(1):26. https://doi.org/10.1186/s41182-016-0024-7.
48. Thompson AM, O'Connor PD, Blaser A, Yardley V, Maes L, Gupta S, et al. Repositioning antitubercular 6-nitro-2,3-dihydroimidazo[2,1-b][1,3]oxazoles for neglected tropical diseases: structure–activity studies on a preclinical candidate for visceral leishmaniasis. *J Med Chem* 2016;**59**(6):2530–50.
49. Alberca LN, Sbaraglini ML, Balcazar D, Fraccaroli L, Carrillo C, Medeiros A, et al. Discovery of novel polyamine analogs with anti-protozoal activity by computer guided drug repositioning. *J Comput Aided Mol Des* 2016;**30**(4):305–21.
50. Knox C, Law V, Jewison T, Liu P, Ly S, Frolkis A, et al. DrugBank 3.0: a comprehensive resource for "Omics" research on drugs. *Nucleic Acids Res* 2011;**39**(1):D1035–41. https://doi.org/10.1093/nar/gkq1126.
51. Novick P, Ortiz O, Poelman J, Abdulhay A, Pande V. SWEETLEAD: an in silico database of approved drugs, regulated chemicals, and herbal isolates for computer-aided drug discovery. *PLoS One* 2013;**8**(11).
52. Rashmi M, Swati D. In silico drug re-purposing against African sleeping sickness using GlcNAc-PI de-*N*-acetylase as an experimental target. *Comput Biol Chem* 2015;**59**:87–94. https://doi.org/10.1016/j.compbiolchem.2015.09.010.
53. Kaiser M, Maes L, Tadoori LP, Spangenberg T, Ioset J-R. Repurposing of the open access malaria box for kinetoplastid diseases identifies novel active scaffolds against trypanosomatids. *J Biomol Screen* 2015;**20**(5):634–45.
54. Ingram-Sieber K, Cowan N, Panic G, Vargas M, Mansour N, Bickle Q, et al. Orally active antischistosomal early leads identified from the open access malaria box. *PLoS Negl Trop Dis* 2014;**8**(1).
55. Bessoff K, Spangenberg T, Foderaro JE, Jumani RS, Ward GE, Huston CD. Identification of cryptosporidium parvum active chemical series by repurposing the open access malaria box. *Antimicrob Agents Chemother* 2014;**58**(5):2731–9.

56. Thomas SM, Purmal A, Pollastri M, Mensa-Wilmot K. Discovery of a carbazole-derived lead drug for human African trypanosomiasis. *Sci Rep* 2016;**6**(1):32083. Available from. http://www.nature.com/articles/srep32083
57. Jones AJ, Kaiser M, Avery VM. Identification and characterization of FTY720 for the treatment of human african trypanosomiasis. *Antimicrob Agents Chemother* 2016;**60**(3):1859–61.
58. Amata E, Bland ND, Hoyt CT, Settimo L, Campbell RK, Pollastri MP. Repurposing human PDE4 inhibitors for neglected tropical diseases: design, synthesis and evaluation of cilomilast analogues as *Trypanosoma brucei* PDEB1 inhibitors. *Bioorg Med Chem Lett* 2014;**24**(17):4084–9. https://doi.org/10.1016/j.bmcl.2014.07.063.
59. Szymański P, Markowicz M, Mikiciuk-Olasik E. Adaptation of high-throughput screening in drug discovery-toxicological screening tests. *Int J Mol Sci* 2012;**13**(1):427–52.
60. Harris CJ, Hill RD, Sheppard DW, Stouten MJ, Stouten PF. The design and application of target-focused compound libraries. *Comb Chem High Throughput Screen* 2011;**14**(6):521–31. https://doi.org/10.2174/138620711795767802.
61. Welsch ME, Snyder SA, Stockwell BR. Privileged scaffolds for library design and drug discovery. *Curr Opin Chem Biol* 2010;**14**(3):347–61.
62. Watanabe Y, Kawaguchi K, Saito S, Okabe T, Yonesu K, Egashira S, et al. An HTRF based high-throughput screening for discovering chemical compounds that inhibit the interaction between *Trypanosoma brucei* Pex5p and Pex14p. *Biochem Biophys Rep* 2016;**6**:260–5.
63. Valenciano AL, Ramsey AC, Santos WL, Mackey ZB. Discovery and antiparasitic activity of AZ960 as a *Trypanosoma brucei* ERK8 inhibitor. *Bioorg Med Chem* 2016;**24**(19):4647–51. https://doi.org/10.1016/j.bmc.2016.07.069.
64. Diaz R, Luengo-Arratta SA, Seixas JD, Amata E, Devine W, Cordon-Obras C, et al. Identification and characterization of hundreds of potent and selective inhibitors of *Trypanosoma brucei* growth from a kinase-targeted library screening campaign. *PLoS Negl Trop Dis* 2014;**8**(10). https://doi.org/10.1371/journal.pntd.0003253.
65. Urbaniak MD, Capes AS, Crossman A, O'Neill S, Thompson S, Gilbert IH, et al. Fragment screening reveals salicylic hydroxamic acid as an inhibitor of *Trypanosoma brucei* GPI GlcNAc-PI de-*N*-acetylase. *Carbohydr Res* 2014;**387**(1):54–8. https://doi.org/10.1016/j.carres.2013.12.016.
66. Benítez D, Medeiros A, Fiestas L, Panozzo-Zenere EA, Maiwald F, Prousis KC, et al. Identification of novel chemical scaffolds inhibiting trypanothione synthetase from oathogenic trypanosomatids. *PLoS Negl Trop Dis* 2016;**10**(4).
67. Ripphausen P, Nisius B, Peltason L, Bajorath J. Quo vadis, virtual screening? a comprehensive survey of prospective applications. *J Med Chem* 2010;**53**(24):8461–7. https://doi.org/10.1021/jm101020z.
68. Zhu T, Cao S, P-C S, Patel R, Shah D, Chokshi HB, et al. Hit identification and optimization in virtual screening: practical recommendations based on a critical literature analysis. *J Med Chem* 2013;**56**(17):6560–72. https://doi.org/10.1021/jm301916b.
69. Tatipaka HB, Gillespie JR, Chatterjee AK, Norcross NR, Hulverson MA, Ranade RM, et al. Substituted 2—phenylimidazopyridines: a new class of drug leads for human African trypanosomiasis. *J Med Chem* 2014;**57**:828–35.
70. Buchynskyy A, Gillespie JR, Hulverson MA, McQueen J, Creason SA, Ranade RM, et al. Discovery of *N*-(2-aminoethyl)-*N*-benzyloxyphenyl benzamides: new potent *Trypanosoma brucei* inhibitors. *Bioorg Med Chem* 2017;**25**(5):1571–84.
71. Huang W, Zhang Z, Barros-Álvarez X, Koh CY, Ranade RM, Gillespie JR, et al. Structure-guided design of novel *Trypanosoma brucei* methionyl-tRNA synthetase inhibitors. *Eur J Med Chem* 2016;**124**:1081–92.
72. Beig M, Oellien F, Garoff L, Noack S, Luise Krauth-Siegel R, Selzer PM. Trypanothione reductase: a target protein for a combined in vitro and in silico screening approach. *PLoS Negl Trop Dis* 2015;**9**(6). https://doi.org/10.1371/journal.pntd.0003773.
73. Cheuka MP, Mayoka G, Mutai P, Chibale K. The role of natural products in drug discovery and development against neglected tropical diseases. *Molecules* 2017;**22**(1):58. https://doi.org/10.3390/molecules22010058.
74. Jain S, Jacob M, Walker L, Tekwani B. Screening North American plant extracts in vitro against *Trypanosoma brucei* for discovery of new antitrypanosomal drug leads. *BMC Complement Altern Med* 2016;**16**(1):131. https://doi.org/10.1186/s12906-016-1122-0.

75. Annang F, Pérez-Moreno G, García-Hernández R, Cordon-Obras C, Martín J, Tormo JR, et al. High-throughput screening platform for natural product-based drug discovery against 3 neglected tropical diseases. *J Biomol Screen* 2015;**20**(1):82–91. https://doi.org/10.1177/1087057114555846.
76. Havrylyuk D, Zimenkovsky B, Karpenko O, Grellier P, Lesyk R. Synthesis of pyrazoline-thiazolidinone hybrids with trypanocidal activity. *Eur J Med Chem* 2014;**85**:245–54. https://doi.org/10.1016/j.ejmech.2014.07.103.
77. Baker N, de Koning HP, Mäser P, Horn D. Drug resistance in African trypanosomiasis: the melarsoprol and pentamidine story. *Trends Parasitol* 2017;**29**(3):110–8. https://doi.org/10.1016/j.pt.2012.12.005.
78. Munday JC, Settimo L, de Koning HP. Transport proteins determine drug sensitivity and resistance in a protozoan parasite, *Trypanosoma brucei*. *Front Pharmacol* 2015;**6**:1–10.
79. Sbaraglini ML, Talevi A. Hybrid compounds as anti-infective agents. *Curr Top Med Chem* 2017; **17**:1080–95. https://doi.org/10.2174/1568026616666160927160912.
80. Taylor MC, Kelly JM. Iron metabolism in trypanosomatids, and its crucial role in infection. *Parasitology* 2010;**137**(6):899–917.
81. Gehrke SS, Pinto EG, Steverding D, Pleban K, Tempone AG, Hider RC, et al. Conjugation to 4-aminoquinoline improves the anti-trypanosomal activity of deferiprone-type iron chelators. *Bioorg Med Chem* 2013;**21**(3):805–13.

CHAPTER 15

A Global Problem of Toxoplasmosis

Rabia Cakir-Koc, Burak Özdemir
Yildiz Technical University, Istanbul, Turkey

1. INTRODUCTION

Toxoplasma gondii is an intracellular parasite and member of the Apicomplexan family. *T. gondii* was first discovered by Nicole and Manceaux in rodents from North Africa in 1908.[1,2] This parasite is a heteroxenous protozoon and can be transmitted by several routes from one host to another.[3] Pets, farm animals, wild animals, and people can be infected by this parasite.[4]

T. gondii has three life forms: tachyzoite, bradyzoite, and oocyst.[5] Tachyzoite and Bradyzoite are found in humans and other animals[6] but oocyte is found only in feline and Felidae.[7] Tachyzoite is an intracellular form with 2–4 μm dimensions.[8] This is an invasive parasite, responsible for acute infections.[9] All mammalian cells that have nuclei can be infected by tachyzoites.[10] Tachyzoites destroy cell membranes through mechanical or chemical ways, penetrating into the cells. They reside in one vacuole after penetrating the cell and then replicate in the cell.[11] Tachyzoites have sensitivity to heat, cold, and gastric fluids.

Bradyzoites are found in tissue cysts.[12] The diameter of the tissue cyst can be changed between 10 and 200 μm; the tissue cyst includes between 50 and 3000 bradyzoites.[11] Cysts can stay in cells belonging to all kind of tissues for many years without an immune response[13] and are responsible for latent infections.[14]

Oocysts are found only in felines.[15] The oocysts have a thick-walled structure with dimensions of 10–12 μm.[11] Tissue cysts and oocysts replicate with enteroepithelial asexual reproduction (schizogony); sexual reproduction happens after the invasion of epithelial cells in a feline's intestine.[16] Oocysts replicate by sporogony and form sporocysts.[17] After that, each sporocyts is replicated four times, forming sporozoites.[18] Infected felines shed 10 million oocysts into the environment. These oocysts are transformed into sporozoite in 1–7 days in humid soil. Sporozoites will have infectious properties in one day or one week.[19] Oocysts are released into the environment in 3–21 days after inoculation.[16]. Oocysts can live for two years in a humid environment, but they are sensitive to heat above 66°C.[20]

Infectious oocysts contaminate soil and water, which infects animals such as birds, sheep, and cows. Most of these animals develop chronic infections during which their

The Microbiology of Central Nervous System Infections
https://doi.org/10.1016/B978-0-12-813806-9.00015-9

tissues include tissue cysts. When infectious spores are uptaken via an oral route, tachyzoites are formed. These tachyzoites are invasive to all organs such as muscles, liver, the central nervous system (CNS), lymph nodes, and the spleen.[4]

1.1 Epidemiology and Transmission

T. gondii is widely spread around the world.[21,22] The prevalence of *T. gondii* varies from region to region depending on climate and seasonal changes.[23] Seroprevalance of *T. gondii is* low in cold, dry climates and higher altitudes.[24] Seroprevalance of *T. gondii* is very high in France, Tahiti, and Africa.[25] In France, 80% of pregnant women have antitoxoplasma antibodies[26]; this rate is 16%–40% in the United States and England. The rate of *T. gondii* seroprevalance varies between 29.8% and 68.9% in Turkey.[27,28]

T. gondii can be transmitted by several ways but the most important transmission route is feline feces. Feline feces can contaminate soil, water, and plants. Animals can be infected mostly by consuming contaminated water and plants. Then, sporozoites transform to tachyzoites and infect all tissues and organs. People can be infected with *T. gondii* by undercooked meat, raw meat, contaminated plants, and water. Children can also be infected by contact with contaminated soil.[29] Contact with pets (especially cats) or livestock is another route of transmission of *T. gondii*.[3,30]

2. DIAGNOSIS OF TOXOPLASMOSIS METHODS

Symptoms of *T. gondii* infection are not specific and change depending on the infected organs. Several methods must be used for supporting clinical symptoms and establishing a final diagnosis. The diagnosis methods of *T. gondii* infection are classified as direct methods and indirect methods.[31]

2.1 Direct Diagnosis Methods

Direct diagnosis methods depend on the life forms of *T. gondii*. Therefore, biopsies of tissues, blood, bodily fluids, and feces are examined for the presence of tachyzoites and bradyzoites.[32] Direct diagnosis methods include isolation of *T. gondii*, histopathological diagnosis, and a polymerase chain reaction (PCR) [31].

Histopathological diagnosis methods determine tachyzoites in samples of patients or detect bradyzoites from tissue cysts in tissue biopsies.[32]

The PCR method is used for determining the DNA of *T. gondii* in tissues and bodily fluids. The specific gene region of *T. gondii* must be selected and amplified.[32]

A direct fluorescent antibody test is useful for detecting the presence of antigens in patients. Antibodies are labeled with a fluorescent dye and samples are treated with fluorescent-labeled antibodies. If antigens are found in the sample, fluorescent-labeled antibodies will bind with these antigens; this binding can be detected by microscope.[33] This method is used to detect the presence of *T. gondii*.[34]

2.2 Indirect Diagnosis Methods

Direct diagnosis methods may not be enough for determining a *T. gondii* infection because they depend on life forms and the route of infection. But the immune system generates specific antibodies against every infection. Indirect diagnosis methods generally aim to determine antibody response in patients.[31] Serological methods are generally used for different infection agents due to ease, specificity, and sensitivity.[35]

The Sabin Feldman dye test is the most specific reference assay. Live parasites are treated with antibodies and complement-like cofactors. Parasites are stained and evaluated after exposure to serum titers. This test has high sensitivity and specificity compared to other tests. For this reason, this test is recommended as a reference test.[36]

The enzyme linked immunosorbent assay (ELISA) is one of the most commonly used indirect diagnosis methods.[37] In this method, a specific antibody response is detected by using enzyme conjugated secondary antibodies, and optical changes are detected.[38] ELISA is commonly used for diagnosis of *T. gondii* because of its easy application, low cost, and high sensitivity.[39]

The indirect fluorescent antibody test (IFAT) aims to determine levels of parasite-specific antibodies. In IFAT, antiserum that is labeled with fluorescent is used against IgG and IgM.[40]

The agglutination test is based on agglutination of beads coated with *T. gondii* antigens during reaction with specific antibodies.[5] The indirect hemagglutination test depends on agglutination of erythrocytes after the interaction of antigen and antibody.[41] The complement fixation test is another method based on the combined reaction of antigen and antibody.[42]

3. TREATMENT OF TOXOPLASMOSIS

Clinical findings and the severity of disease are important parameters in the selection of treatment. Tachyzoites can be destroyed with specific treatment but cystic forms cannot be eradicated. The optimum time of toxoplasmosis treatment is not well known[12] but treatment should be not less than 2–6 weeks until a regression of symptoms. AIDS patients who are infected by *T. gondii* must use pyrimethamine-sulfadoxin and pyrimethamine-clindamycin during their lifetime.[12] The antibody titer is not used as a parameter for evaluating treatment results because increasing antibody titers after treatment does not indicate the success of treatment.

3.1 Pyrimethamine

Pyrimethamine is an antagonist of folic acid; it includes diaminopyrimidine and inhibits dihydrofolate reductase. It shows synergistic effects with sulphanoamids and they are used together. Pyrimethamine is dissolved in lipids and absorbed in the gastrointestinal system.[43] The critical adverse effect is dose-related irreversible suppression of bone marrow. For this reason, patients must be regularly followed and are advised to use 3–5 mg of folic acid two or three times a week to reduce the suppression of bone marrow. Pregnant women should not use pyrimethamine due to its teratogenic effect.[44,45]

3.2 Sulfadiazine, Sulfamerazine, Sulfamethazine

These drugs inhibit dihydrofolate synthetase and antagonize folic acid synthesis. In the United States, only sulfadiazine is in use. They are excreted from kidneys and have adverse side effects such as allergies, suppression of bone marrow, and toxicity to the kidney and liver.[46]

3.3 Clindamycin

Clindamycin is used primary for treatment of ocular toxoplasmosis, according to several studies.[48,49] Clindamycin is stored in retina pigment epithelia and patients take 300 mg of clindamycin four times a day. The use of the drug should be discontinued in the case of diarrhea due to mortality risks. The most important adverse effect is pseudomembranous colitis. Pseudomembranous colitis is developed under action of toxins produced by *Clostridium difficile*. The use of clindamycin prevents the environment for the propagating of the organism.[47,48]

4. VACCINE STUDIES AGAINST TOXOPLASMOSIS

Several researchers are developing new vaccines against toxoplasmosis. There is only one commercial vaccine against toxoplasmosis: ToxoVax, approved for sheep. Inactivated parasites, crude or purified antigens, recombinant proteins, live and attenuated parasites, DNA, and peptides are potential vaccine candidates against toxoplasmosis.[49]

4.1 Inactivated Parasite

Inactivated parasites were first used by Cutchims and Warren in 1956. These researchers immunized guinea pigs with inactivated tachyzoites and obtained protection against infection with live *T. gondii*.[50] In another study, Krahenbuhl et al. in 1972 fixed the whole *T. gondii* tachyzoites with formalin and immunized mice with these tachyzoites. Survival rates of mice immunized with inactivated tachyzoites were increased.[51]

4.2 Crude, Purified Antigens, Recombinant Proteins

Immunization with SAG1, which was purified by affinity and combined with adjuvants, showed high survival rates and significantly reduced cysts in brains after challenge in mouse.[52–54] In 2006 Louronçe et al. purified MIC1 and MIC4 from a parasite extract;

these purified antigens were used to C57BL/6 mice with brain cysts. Results showed that the reduction rate of brain cysts was 68% and the survival rate was 80%.[55]

Recombinant antigens of *T. gondii* are also used for developing new vaccines against toxoplasmosis. Several studies showed that a vaccination with SAG1 recombinant protein provides protection against acute toxoplasmosis and reduces brain cyst formation.[53,54] In 2004 Martin et al. showed the effect of recombinant GRA4 and ROP2 combined with Alum. The result of this research demonstrated that recombinant GRA4 and ROP2 combined with Alum provided protection against brain cyst formation in mice.[56]

4.3 Live and Attenuated Parasite

In 1983, Waldeland and Frenkel suggested that live or attenuated parasites may have better protection than inactivated life forms.[57] Lindsay et al. reported that the RH strain showed protection in pig models; they did not develop any cysts in tissues during the two weeks after vaccination.[58] In 2004 Kringel et al. suggested that RH tachyzoites combined with CpG adjuvant could protect 52% inoculated pigs and provide sterile immunity against tissue cyst formation. OF1 mice were immunized with a *T. gondii* strain that was attenuated by deletion of the MIC1 and MIC3 gene. After inoculation, it was shown that brain cyst formation was reduced by more than 96% and transplacental transmission was also reduced.[59]

4.4 DNA Vaccine

The DNA vaccines are simple rings of DNA containing a gene encoding an antigen and a promoter/terminator to make the gene expressed in mammalian cells.[60] Unlike conventional protein-based vaccines, DNA vaccines are based on bacterial plasmids that encode vaccine antigens driven by efficient eukaryotic promoters.[61,62] Like protein vaccines, DNA vaccines can be delivered through a variety of different routes including intramuscular, subcutaneous, mucosal, or transdermal delivery. However, unlike protein antigens, the DNA vaccine must gain entry to the cytoplasm of cells at the injection site in order to induce antigen expression in vivo, thereby enabling antigen presentation on major histocompatibility molecules (MHCs) for T-cell recognition. DNA vaccines have been successful in a variety of animal models in preventing or treating infectious diseases, cancer, autoimmunities, or allergies.[43,62] DNA-based vaccination is one of the most promising strategies for the development of new generations of effective vaccines against intracellular parasites such as *T. gondii*.

Chen et al. showed that TgCDPK1 has potential as a vaccine candidate for inducing a strong humoral and cellular response against a *T. gondii* infection by using a constructed eukaryotic plasmid, pVAX-CDPK.[63] In another study, a DNA vaccine (pTgGST) encoding *T. gondii* antioxidant glutathione-*S*-transferase (TgGST) constructed using vector pVAX I triggered strong humoral and cellular responses and induced partial protection against acute toxoplasmosis.[64] A DNA vaccine candidate (including twelve T- and B-cell epitopes from SAG1, GRA2, GRA7, and ROP16) combined with genetic adjuvant pEGFP-RANTES (pRANTES) might be new candidates for further vaccine development against *T. gondii* infections.[65] İmmune responses and protective effects of calcium-dependent protein kinase 1 (TgCDPK1), pVAX-CDPK1, in the presence or absence of IL-7 and IL-15 plasmids (pVAX-IL-7-IL-15), against acute and chronic *T. gondii* infection were analyzed. The results indicated that specific humoral and cellular immune responses were elicited by DNA vaccines against acute and chronic *T. gondii* infection in mice.[66] Rahimi et al. demonstrated that a CaPN-based Cocktail DNA(pcROM4 + pcGRA14) vaccine showed the longest survival time. Also, the DNA cocktail vaccine including adjuvant boosted immune responses and increased protective efficacies against acute toxoplasmosis.[67]

4.5 Peptide Vaccines

Peptides that have 20–30 amino acid sequences are also used in vaccines against toxoplasmosis. Immunogenic peptide molecules representing the specific epitope are used as an antigen.[68–72] Since epitopes are the antigenic determinants within larger proteins, these peptides are considered sufficient for activation of the appropriate cellular and humoral responses[73–75] while eliminating allergenic and/or reactogenicity responses.[68] Additionally, peptide vaccines can be used for the induction of broad-spectrum immunity against multiple serological variants (serovars) or strains. On the other hand, owing to the relatively small size of peptides, they are often weakly immunogenic by themselves and therefore require carrier molecules for chemical stability and adjuvant effects.[76,77]

The host defense against *T. gondii* ultimately depends on a protective Th1-based immune response. In the process of immune response, CD8+ T cells play a major role in protecting against *T. gondii* by IFN-γ secretion and also by direct cytotoxicity.[73] Immunization with a multiple antigenic peptide (MAP) vaccine against *T. gondii* including one B-cell and two T-cell epitopes derived from SAG1, GRA4, and GRA1 showed significantly increased humoral and cellular immune responses and prolonged survival time in comparison with unvaccinated controls.[14] Blanchard et al. (2008) identified peptides derived from the GRA6, which played a role in the Ld restricted immune response to *T. gondii*.[6,78] In addition, several studies identified other peptides that could bind to human MHC class I and induce CD8+ T-cell IFN- γ production and protect against parasite infection.

5. CLINICAL PICTURES OF TOXOPLASMOSIS

Toxoplasmosis occurs all over the world in several forms. *T. gondii*, the main causative agent of toxoplasmosis, is responsible for both animal and human infections.[79,80] Toxoplasmosis causes not only health problems but also economic losses.[81] In people, clinical pictures of toxoplasmosis can be grouped as acquired toxoplasmosis, congenital toxoplasmosis, toxoplasmosis in immunodeficient patients, and ocular toxoplasmosis.[40]

5.1 Acquired Toxoplasmosis

The course of acquired toxoplasmosis is generally subclinical and asymptomatic.[82] Ten to twelve percent of people with acute toxoplasmosis have a flu-like syndrome. In these patients, the symptoms are fever, lymphadenopathy, ailment, myalgia, maculopapular skin lesions, and, with less frequency, hepatosplenomegaly and atypical lymphocytes.[82]

5.2 Congenital Toxoplasmosis

Congenital toxoplasmosis occurs in the transmission of the infection to the fetus via the transplacental route. Seroprevalance of congenital toxoplasmosis is 4.4–4.6/1000 and infection is often transmitted to the fetus during primer infection of pregnant women. Also, the fetus is sometimes infected due to immunosuppression situations and steroids. Most of the infected neonatal do not express any symptoms or clinical findings.[83]

Hydrocephalia, chorioretinitis, and cerebral calcification are seen in serious cases.[12,84] Convulsion, rapid increase of the head diameter, bulging fontanelles, and nystagmus are found in neural locations of *T. gondii*. Toxoplasma-related CNS and eye diseases are developed in the first three months whereas full-term infants may have mild forms of disease with lymphadenopathy and hepatosplenomegaly.[84]

5.3 Ocular Toxoplasmosis

Ocular toxoplasmosis is characterized with focal necrotizing retinitis and a focal scar.[85] Most retinal lesions are bilateral in ocular toxoplasmosis.[86] Congenital toxoplamosis also frequently causes bilateral chorioretinitis. Chorioretinitis is the most frequent finding with a rate of 91% and is associated with other lesions with a rate of 71%.[87] The next most frequent findings of ocular toxoplasmosis are microphthalmia and strabismus.[8] Iridocyclitis, cataracts, and glaucoma are formed at the anterior segment of the eye.[88] Active lesions can be of various magnitude but the shape is usually oval.[89] The retina becomes thick in an acute period.[90] The formation of lesions starts from the retina and then inflammation continues to the choroid because *T. gondii* shows affinity for neural tissue. Activation is usually seen only in one eye every time.[91] Hemianopsia can be seen depending on the location of a lesion. Other eye involvements are papillitis, vitreal inflammation, and defect of neural fibers. Vision becomes normal when inflammation regresses but recovery is not complete.[92]

5.4 Toxoplasmosis in Immunodeficient Patients

Toxoplasma meningoencephalitis is mostly seen in immunodeficient patients. Its symptoms are characterized as changes of consciousness, convulsion, paralysis, and emotional dysregulations. Toxoplasma meningoencephalitis is associated with several psychiatric and CNS diseases such as schizophrenia, obsessive-compulsive disorder, major depression, epilepsy, and suicide attempts.

Toxoplasma can rarely infect such organs as the lungs, eyes, and heart.[32,93] The disease can be formed with latent infection consequently. *T. gondii* generates aspheric one or multiple lesions in CNS in patients who have AIDS. Diagnosis in immunodeficient patients is very difficult because of suppressed antibody production.[14]

6. *T. GONDII* AND THE CNS

T. gondii indicates a strong tropism to the CNS, especially hemispheres, basal ganglions, the cerebellum, and the peduncle. In neuropathological studies, *T. gondii* shows *in vitro* selective effect on neurons, glial cells, and especially astrocytes. In postmortem research, abnormal glial cells and, in particular, a decreased number of astrocytes are detected in brains of schizophrenia patients.[94] Toxoplasmosis causes acute necrotizing encephalitis and serious neuropathological problems with the formation of glial nodules, especially in immunodeficient patients.[95]

Another effect of *T. gondii* on the CNS is a change in neurotransmitter pathways. This intracellular parasite affects the synthesis of dopamine, norepinephrine, and other neurotransmitters.[96] The level of neurotransmitters—especially the ones such as dopamine glutamate and gamma amino butyric acid—are abnormal in schizophrenia patients.[97] The relationship between the infection of *T. gondii* and the mechanism of increased dopamine levels is not fully known.[95]

During the toxoplasma infection, astrocytes became activated and this increases the synthesis of kynurenic acid. An increased level of kynurenic acid in the brain inhibits the neurotransmitter receptors of glutamine and nicotine; it is believed that this inhibition causes cognitive symptoms in schizophrenia.[97] In one study, it was emphasized that *T. gondii* causes anomalies in the catecholamine metabolism and these anomalies elicit changes in psychological and motors functions in rodents.[95]

Several researchers mentioned that changes in the level of neurotransmitters cause behavioral changes. In animal models it was shown that an infection of *T. gondii* can cause changes of neurotransmitter levels, which play critical roles on behavioral differences and the pathogenesis of schizophrenia.[98] Congenital toxoplasmosis plays a more critical role on the etiology of schizophrenia than acquired toxoplasmosis.[99] Another study conducted with 875 soldiers who have latent toxoplasmosis showed that toxoplasma-positive soldiers have lower psychomotor performance than noninfected ones.[98]

Another pathway affected by *T. gondii* is the tryptophan metabolic pathway. Tryptophan is an essential amino acid and precursor mediator of serotonin and melatonin.[100] The tachyzoite form of *T. gondii* needs this amino acid for replication. A study emphasized that *T. gondii* has two genes encoding tyrosine hydroxylase, which reduces dopamine synthesis in the brain. These findings show that *T. gondii* can cause behavioral changes in hosts with effects on dopamine and serotonin.[95]

T. gondii is a strong stimulant of the immune system and this parasite directly causes the production of immune system mediators in the brain. Neurodegeneration is caused by neuroinflammation due to released cytokines by cells of the immune system. Tachyzoites cause a more intense inflammatory response related to cytokine in host cells than bradyzoites. Researchers suggested that the propagation of tachyzoites after cysts rupture is associated with the beginning of schizophrenia and other mental diseases.[100,101]

7. CONCLUSION

Toxoplasmosis is one of the most common parasitic diseases in the world because it is transmitted in several ways from one host to another, infecting many mammals. Risk factors, risky regions, and people who are at risk must be determined and an emergency plan should be made against the possibility of a pandemic. Furthermore, people under risk should be educated about sanitary living conditions and environmental hygiene. A diagnosis of toxoplasmosis is very crucial for detecting the stage of the disease and determining treatment methods.

Acquired toxoplasmosis, congenital toxoplasmosis, ocular toxoplasmosis, and toxoplasmosis in immunodeficient patients are clinical types of toxoplasmosis commonly seen in people. Several antibiotics are used for treatment such as pyrimethamine and clindamycine, but these antibiotics have strong side effects and some of them are irreversible. For this reason, in recent years many studies have been undertaken with the goal of developing a vaccine for toxoplasmosis with the help of modern biotechnology. Recombinant protein-based vaccines are one of the studies seeking vaccination against *T. gondii*.[102] On the other hand, peptide-based vaccines have several advantages such as being cheaper compared to others, being manufactured without live organisms, being nontoxic, and having high stability.[103] Another approach for developing new vaccines against *T. gondii* is the DNA vaccine. DNA vaccines have a number of attractive properties that contribute to strong interest in their development.[104] These vaccines have some problems such as degradation and delivery. In recent years, researchers have used nanobiomaterials, which are characterized as small structures for drug delivery systems and tissue engineering because of their properties.[105]

T. gondii indicates a strong tropism to the CNS, especially in hemispheres, basal ganglions, the cerebellum, and the peduncle. This strong tropism causes several diseases such as schizophrenia, epilepsy, obsessive-compulsive disorder, and major depression. As a consequence of these diseases, suicidal attempts and self-destructive behavior are seen. But the pathways that affect these diseases and behaviors are unknown. Determining these pathways is very important for treating these diseases by developing new drugs and vaccines against toxoplasmosis. New treatment approaches against *T. gondii* can also be developed by using nanotechnological methods based on drug encapsulation into polymeric nanoparticles.[106]

To sum up, toxoplasmosis is a very serious parasitic disease that cannot be ignored. This disease has a wide range of symptoms due to its ability to infect any kind of nucleated cells and a very close relation to the CNS. As mentioned above, antibiotics used against toxoplasmosis have strong adverse effects and these effects can be irreversible. For these reasons, new treatment materials and methods such as peptide vaccines, recombinant protein vaccines, and DNA vaccines should be developed by using modern biotechnology methods.

REFERENCES

1. Collard PJ, Collard P. *The development of microbiology*. Great Britain: Cambridge University Press; 1976.
2. Sastry SA, Bhat S. *Essentials of medical parasitology*. India: Jaypee Brothers, Medical Publishers Pvt. Limited; 2014.
3. Tenter AM, Heckeroth AR, Weiss LM. *Toxoplasma gondii*: from animals to humans. *Int J Parasitol* 2000;**30**(12):1217–58.
4. Dubey J, Jones J. *Toxoplasma gondii* infection in humans and animals in the United States. *Int J Parasitol* 2008;**38**(11):1257–78.
5. Weiss LM. *Toxoplasma gondii: the model apicomplexan—perspectives and methods*. USA: Elsevier Science; 2013.
6. Ho-yen DO, Joss AWL. *Human toxoplasmosis*. UK: Oxford University Press; 1992.
7. Leventhal R, Cheadle RF. *Medical parasitology: a self-instructional text*. USA: F.A. Davis; 2011.
8. Kliegman R, Stanton B, Geme JWS, Schor NF, Behrman RE. *Nelson textbook of pediatrics: the field of pediatrics*. USA: Elsevier; 2016.
9. Ryan SJ. *Retina*. USA: Mosby; 2006.
10. Kurjak A, Chervenak FA. *Textbook of perinatal medicine*. 2 Volume Set. UK: Jaypee Brothers, Medical Publishers Pvt. Limited; 2015.
11. Dubey J, Lindsay D, Speer C. Structures of *Toxoplasma gondii* tachyzoites, bradyzoites, and sporozoites and biology and development of tissue cysts. *Clin Microbiol Rev* 1998;**11**(2):267–99.
12. McMillan JA, Feigin RD, DeAngelis C, Jones MD. *Oski's pediatrics: principles & practice*. USA: Lippincott Williams & Wilkins; 2006.
13. Newsome DA. *Retinal dystrophies and degenerations*. USA: Raven Press; 1988.
14. Bennett JE, Dolin R, Blaser MJ. *Principles and practice of infectious diseases*. USA: Elsevier, Health Sciences Division; 2014.
15. Dubey JP. *Toxoplasmosis of animals and humans*. 2nd ed. USA: CRC Press; 2016.
16. Halonen SK, Weiss LM. Neuroparasitology and tropical neurology. In: *Toxoplasmosis*. Netherlands: Elsevier Science; 2013 [chapter 8].
17. Hausmann K, Hülsmann N, Radek R. *Protistology*. Germany: E. Schweizerbart'sche Verlagsbuchhandlung; 2003.
18. Science UoSP-GFiV. *Feline internal medicine: the Valentine Charlton Course for Veterinarians, 16–20 June 1997*. Australia: Post Graduate Committee in Veterinary Science, University of Sydney; 1997.
19. Dubey J. Advances in the life cycle of *Toxoplasma gondii*. *Int J Parasitol* 1998;**28**(7):1019–24.
20. Mai K, Sharman PA, Walker RA, et al. Oocyst wall formation and composition in coccidian parasites. *Mem Inst Oswaldo Cruz* 2009;**104**(2):281–9.
21. Flegr J. Effects of toxoplasma on human behavior. *Schizophr Bull* 2007;**33**(3):757–60.
22. Cumali G, Süleyman Y, Fahri B. *Toxoplasma gondii* antibodies in type1 diabetes mellitus. *Turkiye Klinikleri J Med Sci* 2008;**28**(6):619–21.
23. Shin D-W, Cha D-Y, Hua QJ, Cha G-H, Lee Y-H. Seroprevalence of *Toxoplasma gondii* infection and characteristics of seropositive patients in general hospitals in Daejeon, Korea. *Korean J Parasitol* 2009;**47**(2):125–30.
24. Cevizci S, Tunga Babaoğlu Ü, Güleç Öyekçin D. *Toxoplasma gondii*, Ruh Sağlığı ve Şizofreni. *TAF Prev Med Bull* 2013;**12**(2):199–208.

25. Stray-Pedersen B. *Toxoplasma gondii*—epidemiology. In: *Parasites of the colder climates*. USA: CRC Press; 2003. p. 149 [chapter 14].
26. Topçu AW, Söyletir G, Doğanay M. *İnfeksiyon hastalıkları ve mikrobiyolojisi*. Nobel Tıp; 2002.
27. Altindis M, Tanir H. Gebe kadinlarda *Toxoplasma gondii* ve Sitomegalovirus antikorlari sikliği. *Genel Tip Dergisi* 2002;**12**(1):9–13.
28. Tamer GS, Dundar D, Caliskan E. Seroprevalence of *Toxoplasma gondii*, rubella and cytomegalovirus among pregnant women in western region of Turkey. *Clin Invest Med* 2009;**32**(1):43–7.
29. Hill DE, Dubey JP. Toxoplasma gondii. In: *Biology of foodborne parasites*. USA: CRC Press; 2015. p. 209–22.
30. Macpherson CN. Human behaviour and the epidemiology of parasitic zoonoses. *Int J Parasitol* 2005; **35**(11):1319–31.
31. Silbert-Flagg JA, Sloand ED. *Pediatric Nurse Practitioner Certification Review Guide*. USA: Jones & Bartlett Learning; 2015.
32. Montoya JG. Laboratory diagnosis of *Toxoplasma gondii* infection and toxoplasmosis. *J Infect Dis* 2002;**185**(Supplement 1):S73–82.
33. Kaplan, Medical K. *USMLE step 1 lecture notes 2017: immunology and microbiology*. USA: Kaplan Pub; 2017.
34. Sert M, Koc RC, Kilinc YB. Novel Fitc-labeled Igy antibody: fluorescence imaging *Toxoplasma gondii* in vitro. *Sci Rep* 2017;**7**:852.
35. Budama-Kilinc Y, Cakir-Koc R. Influenza diagnosis with a specific emphasis on the M2e antigen as a diagnostic tool. In: *Steps forwards in diagnosing and controlling influenza*. Rijeka: InTech; 2016.
36. Balows A, Hausler WJJ, Ohashi M, Turano A. *Laboratory diagnosis of infectious diseases: principles and practice*. New York: Springer; 2012.
37. Budama-Kılınc Y, Cakır-Koc R, Badur S. The development of a universally conserved M2e and hemagglutinin peptide-based elisa method against influenza A. *J Clin Anal Med* 2017;**6**(6):1–4.
38. Cakir-Koc R. Production of anti-SAG1 IgY antibody against *Toxoplasma gondii* parasites and evaluation of antibody activity by ELISA method. *Parasitol Res* 2016;**115**(8):2947–52.
39. Petrikovsky BM. *Fetal disorders: diagnosis and management*. Canada: Wiley; 1999.
40. Miller NR, Walsh FB, Hoyt WF. *Walsh and Hoyt's clinical neuro-ophthalmology*. USA: Lippincott Williams & Wilkins; 2005.
41. Wild D. *The immunoassay handbook*. Netherlands: Elsevier; 2005.
42. Lin CC, Verp M, Sabbagha RE. *The high-risk fetus: pathophysiology, diagnosis, and management*. New York: Springer; 2012.
43. Ovalı Ç, Tugal Tutkun İ. *Toksoplazmozis. İstanbul: İstanbul Tıp Fakültesi Göz Hastalıkları AnaBilim Dalı*. http://www.itf.istanbul.edu.tr/goz/Toksoplazmozis.pdf; 2001.
44. Arcangelo VP, Peterson AM. *Pharmacotherapeutics for advanced practice: a practical approach*. USA: Lippincott Williams & Wilkins; 2006.
45. Ambroise-Thomas P, Petersen E. *Congenital toxoplasmosis: scientific background, clinical management and control*. Paris: Springer; 2013.
46. Aminoff MJ. *Neurology and general medicine*. China: Churchill Livingstone; 2008.
47. Scholar EM, Pratt WB. *The antimicrobial drugs*. USA: Oxford University Press; 2000.
48. Mehlhorn H. *Encyclopedic reference of parasitology: diseases, treatment, therapy*. Springer Berlin Heidelberg; 2001.
49. Taweenan W. *Vaccination against toxoplasmosis;* 2011.
50. Cutchins EC, Warren J. Immunity patterns in the guinea pig following toxoplasma infection and vaccination with killed toxoplasma 1. *Am J Trop Med Hyg* 1956;**5**(2):197–209.
51. Cong H, et al. Toxoplasma gondii HLA-B* 0702-restricted GRA7 20–28 peptide with adjuvants and a universal helper T cell epitope elicits CD8+ T cells producing interferon-γ and reduces parasite burden in HLA-B* 0702 mice. *Hum Immunol* 2012;**73**(1):1–10.
52. Cong H, et al. Human immunome, bioinformatic analyses using HLA supermotifs and the parasite genome, binding assays, studies of human T cell responses, and immunization of HLA-A* 1101 transgenic mice including novel adjuvants provide a foundation for HLA-A03 restricted CD8+ T cell epitope based, adjuvanted vaccine protective against *Toxoplasma gondii*. *Immunome Res* 2010;**6**(1):12.
53. Zhang TE, et al. Protective immunity induced by peptides of AMA1, RON2 and RON4 containing T-and B-cell epitopes via an intranasal route against toxoplasmosis in mice. *Parasit Vectors* 2015;**8**(1):15.

54. Debard N, Buzoni-Gatel D, Bout D. Intranasal immunization with SAG1 protein of *Toxoplasma gondii* in association with cholera toxin dramatically reduces development of cerebral cysts after oral infection. *Infect Immun* 1996;**64**(6):2158–66.
55. Lourenço EV, Bernardes ES, Silva NM, Mineo JR, Panunto-Castelo A, Roque-Barreira M-C. Immunization with MIC1 and MIC4 induces protective immunity against *Toxoplasma gondii*. *Microbes Infect* 2006;**8**(5):1244–51.
56. Martin V, Supanitsky A, Echeverria PC, et al. Recombinant GRA4 or ROP2 protein combined with alum or the gra4 gene provides partial protection in chronic murine models of toxoplasmosis. *Clin Diagn Lab Immunol* 2004;**11**(4):704–10.
57. Waldeland H, Frenkel J. Live and killed vaccines against toxoplasmosis in mice. *J Parasitol* 1983;**69**: 60–5.
58. Lindsay DS, Blagburn BL, Dubey J. Safety and results of challenge of weaned pigs given a temperature-sensitive mutant of *Toxoplasma gondii*. *J Parasitol* 1993;**79**:71–6.
59. Kringel H, Dubey J, Beshah E, Hecker R, Urban J. CpG-oligodeoxynucleotides enhance porcine immunity to *Toxoplasma gondii*. *Vet Parasitol* 2004;**123**(1):55–66.
60. Liu M. DNA vaccines: a review. *J Intern Med* 2003;**253**(4):402–10.
61. Rajčáni J, Moško T, Režuchová I. Current developments in viral DNA vaccines: shall they solve the unsolved? *Rev Med Virol* 2005;**15**(5):303–25.
62. Li L, Petrovsky N. Molecular mechanisms for enhanced DNA vaccine immunogenicity. *Expert Rev Vaccines* 2016;**15**(3):313–29.
63. Chen J, Li Z-Y, Huang S-Y, et al. Protective efficacy of *Toxoplasma gondii* calcium-dependent protein kinase 1 (TgCDPK1) adjuvated with recombinant IL-15 and IL-21 against experimental toxoplasmosis in mice. *BMC Infect Dis* 2014;**14**(1):487.
64. Wang S, Hassan IA, Liu X, et al. Immunological changes induced by *Toxoplasma gondii* glutathione-S-transferase (TgGST) delivered as a DNA vaccine. *Res Vet Sci* 2015;**99**:157–64.
65. Cao A, Liu Y, Wang J, et al. *Toxoplasma gondii*: vaccination with a DNA vaccine encoding T-and B-cell epitopes of SAG1, GRA2, GRA7 and ROP16 elicits protection against acute toxoplasmosis in mice. *Vaccine* 2015;**33**(48):6757–62.
66. Chen J, Li Z-Y, Petersen E, Liu W-G, Zhu X-Q. Co-administration of interleukins 7 and 15 with DNA vaccine improves protective immunity against *Toxoplasma gondii*. *Exp Parasitol* 2016;**162**:18–23.
67. Rahimi MT, Sarvi S, Sharif M, et al. Immunological evaluation of a DNA cocktail vaccine with co-delivery of calcium phosphate nanoparticles (CaPNs) against the *Toxoplasma gondii* RH strain in BALB/c mice. *Parasitol Res* 2016;**116**:1–8.
68. Li W, Joshi MD, Singhania S, Ramsey KH, Murthy AK. Peptide vaccine: progress and challenges. *Vaccines* 2014;**2**(3):515–36.
69. Bittle JL, Houghten RA, Alexander H, Shinnick TM, Sutcliffe JG, Lerner RA. Protection against foot-and-mouth disease by immunization with a chemically synthesized peptide predicted from the viral nucleotide sequence. *Nature* 1982;**298**:30–3.
70. Langeveld JP, Casal JI, Cortés E, et al. Effective induction of neutralizing antibodies with the amino terminus of VP2 of canine parvovirus as a synthetic peptide. *Vaccine* 1994;**12**(15):1473–80.
71. Kilinc YB, Akdeste ZM, Koc RC, Bagirova M, Allahverdiyev A. Synthesis and characterization of antigenic influenza A M2e protein peptide-poly (acrylic) acid bioconjugate and determination of toxicity in vitro. *Bioengineered* 2014;**5**(6):357–62.
72. Eroglu BI, Kilinc YB, Mustafaeva Z. Bioconjugation of hepatitis B antigenic peptide with polymeric carriers through various carbodiimide chemistry. *Turk J Biochem* 2011;**36**:222–9.
73. Bijker MS, Melief CJ, Offringa R, Van Der Burg SH. Design and development of synthetic peptide vaccines: past, present and future. *Expert Rev Vaccines* 2007;**6**(4):591–603.
74. Lin SY-H, Cheng C-W, Su EC-Y. Prediction of B-cell epitopes using evolutionary information and propensity scales. *BMC Bioinform* 2013;**14**(2):S10.
75. Apellániz B, Nieva JL. The use of liposomes to shape epitope structure and modulate immunogenic responses of peptide vaccines against HIV MPER. *Adv Protein Chem Struct Biol* 2015;**99**:15–54 [chapter 2].
76. Purcell AW, McCluskey J, Rossjohn J. More than one reason to rethink the use of peptides in vaccine design. *Nat Rev Drug Discov* 2007;**6**(5):404–14.
77. Aguilar J, Rodriguez E. Vaccine adjuvants revisited. *Vaccine* 2007;**25**(19):3752–62.

78. Blanchard N, Gonzalez F, Schaeffer M, et al. Immunodominant, protective response to the parasite *Toxoplasma gondii* requires antigen processing in the endoplasmic reticulum. *Nat Immunol* 2008;**9**(8): 937–44.
79. Arditi A, Horowitz A, Lang MA, Rosenthal B, Seidman K, Stuen C. *Vision rehabilitation.* Netherlands: Taylor & Francis; 2000.
80. Abee CR, Mansfield K, Tardif SD, Morris T. *Nonhuman primates in biomedical research: diseases.* Canada: Elsevier Science; 2012.
81. Sofos JN. Food safety management. In: *Meat and meat products.* USA: Elsevier Science; 2013 [chapter 6].
82. Khurana A, Khurana B. *Comprehensive ophthalmology: with supplementary book—review of ophthalmology.* India: Jaypee Brothers, Medical Publishers Pvt. Limited; 2015.
83. Petersen E, Pollak A, Reiter-Owona I. Recent trends in research on congenital toxoplasmosis. *Int J Parasitol* 2001;**31**(2):115–44.
84. Lopez A, Wilson M. Congenital toxoplasmosis. *Am Fam Physician* 2003;**67**:2131–8.
85. Long SS, Pickering LK, Prober CG. *Principles and practice of pediatric infectious disease.* China: Elsevier, Health Sciences Division; 2012.
86. Bope ET, Kellerman RD. *Conn's current therapy 2015 E-book.* USA: Elsevier Health Sciences; 2014.
87. McAuley JB. Congenital toxoplasmosis. *J Pediatr Infect Dis Soc* 2014;**3**(Suppl. 1):S30–5.
88. Foster CS, Vitale AT. *Diagnosis & treatment of uveitis.* India: Jaypee Brothers, Medical Publishers Pvt. Limited; 2013.
89. Park Y-H, Nam H-W. Clinical features and treatment of ocular toxoplasmosis. *Korean J Parasitol* 2013;**51**(4):393–9.
90. Chinmaya S. Comprehensive notes in ophthalmology. 2011, India: Jaypee Brothers Publishers.
91. Bharadwaj AS, Appukuttan B, Wilmarth PA, et al. Role of the retinal vascular endothelial cell in ocular disease. *Prog Retin Eye Res* 2013;**32**:102–80.
92. Guyer DR. *Retina, vitreous, macula.* Saunders; 1999.
93. Luft BJ, Remington JS. Toxoplasmic encephalitis in AIDS. *Clin Infect Dis* 1992;**15**(2):211–22.
94. Cetinkaya Z, Gecici O, Namli MN. Anti-*Toxoplasma gondii* antibodies in patients with schizophrenia—preliminary findings in a Turkish sample. *Schizophr Bull* 2007;**33**(3):789–91.
95. Yolken R, Dickerson F, Fuller Torrey E. Toxoplasma and schizophrenia. *Parasite Immunol* 2009; **31**(11):706–15.
96. Flegr J. Influence of latent Toxoplasma infection on human personality, physiology and morphology: pros and cons of the Toxoplasma–human model in studying the manipulation hypothesis. *J Exp Biol* 2013;**216**(1):127–33.
97. Webster J, Lamberton P, Donnelly C, Torrey E. Parasites as causative agents of human affective disorders? The impact of anti-psychotic, mood-stabilizer and anti-parasite medication on *Toxoplasma gondii*'s ability to alter host behaviour. *Proc R Soc Lond B: Biol Sci* 2006;**273**(1589):1023–30.
98. Zhu S, Du Y, Li Q, Dong Z. High risk of psychosis may be associated with toxoplasmosis. *Life Sci J* 2007;**4**(4):38–41.
99. Tamer GS, Dundar D, Yalug I, Caliskan S, Yazar S, Aker A. The schizophrenia and *Toxoplasma gondii* connection: infectious, immune or both? *Adv Ther* 2008;**25**(7):703–9.
100. Fabiani S, Pinto B, Bruschi F. Toxoplasmosis and neuropsychiatric diseases: can serological studies establish a clear relationship? *Neurol Sci* 2013;**34**(4):417–25.
101. Jung B-K, Pyo K-H, Shin KY, et al. *Toxoplasma gondii* infection in the brain inhibits neuronal degeneration and learning and memory impairments in a murine model of Alzheimer's disease. *PLoS One* 2012;**7**(3).
102. Cakır-Koc R, Budama-Kılınc Y. Development and characterization of a formulation based covalent conjugation with polyacrylic acid and recombinant major surface antigen (SAG1) of *Toxoplasma gondii. Turkish J Biochem* 2017;**45**.
103. Sergeyev O, Barinsky I. Synthetic peptide vaccines. *Vopr Virusol* 2016;**61**(1):5.
104. Teunissen MBM. *Intradermal immunization.* USA: Springer Berlin Heidelberg; 2011.
105. Budama-Kılınc Y, Ozdemir B, Gozutok K. Peptide-based nanobiomaterials. In: Razavi M, Thakor A, editors. *Nanobiomaterials science, development and evaluation.* UK: Elsevier Science; 2017. p. 135–43.
106. Budama-Kilinc Y, Cakir-Koc R, Horzum-Bayir O. The cytotoxicity, characteristics, and optimization of insulinloaded nanoparticles. *Orbital: Electron J Chem* 2017;**9**(1):66–71.

INDEX

Note: Page numbers followed by *f* indicate figures and *t* indicate tables.

D

E

F

G

H

Q

R

S

V

W

X

Z

FSC
www.fsc.org
MIX
Papier | Fördert
gute Waldnutzung
FSC® C083411

Kolibri 360 GmbH
Ferdinand-Jühlke-Straße 7
99095 Erfurt, Deutschland
produktsicherheit@kolibri360.de